# Photochemical Synthesis

# BEST SYNTHETIC METHODS

Series Editors



R. F. Heck, *Palladium Reagents in Organic Syntheses,* 1985

A. H. Haines, *Methods for the Oxidation of Organic Compounds: Alkanes, Alkenes, Alkynes, and Arenes,* 1985

P. N. Rylander, *Hydrogenation Methods,* 1985

E. W. Colvin, *Silicon Reagents in Organic Synthesis,* 1988

A. Pelter, K. Smith and H. C. Brown, *Borane Reagents*, 1988

B. Wakefield, *Organolithium Methods*, 1988

A. H. Haines, *Methods for the Oxidation of Organic Compounds: Alcohols, Alcohol Derivatives, Alkyl Halides, Nitroalkanes, Alkyl Azides, Carbonyl Compounds, Hydroxyarenes and Aminoarenes*, 1988.

I. Ninomiya and T. Naito, *Photochemical Synthesis*, 1989

In preparation

G. H. Davies, R. H. Green, D. R. Kelly and S. M. Roberts, *Biotransformations in Preparative Organic Chemistry: The Use of Isolated Enzymes and Whole Cell Systems in Synthesis,* 1989

# Photochemical Synthesis

I. Ninomiya and T. Naito

*Laboratory of Medicinal Chemistry*
*Kobe Women's College of Pharmacy*
*Higashinada-Ku, Kobe*
*Japan*

**Academic Press**
*Harcourt Brace Jovanovich, Publishers*
London San Diego New York Berkeley
Boston Sydney Tokyo Toronto

ACADEMIC PRESS LIMITED
24–28 Oval Road
London NW1 7DX

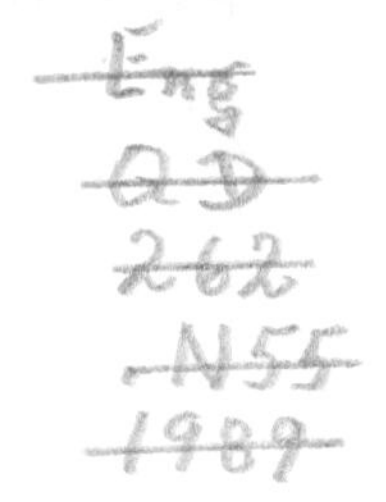


*US Edition published by*
ACADEMIC PRESS INC.
San Diego, CA 92101



This book is a guide providing general information concerning its subject matter; it is not a procedural manual. Synthesis of chemicals is a rapidly changing field. The reader should consult current procedural manuals for state-of-the-art instructions and applicable government safety regulations. The publisher and the author do not accept responsibility for any misuse of this book, including its use as a procedural manual or as a source of specific instructions

**British Library Cataloguing in Publication Data**

Ninomiya, I.
Photochemical synthesis.
1. Organic compounds. Synthesis.
Photochemical techniques
I. Title II. Naito, T. III. Series
547.2

ISBN 0-12-519490-0



Typeset in Great Britain by EJS Chemical Composition, Bath
Printed in Great Britain by St Edmundsbury Press Limited, Bury St Edmunds, Suffolk

# Contents

# Contents

# Foreword

There is a vast and often bewildering array of synthetic methods and reagents available to organic chemists today. Many chemists have their own favoured methods, old and new, for standard transformations, and these can vary considerably from one laboratory to another. New and unfamiliar methods may well allow a particular synthetic step to be done more readily and in higher yield, but there is always some energy barrier associated with their use for the first time. Furthermore, the very wealth of possibilities creates an information-retrieval problem. How can we choose between all the alternatives, and what are their real advantages and limitations? Where can we find the precise experimental details, so often taken for granted by the experts? There is therefore a constant demand for books on synthetic methods, especially the more practical ones like *Organic Syntheses*, *Organic Reactions*, and *Reagents for Organic Synthesis*, which are found in most chemistry laboratories. We are convinced that there is a further need, still largely unfulfilled, for a uniform series of books, each dealing concisely with a particular topic from a *practical* point of view—a need, that is, for books full of preparations, practical hints and detailed examples, all critically assessed, and giving just the information needed to smooth our way painlessly into the unfamiliar territory. Such books would obviously be a great help to research students as well as to established organic chemists.

We have been very fortunate with the highly experienced and expert organic chemists who, agreeing with our objective, have written the first group of volumes in this series, *Best Synthetic Methods*. We shall always be pleased to receive comments from readers and suggestions for future volumes.

A.R.K., O.M.-C., C.W.R.

# Preface

The importance of photochemical reactions in organic chemistry is now well established. Although their role is not as wide-ranging as that of thermal reactions, photochemical reactions have provided a considerable degree of insight into the nature and mechanism of organic reactions.

Organic photochemistry has two aspects: a theoretical one, which is primarily the interest of physical chemists, and the application of photochemical methods to the preparation of organic compounds with specific structures.

Over the past few decades, the growth in interest in photochemistry has led to the production of a vast amount of literature—so much so that even specialists in the field have found some difficulty in following the latest developments, particularly in the area of organic synthesis.

This book is aimed at those organic chemists who are searching for syntheses of specific compounds that have not been possible using established thermal methods.

Using our experience of the application of photochemical methods to organic synthesis over the past two decades, we have attempted to make this book of practical use to organic chemists who are interested in the synthesis of specific structures. We have therefore classified photochemical reactions by the type of structures resulting from them, that is, by functional groups. We hope that this approach will make the book accessible to organic chemists who have no previous experience in photochemistry.

We have been very fortunate in having in our laboratory Drs Toshiko Kiguchi and Okiko Miyata, who are highly experienced and expert synthetic organic chemists with much experience of the applications of organic photochemistry.

ICHIYA NINOMIYA
TAKEAKI NAITO

# Preface

# Detailed Contents

**6. Preparation of Carbocyclic Compounds**

**7. Preparation of Heterocyclic Compounds**

**8. Cage Compounds**



**9. Spiro Compounds**



**10. Equipment for and Techniques in Photochemical Synthesis**

# –1–

# Preparation of Substituted Alkanes

## 1.1 PREPARATION OF ALCOHOLS

Photochemical synthesis of alcohols is accomplished in practice by two methods: (i) photochemical addition of either water or an alcohol to C=C double bonds, which provides the expected homologues of the starting alcohol, and (ii) formation of allylic alcohols via the corresponding hydroperoxides by photooxygenation of unsaturated compounds.

### 1.1.1 Preparation of alcohols by photoaddition of alcohols and water

Alcohols add to C=C double bonds, under ultraviolet irradiation in two ways (i) addition at the carbon atom to which the hydroxyl function is attached; or (ii) addition at the hydroxyl group. In most free-radical and photochemical reactions the first mechanism occurs, although in some cases the second route operates [1]. Methanol, ethanol, propanol and other alcohols react with double bonds under UV irradiation to give 1 : 1 adducts that are higher alcohols [2].

$$R^1CH_2CH(R^2)CH(OH)R^3 \xleftarrow[{[1]}]{h\nu} \left[ R^1CH{=}CHR^2 + R^3CH_2OH \right] \xrightarrow[{[2]}]{h\nu} R^1CH_2CH(OCH_2R^3)R^2$$

The addition reactions of alcohols to unsaturated compounds can be initiated by a variety of photosensitizers. Thus irradiation of ethanol, isopropanol or other alcohols with maleic acid in the presence of benzophenone or anthraquinone as photosensitizers led to the formation of the corresponding 1 : 1 addition products (**1**) [3,4].

$$\mathrm{Me(R)CHOH} + \mathrm{HO_2C{-}CH{=}CH{-}CO_2H} \xrightarrow[\text{sens.}]{h\nu} \underset{(\mathbf{1})}{\mathrm{HO{-}C(R)(Me){-}CH(CO_2H){-}CH_2{-}CO_2H}} \xrightarrow{-H_2O} \text{lactone (5-R-5-Me-4-}CO_2H\text{-tetrahydrofuran-2-one)}$$

The photoaddition of isopropanol to $\alpha,\beta$-unsaturated ketones has also been reported to yield 1 : 1 adducts (**2**), and benzophenone-photosensitized addition of isopropanol to acetylene dicarboxylic acid led to 1 : 2 adducts (**3**) and (**4**) [5, 6].

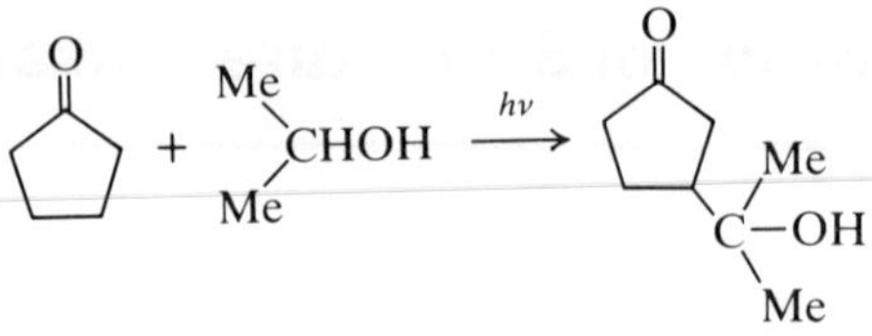

(**2**)

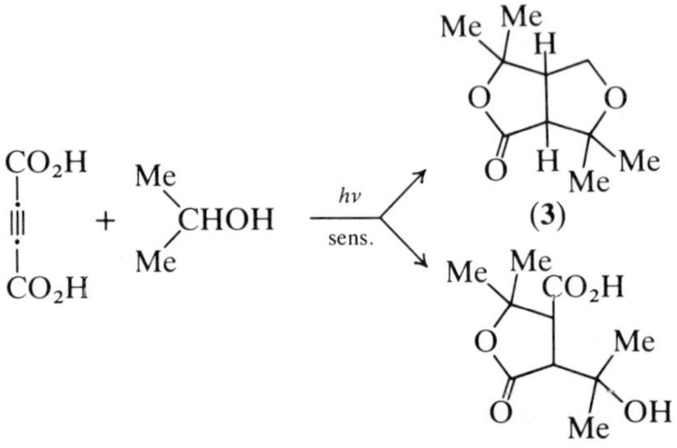

(**3**)

(**4**)

The photochemical addition reactions of water to C=C double bonds, yielding alcohols, are worthy of detailed consideration because of their possible significance in biological processes. Wang *et al.* [7] subjected an aqueous solution of 1,3-dimethyluracil (**5**) to UV irradiation and obtained a photoadduct, 6-hydroxy-1,3-dimethylhydrouracil (**6**) in 60–75% yield. Similarly, 5-fluorouracil (**7**) gave 5-fluoro-6-hydroxyhydrouracil (**8**) [8].

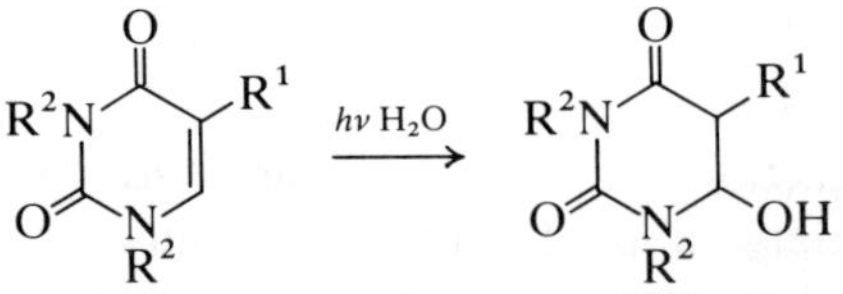

(**5**) $R^1 = H, R^2 = Me$  (**6**) $R^1 = H, R^2 = Me$
(**7**) $R^1 = F, R^2 = H$  (**8**) $R^1 = F, R^2 = H$

During investigations into the photochemical behaviour of some lysergic acid derivatives with the general structure (**9**), Stoll and Schlientz obtained two lumi derivatives (**10**) and (**11**), which are photoadducts with water, by irradiation of (**9**) in dilute acid [9]. The photoaddition of water to other unsaturated ergot alkaloids has been studied systematically by Hellberg [10].

(9) $R = C_{17}H_{20}N_3O_4$ $\xrightarrow[+H_2O]{h\nu}$ (10) + (11)

Other photochemical syntheses of alcohols by photoaddition of water to C=C are shown in the following example. Irradiation of the aqueous solution of octalene (**12**) gave stereoselectively *cis*-decalol (**13**) in 92% yield as well as a small amount of the *trans* isomer (**14**) (8%) [11, 12].

(12) $\xrightarrow[+H_2O,\ \text{xylene}]{h\nu}$ (13) (92%) + (14) (8%)

C=N double bonds can undergo similar photoaddition reactions with alcohols. Irradiation of the carbazolium iodide (**15**) in methanol-acetone yielded the oily 9a-methanol (**16**) [13].

(15) $\xrightarrow[\text{MeOH}]{h\nu}$ (16)

Similarly, when the carbazole (**17**) was irradiated in pure methanol, the oxazolidine derivative (**18**) was formed in 50% yield as a result of the photoaddition of methanol to the C=N bond of oxazoline followed by cyclization [13]. Acid hydrolysis of the product (**18**) gave the methylol (**19**), which on irradiation in methanol regenerated (**18**).

(17) $\xrightarrow[\text{MeOH}]{h\nu}$ (18) $\underset{h\nu,\ \text{MeOH}}{\overset{H^+}{\rightleftarrows}}$ (19)

The transformation of (**17**) to (**18**) was inhibited by oxygen, while that of (**19**) to (**18**) was not. The formation of (**18**) from (**17**) can be explained by the attack of $\cdot CH_2OH$ twice on the starting compound, followed by elimination of water [13].

Photochemical formation of diols (**20**) by irradiation of ketones and aldehydes in alcohols in the presence of titanium tetrachloride is an interesting example of photochemistry in the presence of a metal [14].

$h\nu/TiCl_4$, MeOH → (**20**) +

(54%)

(35%)

(46%) (20%)

(35%)

This reaction proceeds via the following electron-transfer mechanism involving the titanium [15]:

electron transfer → proton transfer → products

**Experimental 1.1** Lumiergotamines [9]

Ergotamine (**9**) (4 g) was dissolved in glacial acetic acid (50 ml) in the cold. The solution was diluted with boiled distilled water (450 ml) that had been saturated with carbon dioxide and was then introduced into a thin-walled glass bottle. After the air had been displaced by carbon dioxide, the vessel was illuminated by exposure to direct sunlight for 30 h (middle of August, Basel). The temperature of the solution did not rise above 30 °C. The liquid became dark brown and no longer showed fluorescence under UV light. It was then made alkaline with concentrated ammonia solution and thoroughly shaken with chloroform. The chloroform extracts were pooled, dried over sodium sulphate and evaporated to dryness.

The products were then isolated by column chromatography on alumina (Merck) (300 g) using 1% ethanolic chloroform as eluent. After elution of the starting ergotamine and by-products, the second fraction gave the brown-red compound

(3.2 g) which was crystallized from methanol to give the crude lumiergotamine-I (**10**) (2.6 g, MP 233–237 °C). Recrystallization from chloroform–methanol gave the pure lumiergotamine-I (**10**) as colourless prisms (MP 247 °C). The third fraction gave the crude lumiergotamine-II (**11**) (205 mg) which was recrystallized from acetone to afford pure material (**11**) as colourless crystals (MP 192 °C).

## 1.1.2 Preparation of alcohols by photooxygenation of C=C double bonds

As described in Section 4.2.2 below, the photochemical synthesis of allylic alcohols via intermediate hydroperoxides is of great importance for synthetic purposes. The photooxygenation of allylic alcohols by the "ene" reaction described in Section 4.2.2 gives allylic hydroperoxides in good yields; these then undergo a variety of transformations: (a) allylic isomerization; (b) dehydration to form $\alpha,\beta$-unsaturated ketones; (c) skeletal changes initiated by migration of groups to oxygen; and (d) reduction to form the allylic alcohols.

$$\overset{*}{C}{=}C{-}CH \xrightarrow{{}^1O_2} \overset{*}{C}(OOH){-}C{=}C \xrightarrow{d} \overset{*}{C}(OH){-}C{=}C$$

$$\xrightarrow{c}\ {-}\overset{*}{C}{=}O + O{=}C{-}CH \qquad \xrightarrow{a}\ \overset{*}{C}{=}C{-}C(OOH) \qquad \xrightarrow{b,\ -H_2O}\ \overset{*}{C}(=O){-}C{=}C$$

Route (d) is illustrated by the synthesis of rose oxide (**24**) [16]. Photooxygenation of citronellol (**21**) followed by reduction of the resulting hydroperoxide afforded two allylic alcohols (**22**) and (**23**), both of which

[1] $^1O_2$ [2] $H_2$ (98%)

OH OH OH OH OH

(**21**) (**22**) (2 : 3) (**23**)

O

(**24**)

were subjected to acid cyclization to give rose oxide (**24**) in quantitative yield. This reaction sequence is known to be applicable on an industrial scale.

Many examples have been reported of the application of photochemical preparation of allylic alcohols to natural-product synthesis—in particular, to terpene and alkaloid syntheses, which have been reviewed [17, 18].

## 1.2 PREPARATION OF ETHERS

Substituted ethers can be prepared by photoaddition of either alcohols or ethers to C═C double bonds.

### 1.2.1 Preparation of ethers by photoaddition of alcohols to C═C double bonds

Among cyclic alkenes, cycloheptenes and cyclohexenes have been the most extensively studied, and are good starting alkenes for the photochemical preparation of ethers. For example, sensitized irradiation of 1-methylcyclohexene (**25a**) or cycloheptene (**25b**) in methanol afforded a mixture of the corresponding exocyclic alkene (**27**) and the methyl ether (**28**) [19, 20]. These products were apparently formed via initial photosensitized *cis–trans* isomerization of the alkene followed by protonation of the resulting *trans*-cycloalkene by the solvent as a ground-state reaction. The driving force for protonation by such a weak acid as methanol is presumably the accompanying relief of strain in the cycloalkene. The resulting carbocation (**26**) undergoes competing nucleophilic trapping to afford the ether (**28**) and deprotonation to afford the exocyclic alkene (**27**). Any accompanying deprotonation leading to the preferential formation of the endocyclic alkene simply regenerates the starting alkene (**25**), which is then recycled.

$-H^+$

$(CH_2)_{n-2}$ $\underset{}{\overset{h\nu}{\rightleftharpoons}}$ [*trans*] $\xrightarrow{ROH}$ $\oplus$ H $(CH_2)_{n-2}$ $\xrightarrow{-H^+}$ H $(CH_2)_{n-2}$

(**25abc**) (**26abc**)

(**a**) $n = 6$, (**b**) $n = 7$, (**c**) $n = 8$

(**27a**) $n = 6$ (trace)
(**27b**) $n = 7$ (38%)

$\xrightarrow{ROH}$ OR H $(CH_2)_{n-2}$

(**28a**) $n = 6$ (93%)
(**28b**) $n = 7$ (51%)

Synthetic application of this photoprotonation reaction to the synthesis of ethers offers advantages over traditional ground-state methods of protonation since acid-catalysed addition of alcohols to double bonds is irreversible and does not go to completion, while the photochemical reaction proceeds until the one unsaturated compound has been consumed. In a preparative procedure, irradiation of cyclohexene in methanol containing 0.6% $H_2SO_4$ afforded the ether (**29**) in 70% isolated yield [21].

$h\nu$, MeOH—$H^+$

(**29**) (70%)

The same type of addition reaction of alcohols, usually methanol, to pyrimidine bases, leading to the formation of an ether, is a well-known reaction in photobiology [22].

$h\nu$, MeOH

## 1.2.2 Preparation of ethers by photoaddition of ethers to C═C double bonds

Photochemical addition of ethers to C═C double bonds provides a useful method for the synthesis of homologues of ethers possessing a potentially reactive C—H bond α to the ether linkage. These hydrogens can be ejected from the molecule [23], leading to ether free radicals, by direct irradiation of the ether, or abstracted by photoactivated ketones such as acetone, acetophenone or benzophenone [24, 25]. The photoaddition of

$$R^1CH_2OR^2 + R^3CH{=}CH_2 \xrightarrow[\text{acetone}]{h\nu} R^3CH_2CH_2CH(R^1)OR^2$$

+ RCH═CH$_2$ → ($h\nu$, acetone or acetophenone)

+ (CO$_2$Et)CH═CH(CO$_2$Et) → ($h\nu$, acetone or acetophenone)

X = [(CH$_2$)$_n$: $n$ = 0 or 1] or oxygen

cyclic ethers to C=C double bonds can be induced directly by light or by irradiation together with a ketonic compound with high chemical yields. The products are the $\alpha$-alkylated ethers [26–28].

Diethyl maleate or fumarate gave the 1 : 1 adduct with 1,3-dioxolane in over 90% chemical yield via the following route [29, 30]:

$$\text{(CHCO}_2\text{Et)}_2 + \text{1,3-dioxolane} \xrightarrow[\text{ketone}]{h\nu} \text{2-(1,3-dioxolanyl)–CH(CO}_2\text{Et)CH}_2\text{CO}_2\text{Et}$$

$$\text{1,3-dioxolane} \xrightarrow[\text{ketone}]{h\nu} \text{1,3-dioxolan-2-yl }\dot{\text{C}}\text{H}$$

$$\text{1,3-dioxolan-2-yl }\dot{\text{C}}\text{H} + \text{CH}_2\text{=CHR} \longrightarrow \text{(1,3-dioxolanyl)–CH}_2\dot{\text{C}}\text{HR} \longrightarrow \text{product}$$

**Experimental 1.2** Irradiation of 1,3-dioxolane and diethyl maleate [30]

A mixture of diethyl maleate (1 g), 1,3-dioxolane (110 ml) and acetone (10 ml) was irradiated for 1 h. A solution of diethyl maleate (2.65 g) in acetone (5 ml) was then added in four equal portions at 1 h intervals and the mixture was irradiated until consumption of diethyl maleate was complete (*ca.* two more hours after the last addition). Excess reagents were removed by distillation under reduced pressure and the residue (6 g) was chromatographed on silica gel (0.05–0.02 mm, Merck). Acetone–petroleum ether (1 : 9) eluted diethyl(1,3-dioxolanyl)-2-succinate (4.59 g, 87.5%), BP 120–125 °C (1.5 mm Hg).

### 1.2.3 Preparation of ethers by photolysis of alkyl hypoiodites

The "hypoiodite reaction" [31] — oxidation of non-activated methylene groups with lead tetraacetate and iodine in organic solvents—proceeds smoothly to give the corresponding iodoalcohols.

$$\text{Me}\ \ \text{OH} \longrightarrow [\text{Me}\ \ \text{OI}] \xrightarrow{h\nu} \text{ICH}_2\ \ \text{OH}$$

When the 6$\beta$-hydroxyandrostan-17-one (**30**) was irradiated with strong visible light in order to accelerate the reaction, the cyclic ether (**31**) was obtained in excellent yield [32].

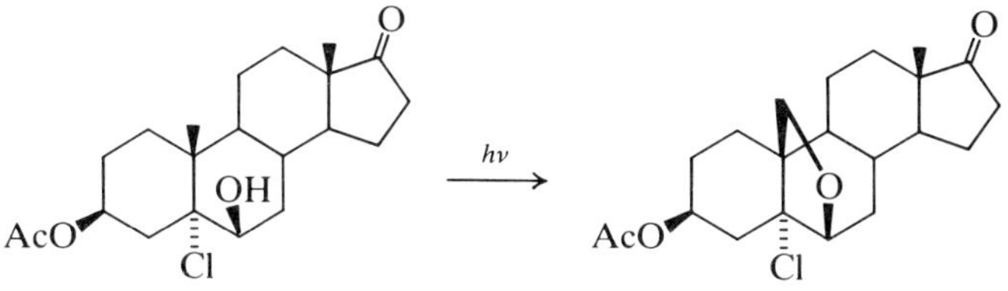

(30) $\xrightarrow{h\nu}$ (31)

**Experimental 1.3** 3β-Acetoxy-5-chloro-6β,19-oxido-5α-androstan-17-one [32]

A suspension of lead tetraacetate (90 g) (thoroughly freed from acetic acid under high vacuum) and $CaCO_3$ (30 g) in cyclohexane (4 l), briefly heated to 80 °C, was mixed with iodine (20 g) and (**30**) (15 g). The mixture was then boiled under reflux with stirring while being irradiated with a 500 W lamp for 60 min. After cooling, the still lightly coloured mixture was filtered through celite and the residue was thoroughly washed with ether. The filtrate was extracted by shaking with 10% sodium thiosulphate (1 l) and with water, dried over sodium sulphate and evaporated under vacuum (water pump). The crystalline crude product was recrystallized from ether–methanol to give pure (**31**) (11.48 g), MP 181–182 °C. A further crop of impure (**31**) (1.35 g) was obtained from the mother liquor.

## 1.3 PREPARATION OF ESTERS

As described in Sections 1.1 and 1.2, some esters have been prepared by photochemical addition of carboxylic acids to C=C double bonds via photoprotonation followed by addition of the acid used as solvent. Irradiation of 2-cyclohexenyl acetate (**32**) in acetic acid containing xylene as sensitizer gave a 1 : 1 mixture of the *cis*- (**33**) and *trans*- (**34**) diacetates [33]. In general, initial photoprotonation occurs at a position near to the substituted carbon as shown in the following examples.

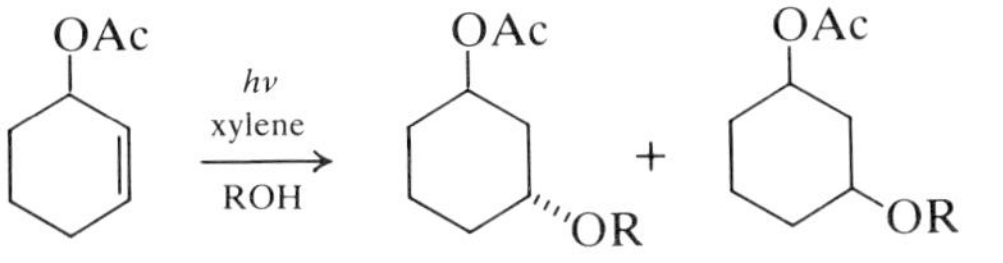

(32) $\xrightarrow[\text{ROH}]{h\nu,\ \text{xylene}}$ (33) + (34)

R = Ac (70%)
R = Et (53%)
R = Me (51%)

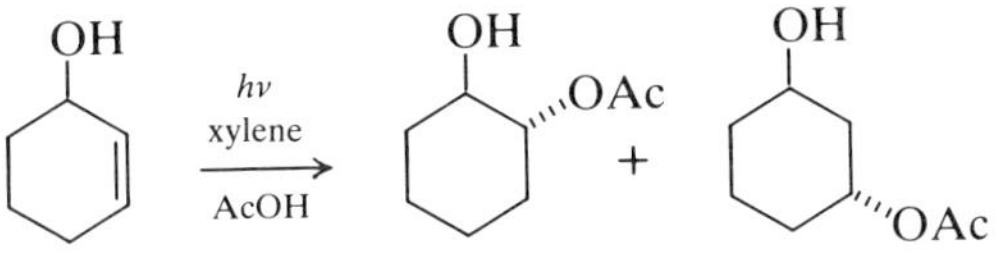

$\xrightarrow[\text{AcOH}]{h\nu,\ \text{xylene}}$ (1 : 10)

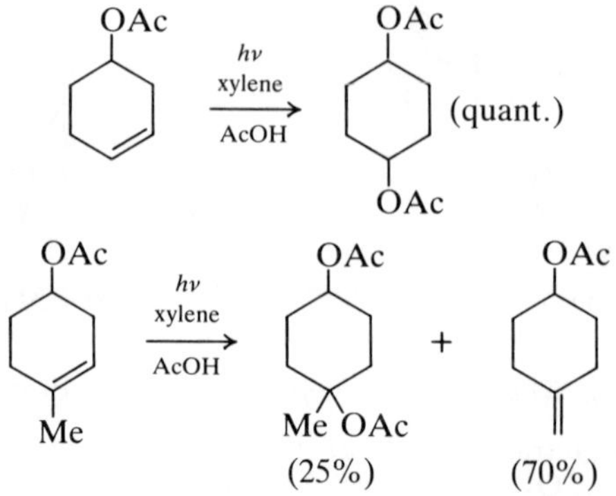

Similarly, the phenylcycloalkenes (**35**) gave acetates and propionates upon irradiation by a high-pressure mercury lamp in the presence of acetophenone [34, 35].

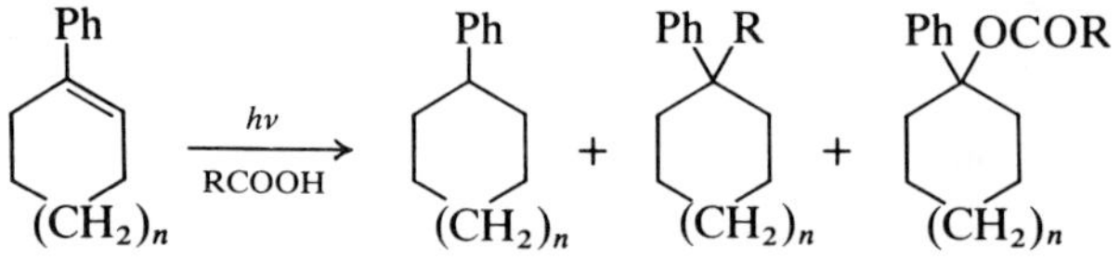

| R | $n$ | % | % | % |
|---|---|---|---|---|
| Me | 1 | 2 | 3 | 31 |
| | 2 | 5 | 4 | 41 |
| Et | 1 | 4 | 7 | 44 |
| | 2 | 3 | 5 | 40 |

TABLE 1.1

Irradiation of acetals

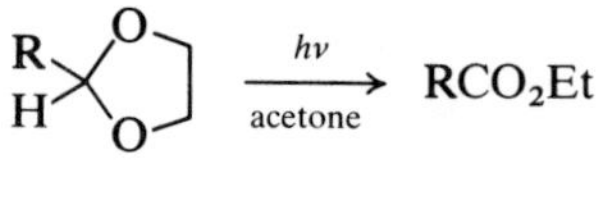

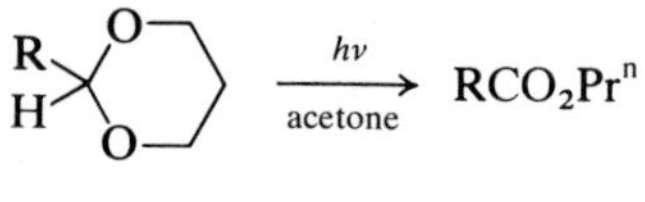

| Acetal ($\rightarrow RCO_2Et$, $h\nu$, acetone) | Yield % | Acetal ($\rightarrow RCO_2Pr^n$, $h\nu$, acetone) | Yield % |
|---|---|---|---|
| 2-Pentyl-1,3-dioxolane | 36 | 2-Heptyl-1,3-dioxane | 23 |
| 2-Heptyl-1,3-dioxolane | 55 | 2-Phenylethyl-1,3-dioxane | 14 |
| 2-Nonyl-1,3-dioxolane | 33 | | |
| 2-Benzyl-1,3-dioxolane | 35 | | |
| 2-Phenylethyl-1,3-dioxolane | 30 | | |

Another potential photochemical synthesis of esters involves photoisomerization of cyclic acetals. Elad and Youssefyeh [36] reported the acetone-sensitized photochemical conversion of cyclic acetals to open-chain carboxylic acid esters (Table 1.1).

**Experimental 1.4** *cis*- and *trans*-1,3-Diacetoxycyclohexanes [33]

Irradiation of a solution of 2-cyclohexenyl acetate (5 g) in acetic acid (300 ml) and xylene (10 ml) in a quartz vessel was effected by means of an external 200 W high-pressure mercury arc for 70 h to afford a *cis/trans* mixture (*ca.* 1:1) of 1,3-diacetoxycyclohexanes (**33**) and (**34**) (70%).

## 1.4 PREPARATION OF HALIDES

There are two photochemical syntheses of halides: (i) photochemical addition of either a halogen molecule or hydrogen halide to C=C double bonds, and (ii) photolysis of acyl hypoiodites.

### 1.4.1 Preparation of halides by photochemical addition of a halogen molecule or hydrogen halide

In general terms the photochemical addition of HX to double bonds can be summarized as follows:

$$HX \xrightarrow{h\nu} \dot{H} + \dot{X} \quad (1)$$

$$RCH{=}CH_2 + \dot{X} \longrightarrow R\dot{C}HCH_2X \quad (2)$$

$$R\dot{C}HCH_2X + HX \longrightarrow RCH_2CH_2X + \dot{X} \quad (3)$$

$$R\dot{C}HCH_2X + RCH{=}CH_2 \longrightarrow R\dot{C}HCH_2CHRCH_2X, \text{etc.} \quad (4)$$

$$2\dot{X} \rightarrow XX \quad (5)$$

Many photochemical addition reactions are free-radical chain reactions. These involve a free radical that adds to the double bond [reaction (2)] to produce a new free-radical, which subsequently reacts by radical transfer with another species present in the system [reaction (3)].

Synthetically useful addition reactions of halogens with double bonds usually involve chlorine and bromine rather than iodine. In general, the reactivity of the halogens in substitution or addition decreases from chlorine to iodine. The reaction products from iodination and some bromination reactions suffer from complicated decomposition side-reactions. Chlorine atoms are so reactive that they give apparent addition rather than substitution products with some aromatic-hydrocarbon derivatives [37]. Some examples of photoaddition of chlorine and bromine to C=C double bonds are shown in Tables 1.2 and 1.3.

TABLE 1.2

Addition of chlorine to unsaturated compounds

| Substrate | Product | Yield (%) |
|---|---|---|
| $CF_2{=}CH_2$ | $ClCF_2CH_2Cl$ | 98 |
| Trifluoropropene | $CF_3CHClCH_2Cl$ | 80 |
| $CF_2{=}CHCH{=}CF_2$ | $CF_2ClCHClCHClCClF_2$ | 33 |
| Perfluorocyclohexene | 1,2-Dichlorodecafluorocyclohexane | 81 |

TABLE 1.3

Addition of bromine to unsaturated compounds

| Substrate | Product | Yield (%) |
|---|---|---|
| $CH_2CHCl$ | $CH_2BrCHBrCl$ | — |
| $CF_2{=}CHF$ | $CF_2BrCHBrF$ | 82 |
| Perfluorocyclohexene | 1,2-Dibromodecafluorocyclohexane | 83 |
| $CF_2{=}CFCF{=}CF_2$ | $CF_2BrCFBrCFBrCF_2Br$ | Quant. |
| $Me_3CC{\equiv}CH$ | $Me_3CCBr{=}CHBr$ | 90 |
| $HC{\equiv}CCMe_2OCOMe$ | $BrCH{=}CBrCMe_2OCOMe$ | 77 |

The photoaddition of hydrogen bromide to a variety of unsaturated compounds is important because of its synthetic applications in the preparation of alkyl and alkenyl halides.

Since hydrogen bromide absorbs light of wavelength shorter than 290 nm, reactions involving addition of hydrogen bromide to double bonds must be induced by light of such wavelength or by light of longer wavelength in the presence of a photosensitizer [38]. The wavelength employed is a determining factor in these reactions. The addition of hydrogen bromide to propene leads to n-propyl bromide and similarly to higher terminal alkenes, all these being anti-Markovnikov additions.

$$RCH{=}CH_2 + HBr \xrightarrow{h\nu} RCH_2CH_2Br$$

The orientation in addition of hydrogen bromide is generally that which would be predicted on the basis of intermediate free-radical stability. The addition of hydrogen bromide to a C=C double bond is in fact controlled by the initial addition of the bromine atoms produced by the photochemical decomposition of the hydrogen bromide. Some examples are shown in Table 1.4 [39, 40].

TABLE 1.4

Addition of hydrogen bromide to unsaturated compounds

| Substrate | Product | Yield (%) |
|---|---|---|
| $CH_2CCl_2$ | 1-Bromo-2,2-dichloroethane | 62 |
| Trifluoropropene | $CF_3CH_2CH_2Br$ | 90 |
| 1,5-Hexadiene | 1,6-Dibromohexane | Quant. |
| CH≡CH | Vinyl bromide | 80 |
| MeC≡CBr | MeCBr═CHBr | 92 |

**Experimental 1.5** Addition of hydrogen bromide to 1,1-dichloroethylene [39]

1,1-Dichloroethylene (20 ml, 24.2 g) was distilled on a vacuum line into a 500 ml round-bottomed Pyrex flask containing a magnetic stirrer bar. Hydrogen bromide (6.0 ml, 16.6 g) was distilled into the line until the pressure was nearly one atmosphere. Absorption of hydrogen bromide took place only after the reaction flask was irradiated with UV light, and then only at the rate of about 1 ml of liquid hydrogen bromide (measured at −78 °C) per hour. After 18 h of irradiation, absorption ceased. Unreacted hydrogen bromide and dichloroethylene were then pumped off. The reaction mixture was diluted with methylene chloride (30 ml), washed with 5% sodium bicarbonate and dried over anhydrous potassium carbonate. The product was distilled in a column filled with glass helices. After the methylene chloride had been distilled off, the temperature rose rapidly to 134 °C. The yield of product, BP 134–144 °C, was 22.4 g (61.5%).

### 1.4.2 Preparation of iodides by photochemical replacement of the carboxyl group by iodine

Barton and Serebryakov [41] have found that the following photochemical decarboxylation may be effected for the preparation of alkyl iodides; the reaction involves the formation of an intermediate acyl hypoiodite (RCOOI):

$$\mathrm{RCOOH} \xrightarrow[\mathrm{Pb(OAc)_4 + I_2}]{h\nu} \mathrm{RI}$$

The carboxylic acid and iodine were added to a 5% w/v suspension of lead tetraacetate in refluxing carbon tetrachloride that was illuminated by a tungsten lamp until the iodine colour persisted. The reaction proceeds more slowly and less cleanly (lower yield) in the dark. This reaction is an excellent method for the preparation of alkyl iodides from carboxylic acids. Illustrative examples are listed in Table 1.5.

TABLE 1.5

Photodecarboxylation reaction

| Substrate | Product | Yield (%) |
|---|---|---|
| 12-Acetoxystearic acid | 11-Acetoxy-1-iodoheptadecane | 82 |
| Cyclohexanecarboxylic acid | Iodocyclohexane | 91 |
| Benzoic acid | Iodobenzene | 56 |
| Adipic acid | 1,4-Diiodobutane | 33 |

## REFERENCES


1. J. A. Marshall and R.D. Carrol, *J. Am. Chem. Soc.* **88**, 4092 (1966).
2. W. H. Urry, F. W. Stacey, E. S. Huyser and O. O. Juveland, *J. Am. Chem. Soc.* **76**, 450 (1954).
3. R. Dulon, M. Vilkas and M. Pfau, *C.R. Acad. Sci. Paris* **249**, 429 (1959).
4. G. O. Schenck, G. Koltzenburg and H. Grossmann, *Angew. Chem.* **69**, 177 (1957).
5. M. Pfau, R. Dulon and M. Vilas, *C.R. Acad. Sci. Paris* **254**, 1817 (1962).
6. G. O. Schenck and R. Steinmetz, *Naturwissenschaften* **47**, 514 (1960).
7. S. Y. Wang, M. Apicella and B. S. Stone, *J. Am. Chem. Soc.* **78**, 4180 (1956).
8. H. A. Lozeron, M. P. Gordon, T. Gabriel, W. Tantz and R. Duschinsky, *Biochemistry* **3**, 1844 (1964).
9. A. Stoll and W. Schlientz, *Helv. Chim. Acta* **38**, 585 (1955).
10. H. Hellberg, *Acta Chem. Scand.* **16**, 1363 (1962).
11. J. A. Marshall and A. R. Hochstetler, *J. Org. Chem.* **31**, 1020 (1966).
12. J. A. Marshall and M. J. Wurth, *J. Am. Chem. Soc.* **89**, 6788 (1967).
13. P. Cerutti and H. Schmid, *Helv. Chim. Acta* **45**, 1992 (1962).
14. T. Sato, H. Kaneko and S. Yamaguchi, *J. Org. Chem.* **45**, 3778 (1980).
15. T. Sato, G. Izumi and T. Imamura, *J. Chem. Soc. Perkin Trans. 1*, **1976**, 788.
16. G. Ohloff, E. Klein and G. O. Schenck, *Angew. Chem.* **73**, 578 (1961).
17. M. Matsumoto and H. Kondo, *J. Synth. Org. Chem. Jpn* **35**, 188 (1977).
18. R. W. Denny and A. Nickon, in *Organic Reactions*, Vol. 20 (ed. W. G. Dauben), p. 133. Wiley, New York, 1973.
19. P. J. Kropp and H.J. Krauss, *J. Am. Chem. Soc.* **89**, 5199 (1967).
20. P. J. Kropp, *J. Am. Chem. Soc.* **88**, 4091 (1966).
21. F. P. Tise and P. J. Kropp, Unpublished result.
22. S. Y. Wang, *Fed. Proc.* **24**, S-71 (1965).
23. K. Pfordte, *Liebig's Ann. Chem.* **625**, 30 (1959).
24. K. Shima and S. Tsutsumi, *Bull. Chem. Soc. Jpn* **36**, 121 (1963).
25. G. O. Schenck, H. D. Becker, K. H. Schulte-Elte and C. H. Krauch, *Chem. Ber.* **96**, 509 (1963).
26. I. Rosenthal and D. Elad, *Tetrahedron* **23**, 3193 (1967).
27. D. Elad and R. D. Youssefyeh, *J. Org. Chem.* **29**, 2031 (1964).
28. R. L. Jacobs and G. G. Ecke, *J. Org. Chem.* **28**, 3036 (1963).
29. D. Alad and I. Rosenthal, *Chem. Commun.* **1966**, 684.
30. I. Rosenthal and D. Elad, *J. Org. Chem.* **33**, 805 (1968).
31. K. Heusler and J. Kalvoda, *Angew. Chem. Int. Ed. Engl.* **3**, 525 (1964).

32. H. Uberwasser, K. Heusler, J. Kalvoda, C. Meystre, P. Wieland, G. Anner and A. Wettstein, *Helv. Chim. Acta* **46**, 344 (1963).
33. T. Okada, V. Shibata, M. Kawanishi and H. Nozaki, *Tetrahedron Lett.* **1970**, 859.
34. S. Fujita, N. Nomi and H. Nozaki, *Tetrahedron Lett.* **1969**, 3557.
35. T. Nylund and H. Morrison, *J. Am. Chem. Soc.* **100**, 7364 (1978).
36. D. Elad and R. D. Youssefyeh, *Tetrahedron Lett.* **1963**, 2189.
37. G. G. Ecke. L. R. Buzbee and A. J. Kobsa, *J. Am. Chem. Soc.* **78**, 79 (1956).
38. W. E. Vaughan, F. F. Rust and T. W. Evans, *J. Org. Chem.* **7**, 477 (1942).
39. T. E. Francis and L. C. Leitch. *Can. J. Chem.* **35**, 500 (1957).
40. A. L. Henne and M. Nager, *J. Am. Chem. Soc.* **73**, 5527 (1951).
41. D. H. R. Barton and E. P. Serebryakov, *Proc. Chem. Soc.* **1962**, 309.

# –2–

# Preparation of Carboxylic Acid Derivatives

## 2.1 PREPARATION OF CARBOXYLIC ACIDS AND ESTERS

2,4-Cyclohexadienones (**1**) are very reactive under irradiation to give a mixture of *cis*-dienylketene (**2**) and the *trans* isomer (**3**) [1]. When irradiation was carried out in the presence of a strong nucleophile the intermediate ketene (**3**) reacted with the nucleophile to give the dienes (**4**); good yields of dienyl carboxylic acids, esters and amides were obtained for YH=$H_2O$, ROH and $NH_2$ [2].

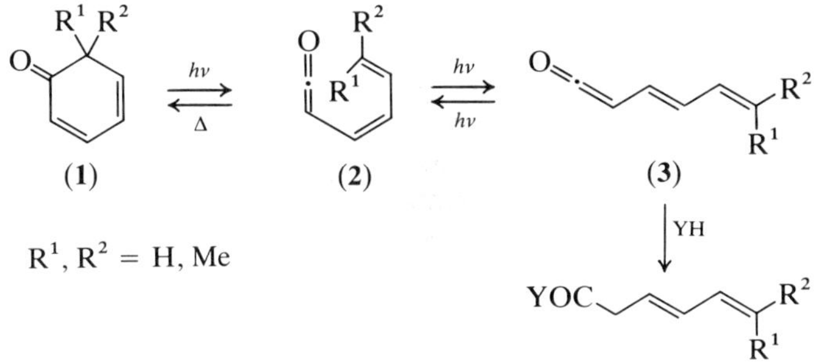

O
R¹ R²
hν
Δ
hν
hν
YH
YOC


(**1**) (**2**) (**3**) (**4**)

$R^1, R^2$ = H, Me

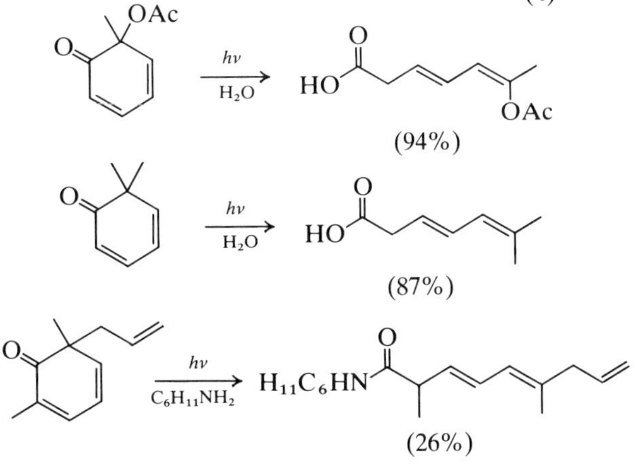

OAc
hν
$H_2O$
HO
(94%)
hν
$H_2O$
HO
(87%)
hν
$C_6H_{11}NH_2$
$H_{11}C_6HN$
(26%)

Analogously, lumisantonin (**5**) [3, 4] and the santodienolide (**6**) [5] gave the corresponding acids, photosantonic acid (**7**) and (**8**) respectively, upon irradiation in the presence of water.

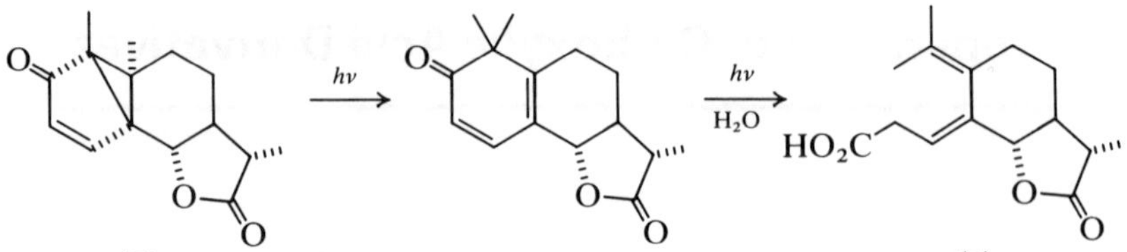

(**5**) $\xrightarrow{h\nu}$ $\xrightarrow[H_2O]{h\nu}$ (**7**)

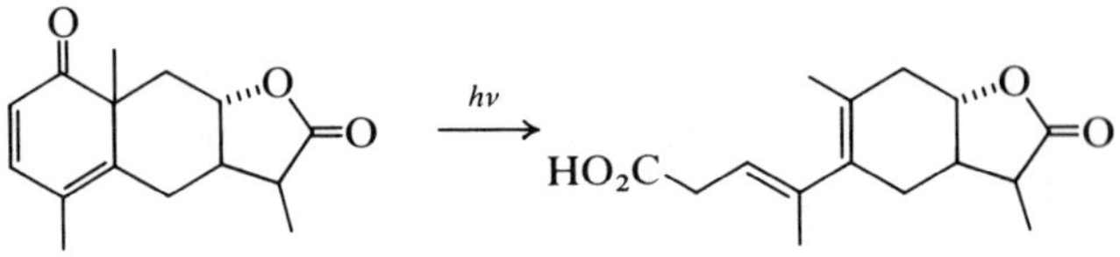

(**6**) $\xrightarrow{h\nu}$ (**8**)

The photochemical preparation of esters has been described in Section 1.3.

**Experimental 2.1** Photosantonic acid [3]

Lumisantonin (**5**) (500 mg) was dissolved in acetic acid (12 ml) and the mixture was then diluted with water (14 ml). Irradiation of the mixture at −5 to +5 °C for 1.5 h with a 125 W Crompton bare-arc mercury lamp, followed by isolation of the acidic portion and crystallization from chloroform–light petroleum, afforded photosantonic acid (**7**) (350 mg).

## 2.2 PREPARATION OF LACTONES

Lactones are photochemically prepared as secondary products by lactonization of the hydroxy-esters that result from the photochemical addition of alcohols to unsaturated carboxylic acids. Photolysis of maleic acid in isopropanol in the presence of benzophenone as a sensitizer gave the lactone (**9**) in 96% yield [6]. The triplet sensitizer benzophenone abstracts the hydrogen of the alcohol to give a ketyl radical, which reacts with maleic acid to give the hydroxy-acid (**10**); finally the lactone (**9**) is formed.

Barton and Beckwith [7] have reported the photochemical synthesis of lactones by photolysis of *N*-iodoamides (**11**). Irradiation of acid amides possessing $\gamma$-hydrogen together with lead tetraacetate and iodine, followed by alkaline hydrolysis of the reaction product, afforded the $\gamma$-lactones (**13**) [7, 8].

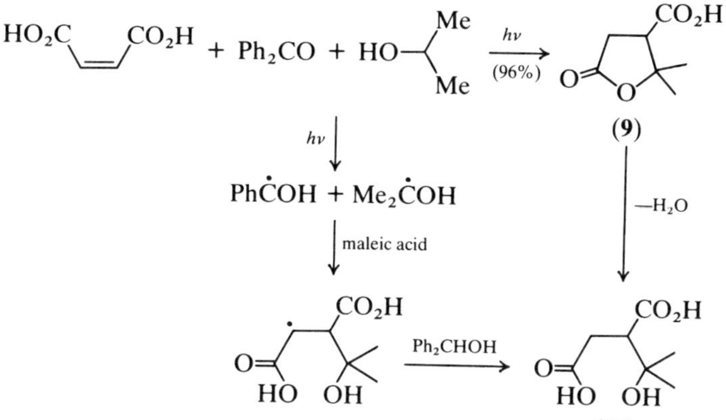

$HO_2C$ $CO_2H$ + $Ph_2CO$ + HO–CH(Me)Me $\xrightarrow[(96\%)]{h\nu}$ $CO_2H$ (**9**)

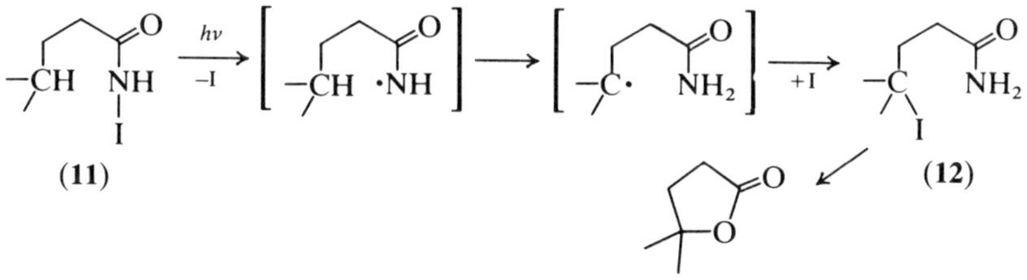

$h\nu$ ↓

$Ph\dot{C}OH + Me_2\dot{C}OH$

↓ maleic acid

$CO_2H$ O= HO OH $\xrightarrow{Ph_2CHOH}$ $CO_2H$ O= HO OH (**10**)

(**9**) $\xrightarrow{-H_2O}$ (**10**)

–CH NH(I) =O (**11**) $\xrightarrow[-I]{h\nu}$ [ –CH •NH =O ] ⟶ [ –C• $NH_2$ =O ] $\xrightarrow{+I}$ –C(I) $NH_2$ =O (**12**) ⟶ =O O (**13**)

The mechanism of this reaction involves the formation of an *N*-iodoamide (**11**), photolysis of (**11**), intramolecular hydrogen transfer, and finally coupling of the resulting radical with iodine. Hydrolysis of the $\gamma$-iodoamide (**12**) thus formed finally gives a $\gamma$-lactone via an intermediate $\gamma$-imino-lactone. For example, the $\gamma$-lactone (**15**) was prepared from the corresponding amide (**14**) [7]:

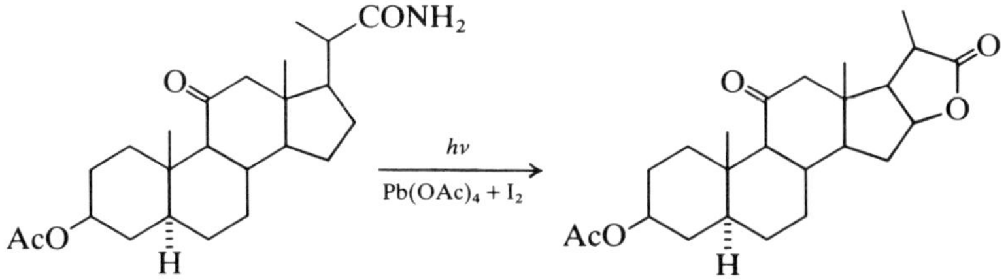

$CONH_2$ O AcO H (**14**) $\xrightarrow[Pb(OAc)_4 + I_2]{h\nu}$ O O =O AcO H (**15**)

**Experimental 2.2** 3$\beta$-Acetoxy-16$\beta$-hydroxy-11-oxo-5$\alpha$-pregnane-20-carboxylic acid lactone

The reaction of 3$\beta$-acetoxy-11-oxo-5$\alpha$-pregnane-20-carboxamide (**14**) (1 mol) with lead tetraacetate (3 mol) and iodine (4 mol) in chloroform at 15 °C for 5 h, under irradiation with UV light through Pyrex, afforded, after alkaline hydrolysis and acetylation of the crude product, 3$\beta$-acetoxy-16$\beta$-hydroxy-11-oxo-5$\alpha$-pregnane-20-carboxylic acid lactone (**15**) (0.55 mol, 55%), MP 265–267 °C.

## 2.3 PREPARATION OF AMIDES AND LACTAMS

Carboxylic amides have been prepared by photochemical addition of formamide to C=C double bonds [9]. The products are the higher homologous amides, and this method could be a synthetic tool for their preparation from unsaturated compounds, as well as the preparation of the corresponding carboxylic acids. The reaction can be induced directly by light (220–250 nm) or initiated photochemically by acetone (>290 nm) with higher chemical yields [10]. It has been found to be applicable to a variety of unsaturated compounds, including terminal, non-terminal and cyclic alkenes, and $\alpha,\beta$-unsaturated esters and acetylenes. This reaction with C=C double bonds is summarized in Table 2.1 [10].

TABLE 2.1

Photoamidation of C=C double bonds

$$R^1{-}CH{=}CH_2 + HCONHR^2 \xrightarrow{h\nu} R^1CH_2CH_2CONHR^2$$

| Substrate | Product | Yield (%) |
|---|---|---|
| 1-Hexene | Heptanamide | 50 |
| 1-Heptene | Octanamide | 61 |
| 1-Octene | Nonanamide | 51 |
| 1-Decene | Undecanamide | 67 |
| Methyl-4-pentenoate | Methyl-5-carbamoylpentanoate | 58 |

When aromatic hydrocarbons were used as the "unsaturated" compounds is this reaction, photosubstitution proceeded to give the corresponding amides [11].

$$C_6H_6 + HCONH_2 \xrightarrow[\text{acetone}]{h\nu} C_6H_5CONH_2 \quad 15\%$$

$$C_{10}H_8 + HCONH_2 \xrightarrow[\text{acetone}]{h\nu} C_{10}H_7CONH_2 \quad 20\%$$

$$C_6H_5Me + HCONH_2 \xrightarrow[\text{acetone}]{h\nu} C_6H_5CH_2CONH_2 \quad 23\%$$

Some amides have been prepared by photolysis of diazoketones in the presence of amines [12]. The diazoketones (**16**) were prepared from the acid and photochemically converted into the *N*-methylanilides of homologous acids in the presence of *N*-methylaniline as a result of the Wolff rearrangement of (**16**) in the Arndt–Eistert reaction.

$$RCO_2H \longrightarrow \underset{(\mathbf{16})}{RC(=O)-CHN_2} \xrightarrow[\text{PhNHMe}]{h\nu} RCH_2-C(=O)-N(Me)-Ph \xrightarrow{LiAlH_4} RCH_2CHO$$

Another photochemical preparation of amides involves the photoisomerization of nitrogen compounds. Aldoximes and nitrones are good precursors for this reaction. Irradiation of an arylaldoxime in a variety of solvents leads to the formation of amides. Thus benzaldehyde oxime (**17**) gave benzamides (**18**) in 41% yield upon irradiation in acetic acid [13].

$$\underset{(\mathbf{17})}{C_6H_5CH=NOH} \xrightarrow{h\nu} \underset{(\mathbf{18})}{C_6H_5CONH_2}$$

Analogously, nitrones undergo a variety of photochemical rearrangements involving intermediate oxaziridine formation to give finally amides. Irradiation of $\alpha$-benzoyl-$\alpha$,*N*-diphenylnitrone (**19**) in anhydrous ether or benzene gave an almost quantitative yield of *N*-phenyldibenzamide (**21**) via the intermediate oxaziridine (**20**) [14]. Diaryloxaziridines are very unstable and break down thermally. When *N*,$\alpha$-diphenylnitrone (**22**) was irradiated in absolute ethanol, a 53% yield of *N*,*N*-diphenylformamide (**25**) was

$$\mathbf{(19)}/\mathbf{(22)} \underset{\Delta}{\overset{h\nu}{\rightleftarrows}} [\mathbf{(20)}/\mathbf{(23)}] \xrightarrow{\Delta} \mathbf{(21)}/\mathbf{(24)}; \quad \mathbf{(25)}$$

(**19**) R = COPh (**20**) R = COPh (**21**) R = COPh (**25**) R = H
(**22**) R = H (**23**) R = H (**24**) R = H

obtained in addition to 5% of benzanilide (**24**) [15]. When acetone was used as solvent, the ratio between the two products was reversed, (**24**) being the main product (75%) accompanied by (**25**) (5%).

Related to this photochemical preparation of amides from nitrones are many examples of the photochemical synthesis of lactams such as quinolones and isoquinolones by photolysis of the corresponding *N*-oxides [16, 17]. However, these syntheses are beyond the scope of this book since they are photochemical transformation reactions of heterocyclic compounds.

## REFERENCES

1. J. Griffiths and H. Hart, *J. Am. Chem. Soc.* **90**, 5296 (1968).
2. D. H. R. Barton and G. Quinkert, *J. Chem. Soc.* **1960**, 1.
3. O. L. Chapman and L. F. Englert, *J. Am. Chem. Soc.* **85**, 3028 (1963).
4. M. H. Fisch and J. H. Richards, *J. Am. Chem. Soc.* **85**, 3029 (1963).
5. W. G. Dauben, D. A. Lightner and W. K. Hayes, *J. Org. Chem.* **27**, 1897 (1962).
6. G. O. Schenck and R. Steinmetz, *Naturwissenschaften* **22**, 514 (1968).
7. D. H. R. Barton and A. J. Beckwith, *Proc. Chem. Soc.* **1963**, 335.
8. D. H. R. Barton, A. J. Beckwith and A. Goosen, *J. Chem. Soc.* **1965**, 181.
9. D. Elad, *Chem. Ind.* **1962**, 362.
10. D. Elad and J. Rokach, *J. Org. Chem.* **29**, 1885 (1964).
11. D. Elad, *Tetrahedron Lett.* **1963**, 77.
12. F. Weygand and H. T. Bestmann, *Chem. Ber.* **92**, 528 (1959).
13. J. H. Amin and P. de Mayo, *Tetrahedron Lett.* **1963**, 1585.
14. A. Padwa, *J. Am. Chem. Soc.* **87**, 4365 (1965).
15. J. S. Splitter and M. Calvin, *J. Org. Chem.* **30**, 3427 (1965).
16. C. Kaneko, A. Yamamoto and M. Gumi, *Heterocycles* **12**, 227 (1979) and references therein.
17. T. Tsuchiya, *Yakugaku Zasshi* **103**, 373 (1983).

# –3–

# Preparation of Aromatic Hydroxy Ketones by Photo-Fries Rearrangement

As special aromatic ketones, *o*- or *p*-hydroxyaromatic ketones can be photochemically prepared by the "photo-Fries reaction" from readily available acyloxybenzenes [1]. The scope of the photo-Fries reaction has been thoroughly investigated and many examples are known. The following reactions are typical:

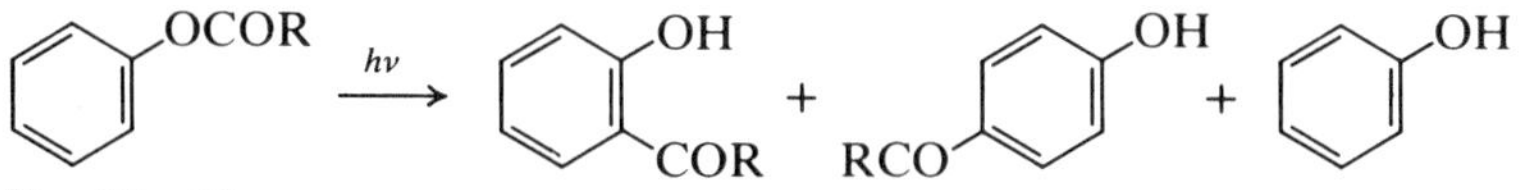

R = Me, Ph

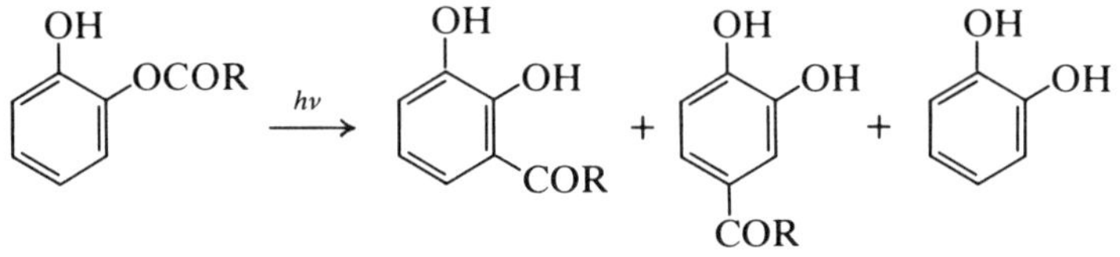

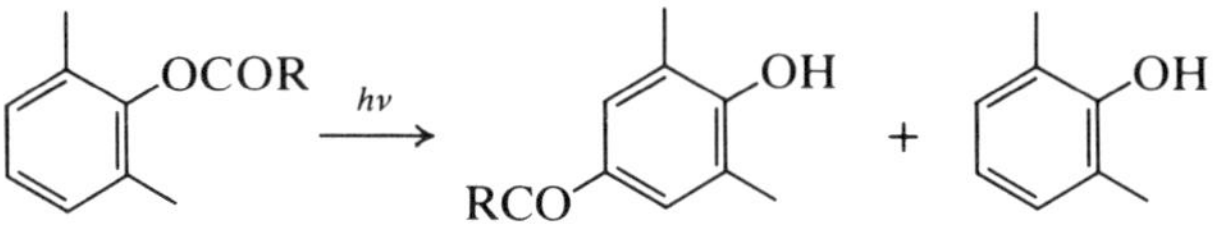

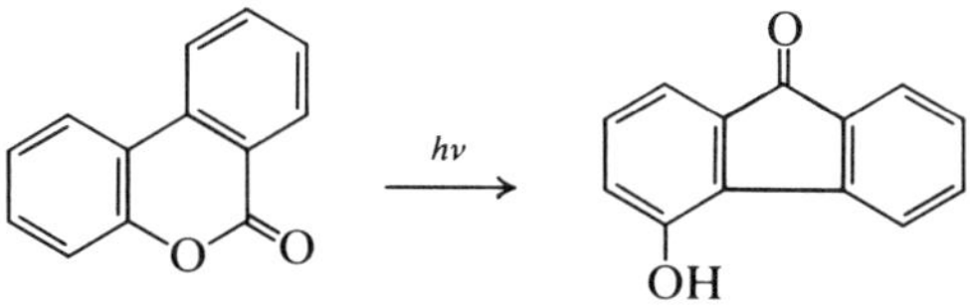

In many cases nearly quantitative total yields of various products are achieved, i.e. of the phenol and the *o*- and *p*-hydroxyketones, with the yields of each individual product varying by up to 50%. The ratio of *ortho* to *para* isomers thus formed also varies, but the *para* isomer usually predominates. Exceptional cases are known where only the *para* isomer appears in the reaction mixture. Since there are too many examples of photo-Fries reactions to describe each case, a summary of typical reactions is given in Table 3.1.

TABLE 3.1

Photo-Fries reaction

| Substrate ArOOCX | | | | | |
|---|---|---|---|---|---|
| Ar | X | *o*-Isomer (%) | *p*-Isomer (%) | Phenol (%) | Recovered starting material (%) |
| Ph | Me | 19 | 15 | 28 | 0 |
| Ph | Ph | 20 | 28 | 14 | — |
| *p*-Bu$^t$C$_6$H$_4$ | Ph | 54 | 0 | — | 17 |
| *p*-Bu$^t$C$_6$H$_4$ | *p*-Bu$^t$C$_6$H$_4$ | 60 | — | — | 21 |
| *p*-Bu$^t$C$_6$H$_4$ | *p*-ClC$_6$H$_4$ | 74 | — | — | 35 |
| *p*-Bu$^t$C$_6$H$_4$ | 3,4-(Cl)$_2$C$_6$H$_3$ | 58 | — | — | 37 |
| *p*-Bu$^t$C$_6$H$_4$ | *p*-(NC)C$_6$H$_4$ | 71 | — | — | 32 |
| *p*-Bu$^t$C$_6$H$_4$ | –Naphthyl | 67 | — | — | 34 |

## 3.1 SCOPE AND LIMITATIONS

The photo-Fries rearrangement has provided a convenient method for the synthesis of *o*- and *p*-hydroxyaromatic ketones. The yields are quite respectable in many cases. The ratio of *ortho*- and *para*-substituted products formed can be controlled within limits by varying the temperature. Generally the *para* isomer predominates at low temperatures, whereas high temperatures lead to the formation of a greater percentage of the *ortho* isomer.

Alkyl-substituted aromatics, in particular polyalkyl aromatics, undergo alkyl elimination, transfer and migration under the acidic conditions.

## 3.2 REACTION MECHANISM

Two mechanisms have been proposed for the photo-Fries rearrangements, both of which are consistent with the intramolecular nature of the reaction. Anderson and Rees [2, 3] proposed cyclic intermediates:

O$_2$CR $\xrightarrow{h\nu}$ (1) + (2)

(1) → *o*-hydroxyketone

(2) → *p*-hydroxyketone

Intermediate (**1**) is strained owing to ring size and loses aromaticity, thus diminishing the attraction of this mechanism. On the other hand, intermediate (**2**) has no ring strain, but the carbon atoms are displaced out of the plane of the aromatic ring. Another mechanism proposed by Kobsa [4] involves homolytic cleavage of the carbonyl-ether oxygen of the ester, with subsequent rearrangement or dissociation:

$O_2CR$ —hν→ [ Ȯ •COR —a→ O, COR, H + O, COR, H ]

solvent | b → phenol; polymer; o-hydroxyketone; p-hydroxyketone

The primary process is the absorption of a photon by the ester, followed by dissociation of the latter into radicals within a solvent cage. Recombination of radicals before diffusion from the cage allows the reformation of the starting compound and the rearranged products. Diffusion from the solvent cage permits the formation of cleavage products. The phenol radical abstracts a hydrogen atom from the solvent and forms the phenolic products.

## 3.3 SUBSTITUENT EFFECTS

When alkyl groups are present in the *ortho* and *para* positions of the phenol portion of the esters, they effectively prevent further attack at these positions by rearranging groups [4]. For example, irradiation of *p*-t-butylphenyl benzoate (**3**) gave the *o*-hydroxyketone (**4**) as the sole product. In contrast with this photochemical reaction, treatment of 4-t-butylphenyl benzoate (**3**) with $AlCl_3$ gave 4-hydroxybenzophenone (**5**) as the major product.

$Bu^t$—C₆H₄—$O_2C$—C₆H₅ (**3**) —hν→ OH, —CO—, $Bu^t$ (**4**)

(**3**) —$AlCl_3$→ HO—C₆H₄—CO—C₆H₅ (**5**)

In all known photo-Fries rearrangements of alkyl-substituted phenol esters, the alkyl groups remain intact and undisturbed.

## 3.4 SOLVENT EFFECTS

Solvent effects remain to be studied in detail. Variation of the solvent in the series dioxane, cyclohexane, isopropanol and benzene for the irradiation of *p*-tolylphenyl benzoate caused respective increases in yield of the order of 20, 25 and 53% of rearranged product [5]. Although the preliminary results of two other reactions indicate that benzene is the preferred solvent for these reactions, in general, protic solvents such as t-butanol are generally recommended for the rearrangement, and in cyclohexane the formation of phenol predominates.

## 3.5 WAVELENGTH EFFECTS

Variation of the alcohol portion of the ester dramatically affects the wavelength of absorption and hence the formation of products. For example, β-napthylfluorene-9-carboxylate was readily photolysed in Pyrex vessels, whereas the corresponding *p*-cresol derivative was unaffected under the same conditions. Furthermore, methyl and cholesteryl fluorene-9-carboxylates were stable in both Pyrex and quartz vessels. Thus it is clear that the photochemical reaction depends upon light absorption by the aryloxy moiety.

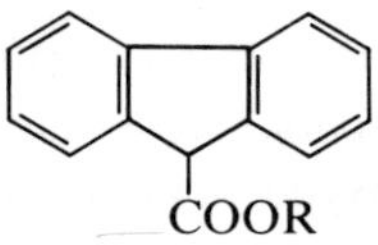

COOR

R = β-napthyl
*p*-methylphenyl
methyl
cholesteryl

## 3.6 SYNTHETIC APPLICATIONS

In the total synthesis of (±)-griseofulvin (**7**) the photo-Fries rearrangement was employed as a key step for the construction of the benzophenone system (**6**) [6].

Ishii *et al.* [7] have established the stereostructure of a new coumarin, rutaetin methyl ether (**8**), by preparing its degradation product (**9**) using the photo-Fries rearrangement of the diacetate (**10**) as a key step.

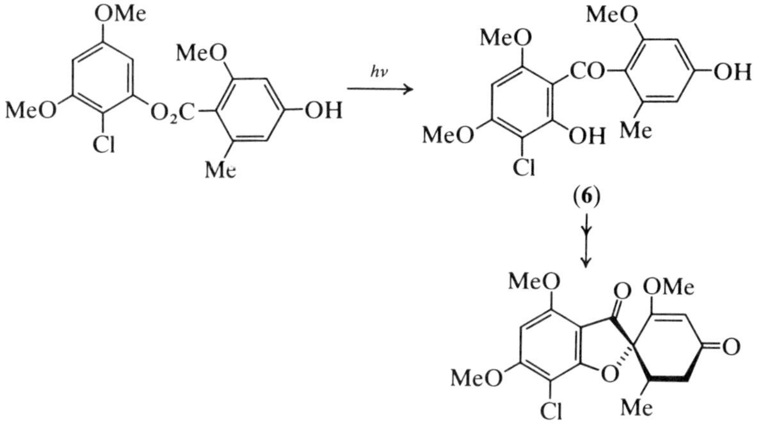

OMe MeO MeO $O_2C$ OH Cl Me $h\nu$ MeO MeO CO OH MeO OH Me Cl

(6)

MeO O OMe O MeO O Cl Me

(7)

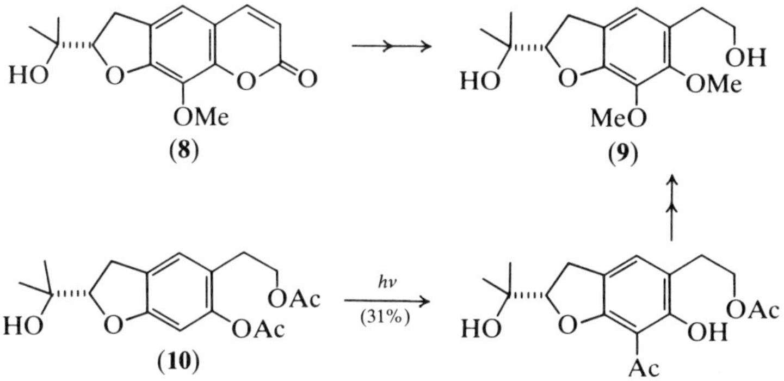

HO O O O OMe

(8)

HO O OH OMe MeO

(9)

HO O OAc OAc

(10)

$h\nu$ (31%)

HO O OAc OH Ac

## 3.7 RELATED PHOTOCHEMICAL REARRANGEMENTS

### 3.7.1 Carbonate esters

Irradiation of the carbonate ester (**11**) in either acetic acid, alcohol or ether led to a photo-Fries rearrangement giving a mixture of two hydroxyesters (**12**) and (**13**) and phenol [8, 9].

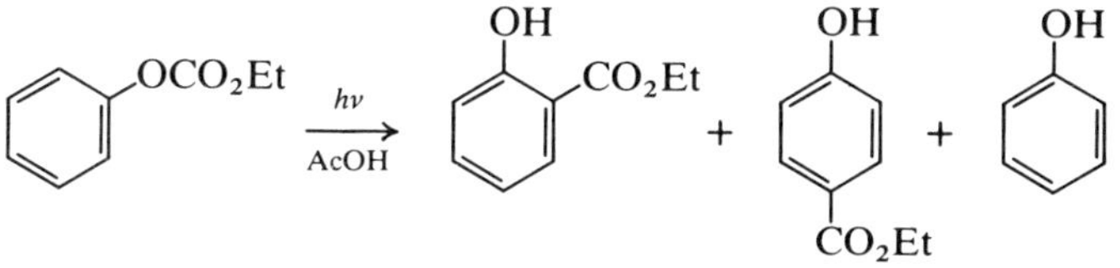

$OCO_2Et$ $\xrightarrow[AcOH]{h\nu}$ OH $CO_2Et$ + OH $CO_2Et$ + OH

(**11**) (**12**) (16%) (**13**) (25%) (7–8%)

### 3.7.2 Vinyl esters

Substitution of the vinyl group for a phenyl group in the alcohol portion of the ester in the previously mentioned photo-Fries rearrangement has given fruitful results in the preparation of diketones, although the yields are not so high [10, 11]. See Table 3.2.

TABLE 3.2

Vinyl ester photochemical reactions

$$ArCO_2\overset{|}{C}=\underset{|}{\overset{|}{C}} \xrightarrow{h\nu} -\overset{O}{\overset{\|}{C}}-\underset{|}{\overset{|}{C}}-COAr$$

$$PhCO_2CH=CH_2 \xrightarrow{h\nu} OHCCH_2COPh \xrightarrow[-CO]{h\nu} MeCOPh$$

| Substrate | Products (Yield, %) |
|---|---|
| Vinyl benzoate | Benzoic acid (10–15%), acetophenone (2–4%) |
| Isopropenyl benzoate | α-Benzoylacetone |
| Cyclohexen-1-yl benzoate | 1-Benzoylhex-5-en-2-one |
| 6-Methylcyclohex-1-yl benzoate | 1-Benzoyl-3-methylhex-5-en-2-one |
| Cholest-2-en-3-yl benzoate | 2-Benzoylcholestan-3-one |
| Androst-16-en-3β,17β-diol-3-acetate-17 benzoate | 16-Benzoylandrostan-3β-ol-17-one acetate |

### 3.7.3 Anilides

Owing to their structural similarity to phenyl esters, the photochemistry of anilides (**14**) has been studied and now provides a synthetic method for the aminoaromatic ketones (**15**) and (**16**) [12, 13]. (See Table 3.3.)

Acet-, propion-, butyr- and benz-anilide all produced *o*- and *p*-aminoketones on irradiation. The yields of the two isomers, as in the ester irradiation, are nearly equal in proportion in the reactions studied so far.

NHCOR (**14**) $\xrightarrow{h\nu}$ RCO, NH$_2$ (**15**) + NH$_2$, COR (**16**) + NH$_2$

TABLE 3.3
Anilide photochemical reactions

| | *o*-Isomer (**15**) (%) | *p*-Isomer (**16**) (%) | Aniline (%) | Recovered starting material (%) |
|---|---|---|---|---|
| Acetanilide (EtOH) | 20 | 25 | 18 | 38 |
| Propionanilide (EtOH) | 22 | 25 | 17 | 10 |
| Butyranilide (EtOH) | 17 | 23 | 20 | 0 |
| Benzanilide (EtOH) | 14 | 12 | Trace | 68 |

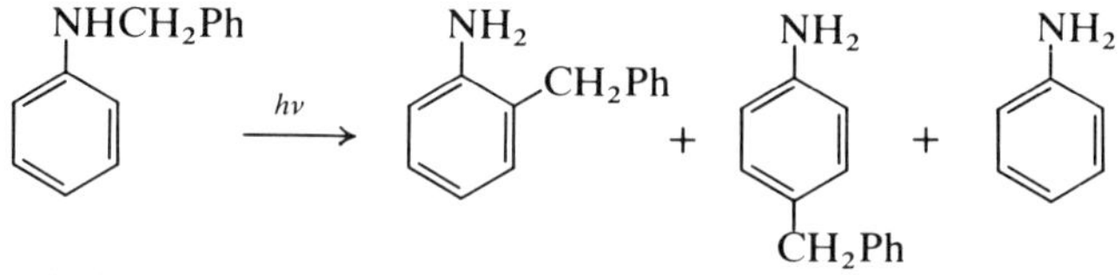


(**17**) (29%) (14%) (6%)

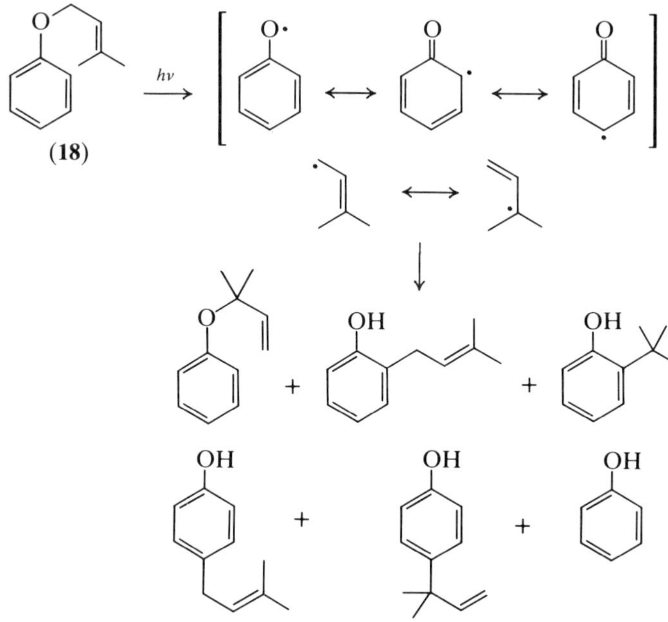


(**18**)

Finally, as a related photochemical rearrangement of the anilide series, *N*-benzylaniline (**17**) and the allyl aryl ether (**18**) were both subjected to irradiation to give a mixture of the photorearrangement products [14].

**Experimental 3.1** 5-t-Butyl-2-hydroxybenzophenone [4]

A solution of *p*-t-butylphenyl benzoate (**3**) (12.72 g, 0.05 mol) in benzene (1 l) was irradiated internally through a filter solution, containing nickel sulphate hexahydrate (600 g $l^{-1}$) and copper sulphate heptahydrate (100 g $l^{-1}$), with a mercury arc (Hanovia Chemical and Manufacturing Co.) until 0.316 einstein of energy was absorbed. After irradiation, the solvent was evaporated under reduced pressure at room temperature and the crude mixture was chromatographed on a 400 × 50 mm column of alumina (Woelm, neutral, activity grade III). Elution with 3 l of benzene–petroleum ether (1 : 4) gave 2.10 g of white crystals which were identified as unchanged material by their IR spectrum. Further elution with ethyl ether yielded 5.78 g (45%) (54% based on the consumed starting material) of a yellow 5-t-butyl-2-hydroxybenzophenone (**4**) which, after distillation, had a melting point of 67 °C.

## REFERENCES

1. V. I. Stenberg, in *Organic Photochemistry* (ed. O. L. Chapman), Vol. 1, p. 127. Marcel Dekker, New York, 1967.
2. J. C. Anderson and C. B. Rees, *Proc. Chem. Soc.* **1960**, 217.
3. J. C. Anderson and C. B. Rees, *J. Chem. Soc.* **1963**, 1781.
4. H. Kobsa, *J. Org. Chem.* **27**, 2293 (1962) and references therein.
5. R. A. Finnegam and D. Knutson, *Chem. Ind.* **1965**, 1837.
6. D. Taub, C. H. Kuo, H. L. Slates and N. L. Wendler, *Tetrahedron* **19**, 1 (1963).
7. H. Ishii, F. Sekiguchi and T. Ishikawa, *Tetrahedron* **37**, 285 (1981).
8. C. Pac and S. Tsutsumi, *Bull. Chem. Soc. Jpn* **37**, 1392 (1964).
9. W. H. Horspool and P. L. Pauson, *J. Chem. Soc.* **1965**, 5162.
10. R. A. Finnegan and A. W. Hagen, *Tetrahedron Lett.* **1963**, 365.
11. M. Feldkinel-Gorodetsky and Y. Mazur, *Tetrahedron Lett.* **1963**, 369.
12. D. Elad, *Tetrahedron Lett.* **1963**, 873.
13. D. Elad, D. V. Rao and V. I. Stenberg, *J. Org. Chem.* **30**, 3252 (1965).
14. J. E. Herweh and C. E. Hoyle, *J. Org. Chem.* **45**, 2195 (1980).

# – 4 –

# Preparation of Olefins, Peroxides and Hydroperoxides

## 4.1 PREPARATION OF OLEFINS

Although there is no photochemical reaction that forms olefins directly, some compounds such as *cis*-olefins that are unobtainable by thermal reactions can be prepared indirectly from the *trans* isomers by photochemical isomerization.

### 4.1.1 Preparation of *cis*-stilbenes

Upon irradiation in solution, *trans*-stilbene is known to isomerize to the corresponding *cis*-isomer and vice versa. Under prolonged irradiation by light of 313 nm wavelength an equilibrium mixture of 93% of *cis* and 7% of *trans* isomers was obtained from both *cis*- and *trans*-stilbenes [1].

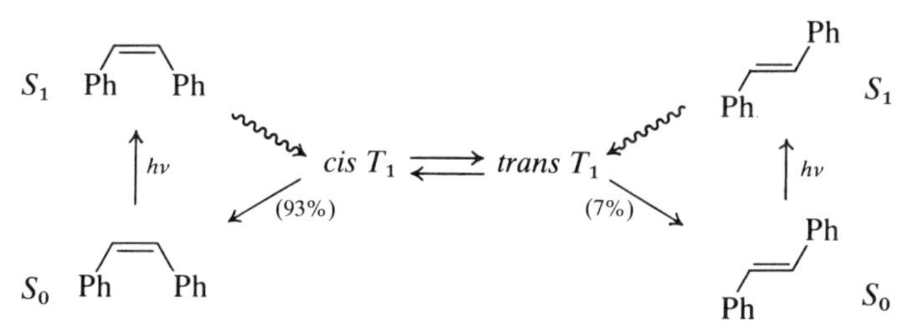


Direct irradiation by 366 nm light does not lead to photochemical isomerization, but in the presence of a small amount of sensitizer such as benzophenone the isomerization proceeded very smoothly to give a mixture of *cis*- and *trans*-stilbenes in a ratio depending upon the triplet energy of the sensitizer used [2].

When the photoisomerization of *trans*-stilbenes was carried out in the presence of an oxidant such as iodine, oxidative photocyclization of the resulting *cis*-stilbenes proceeded smoothly to give phenanthrenes as described in Section 6.4.1.1 [3–5].

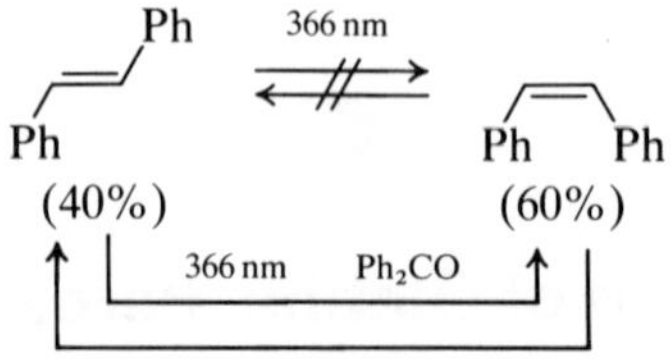

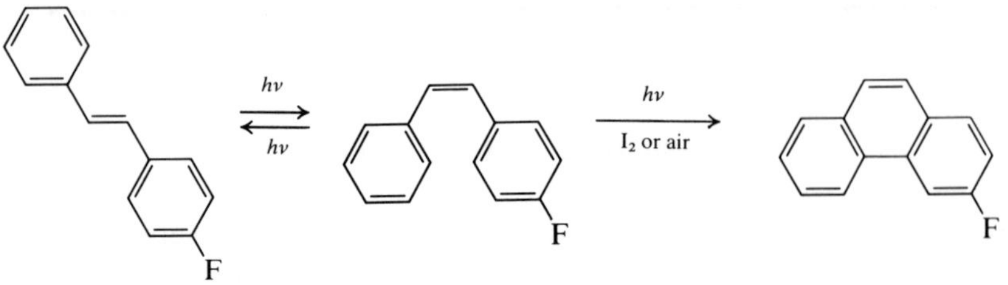

Ph 366 nm Ph Ph
(40%) (60%)
366 nm $Ph_2CO$

hν hν hν $I_2$ or air
F F F

### 4.1.2 Preparation of cyclic *trans*-olefins

Small- and medium-sized cyclic olefins exist in the *cis* form because the *trans* form is unstable owing to the ring strain. However, larger-sized (>8-membered) rings can exist predominantly in the *trans*-olefinic form. Thus *trans*-cyclooctene (**2**) was prepared by irradiation of the *cis* isomer (**1**) in the presence of xylene in over 97% purity [3–5]. Similarly, *cis*,*trans*-cyclodeca-1,5-diene (**3**) and *cis*,*trans*,*trans*-1,5,9-cyclodecatriene (**5**) gave the corresponding *trans* isomers (**4**) and (**6**) and (**7**) respectively [6, 7]. In the latter case, a systematic study by using sensitizers was carried out as shown in Table 4.1.

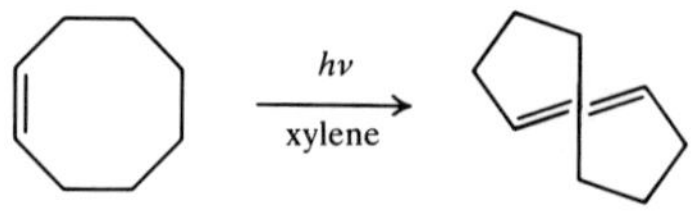

hν xylene

(**1**) (**2**)

(**3**) (1 : 4) (**4**)

hν hν +

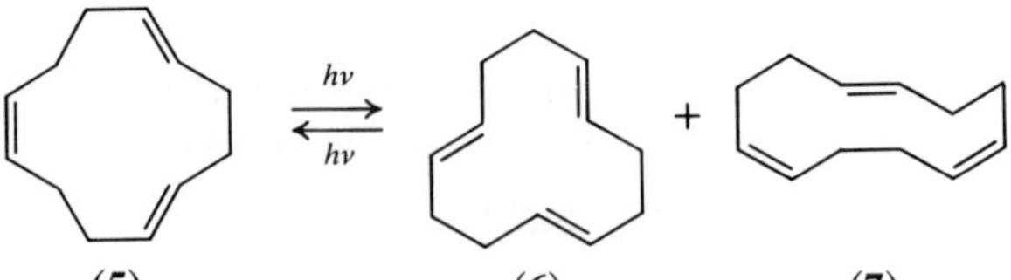

(**5**) (**6**) (**7**)

TABLE 4.1
Isomerization of cyclodecatriene (**5**)

| Sensitizer | (**5**) (%) | (**6**) (%) | (**7**) (%) |
|---|---|---|---|
| Dark | 97 | 3 | 0 |
| No sensitizer | 93 | 4 | 3 |
| Acetone | 41 | 12 | 47 |
| Cyclododecanone | 65 | 9 | 26 |
| Acetophenone | 52 | 39 | 9 |
| Triphenylene | 97 | 3 | 0 |

Irradiation of a pentane solution of *cis,cis*-1,3-cyclooctadiene (**8**) in the presence of acetophenone at 10–15 °C followed by treatment with 20% silver nitrate gave the silver complex of *cis,trans*-1,3-cyclooctadiene (**9**) in 80% yield, which was treated with concentrated ammonium hydroxide to afford the *cis,trans*-olefin (**9**) in 60–70% yields [8].

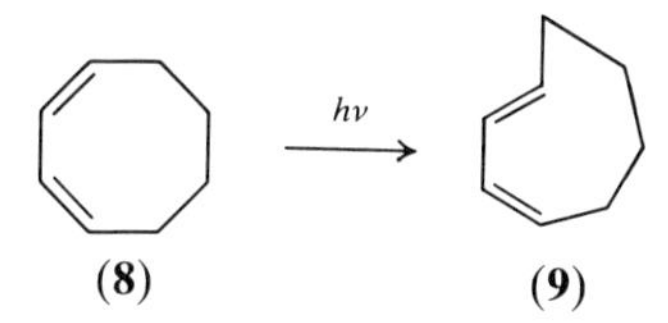


### 4.1.3 Preparation of cyclic *trans*-enones

Although cyclopentanone and cyclohexenone did not undergo photoisomerization, irradiation of *cis*-cyclooctenone (**10**) at −78 °C gave a mixture of *cis*- and *trans*-enones (**11**) [9]. However, cycloheptenone (**12**) gave a mixture of the photoisomerized *trans*-enone (**13**), other photoproducts such

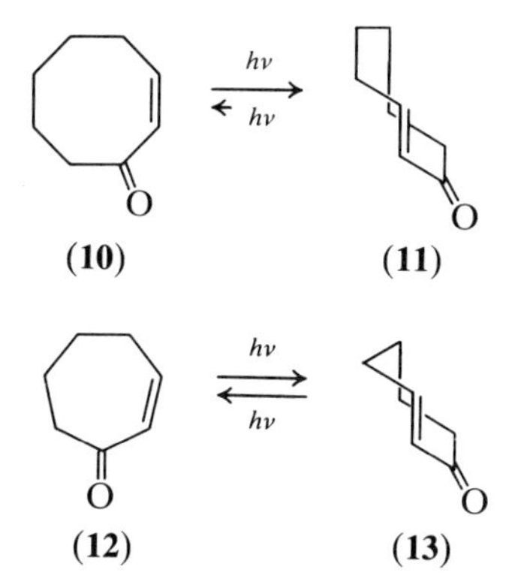

as photodimerized products, and adducts with other molecules as a result of the photochemical reactivity of the formed *trans*-enone (**13**) [10].

### 4.1.4 Synthetic applications

As examples of synthetic applications of photochemical isomerization of cyclic olefins, humulene (**15**) and isocaryophyllene (**17**) were synthesized from the corresponding olefins (**14**) and (**16**) [11].

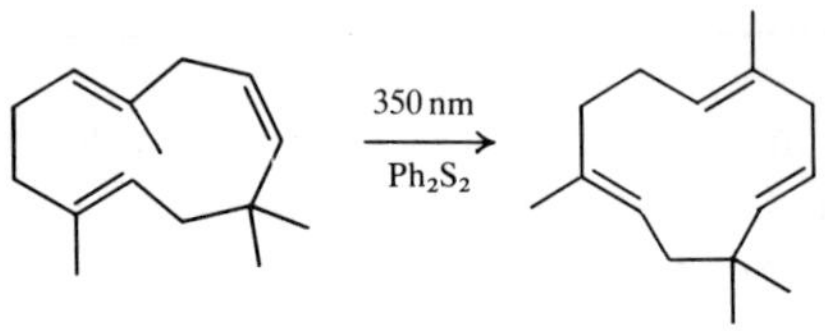

350 nm, $Ph_2S_2$

(**14**) (**15**)

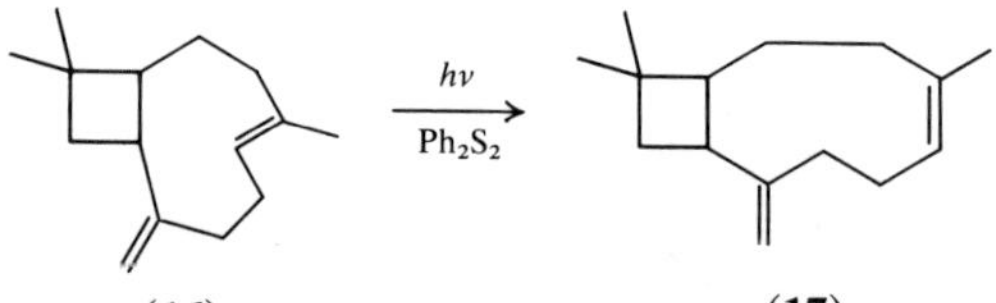

$h\nu$, $Ph_2S_2$

(**16**) (**17**)

*cis–trans* Photochemical isomerization has been of considerable use in the synthesis of various types of unsaturated compounds such as ionone (**18**) [12] and has also played an important role in the photochemistry of the visual process and in the chemistry of vitamin A (**19**), as exemplified by the formation of all-*trans*-rhodopsin by regioselective photochemical isomerization of 11-*cis*-rhodopsin [13].

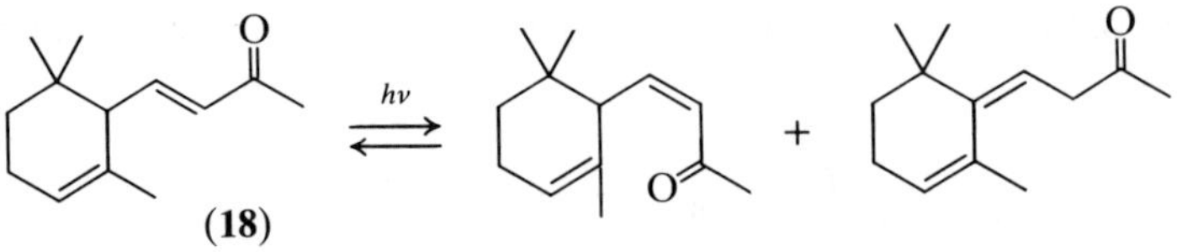

$h\nu$

(**18**)

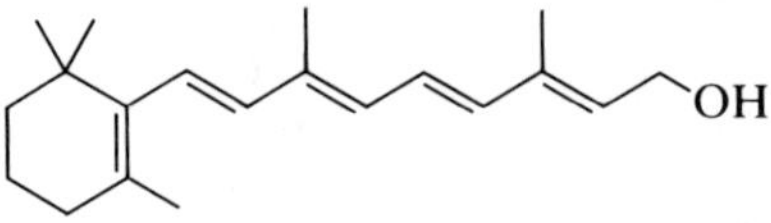

(**19**)
vitamin A

Another example of photochemical synthesis by contrathermodynamic isomerization, is the formation of the exocyclic double bond (**21**) by irradiation of 1-methylcyclohexene (**20**) in non-nucleophilic protic media [14].

**Experimental 4.1** Photoisomerization of (–)-caryophyllene to (–)-isocaryophyllene in the presence of diphenyl sulphide [11]

A mixed solution of (−)-caryophyllene (**16**) (100 g) and diphenyl sulphide (1 g) in absolute benzene (170 ml) was irradiated internally with a high pressure mercury lamp (Philips, HPK, 125 W) in a water-cooled (17–19 °C) immersion apparatus through which nitrogen was bubbled. Gas chromatography (AEROGRAPH, Model 700 Autoprep, 5 m × 5 mm glass column; 30% Carbowax 20M on Chromosorb) showed that the starting caryophyllene (**16**) disappeared at the rate of 3.5 g $min^{-1}$. After 120 min, (**16**) ($R_t$ = 5.1 min) was converted completely into (**17**) ($R_t$ = 4.6 min). Distillation of the reaction mixture to remove the excess of diphenyldisulphide gave the crude product (**17**) which was dissolved in n-hexane and stirred in the presence of alumina (activity 1) (10 g) at room temperature overnight. Filtration and distillation of the residue obtained afforded pure (−)-isocaryophyllene (**17**) (94.8 g), BP 188 °C at 12 mmHg.

## 4.2 PREPARATION OF PEROXIDES AND HYDROPEROXIDES

As is evident from biochemistry, hydroperoxides and peroxides are attractive synthetic intermediates for various types of compounds. Photochemical preparations of hydroperoxides and peroxides have been accomplished by two oxidative methods: (i) by unsensitized photochemical oxidation by the direct action of oxygen; and (ii) by photosensitized oxygenation ($^1O_2$). The former method, however, is of little practical importance.

There is an excellent review of organic synthesis by photosensitized oxygenation [15].

### 4.2.1 Preparation of peroxides

Irradiated cyclic 1,3-dienes readily add one molecule of oxygen, although extra sensitizer is generally necessary to effect the oxygenation. The reaction may formally be regarded as a diene synthesis with oxygen as dienophile [16].

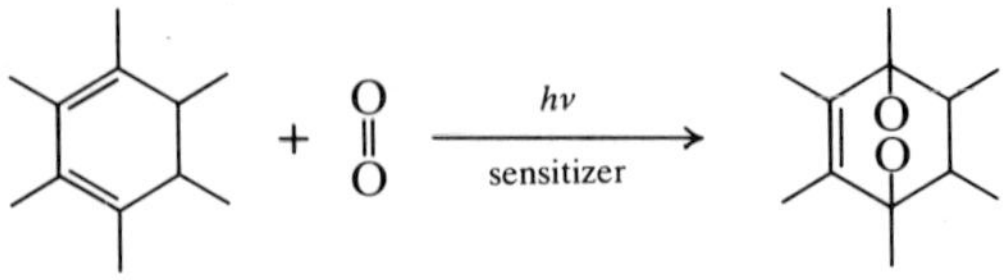

hv

sensitizer

### 4.2.1.1 Preparation of endoperoxides from alicyclic 1,3-dienes

Schenck and Ziegler [17] have succeeded in preparing ascaridole (**23**) from $\alpha$-terpinene (**22**) by irradiation of the 1,3-diene (**22**) in the presence of oxygen and chlorophyll as sensitizer. Similarly, various types of 1,3-dienes (**24**), (**26**) and (**28**) are oxidized by singlet oxygen to give the endoperoxides (**25**), (**27**) and (**29**) as shown.

In the steroid field, photochemical preparation of endoperoxides is a potential synthetic method for the following modified steroidal derivatives [21]:

### 4.2.1.2 Preparation of endoperoxides from heterocyclic cyclopentadienes

Photooxidation of heterocyclic cyclopentadienes containing nitrogen (pyrrole) and oxygen (furan) have received much attention. The latter case

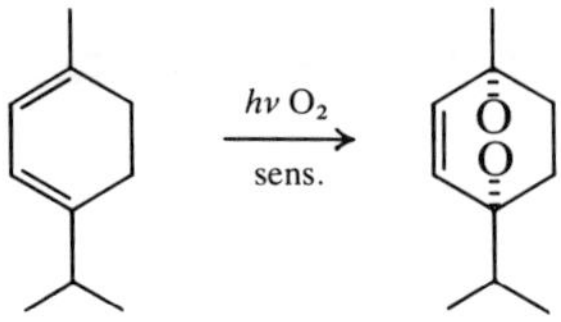

hv $O_2$ sens.

**(22)** **(23)**

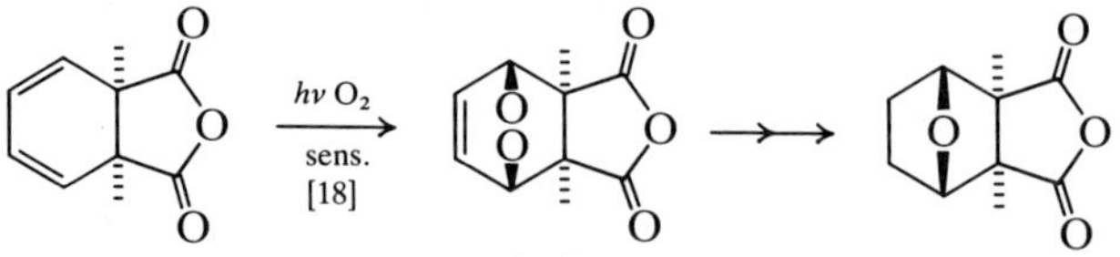

hv $O_2$ sens. [18]

**(24)** **(25)**

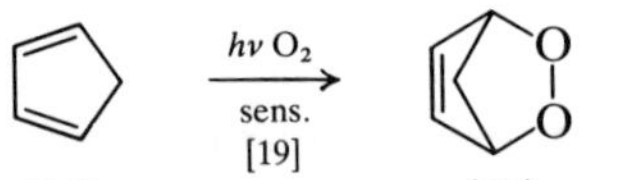

hv $O_2$ sens. [19]

**(26)** **(27)**

hv $O_2$ sens. [20]

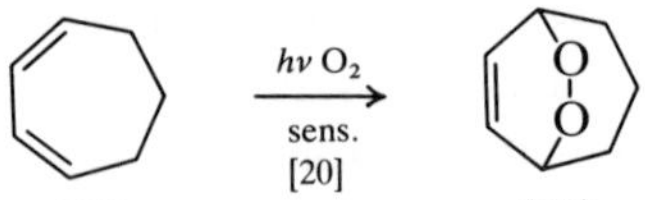

**(28)** **(29)**

is of particular significance since the endoperoxides are good synthetic precursors of 1,4-diketo compounds. Irradiation of an aerated dilute solution of pyrrole containing dyes gave (**31**), probably via the endoperoxide (**30**). Manganese dioxide oxidation of (**31**) gave maleimide (**32**) [22].

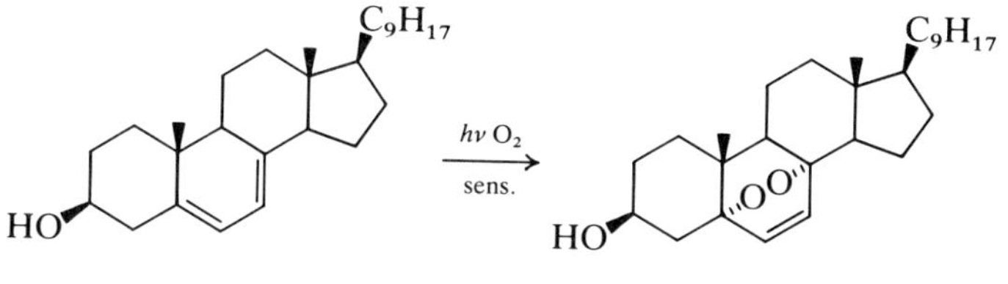

$C_9H_{17}$ HO $\xrightarrow[\text{sens.}]{h\nu\ O_2}$ $C_9H_{17}$ HO

Recent work on the photosensitized oxidation of furan derivatives in solution has revealed that an ozonide-type peroxide is initially formed in what may be regarded as a diene synthesis with oxygen [23]. The endoperoxides (**33**) are stable only at sufficiently low temperatures ($<-90\,°C$). Reduction of the primary peroxide (**33**) leads to ketones or aldehydes (**34**) as shown.

In summary, the endoperoxides formed from furan derivatives are good synthetic precursors for 1,4-diketo compounds and their congeners as shown in the following two examples [24, 25]:

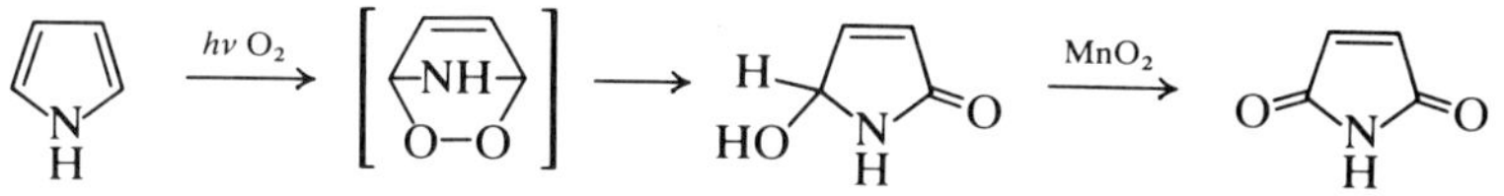

(**30**) (**31**) (**32**)

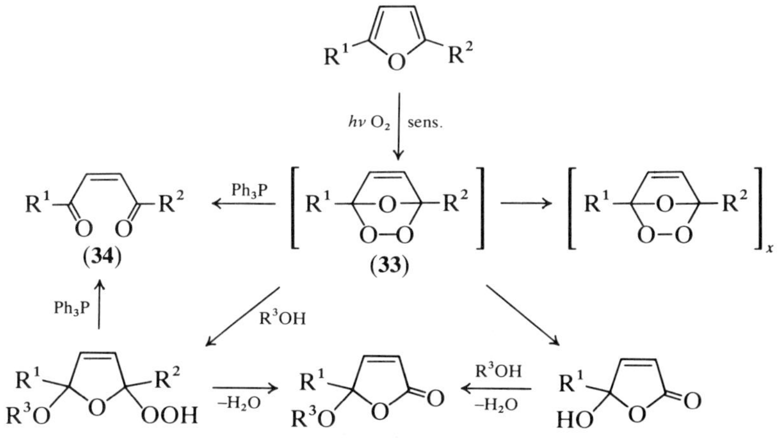

(**34**) (**33**)

(**a**) $R^1 = R^2 = H$; (**b**) $R^1 = Me$, $R^2 = H$; (**c**) $R^1 = R^3 = Me$

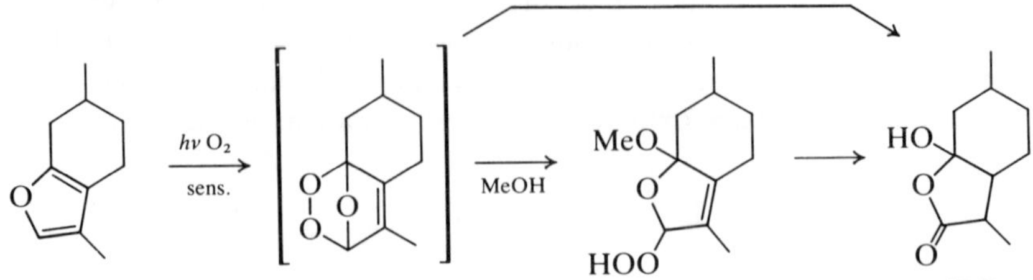

$h\nu$ $O_2$, sens. ; MeOH

[24]

$h\nu$ $O_2$

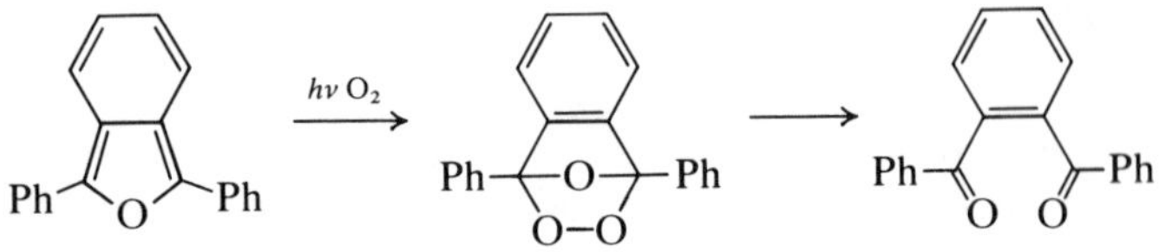

[25]

**Experimental 4.2** 2-Hydroperoxy-5-methoxy-2,5-dimethyl-2,5-dihydrofuran [23]

A solution of 2,5-dimethylfuran (3 g, 0.031 mol) and rose bengal (0.03 g, $3 \times 10^{-5}$ mol) in methanol (250 ml) was irradiated internally with a Sylvania 625-W Sun Gun lamp in a water-cooled immersion apparatus through which oxygen was recirculated. After 5 min, 686 ml (over 90%) of oxygen had been taken up. The solvent was removed with a rotary evaporator, the residue was washed successively with cold diethyl ether and petroleum ether, and the solid residue was sublimed at 63 °C/0.15 mmHg to give 2-hydroperoxy-5-methoxy-2,5-dimethyldihydrofuran (3.6 g, 72%), MP 75–76 °C after recrystallization from diethyl ether.

### 4.2.2 Preparation of hydroperoxides

Preparation of allylic hydroperoxides from unsaturated compounds by reaction with singlet oxygen is called the "ene" reaction and is of significance for synthetic purposes [26, 27]. The ene reaction occurs widely in oxidation by singlet oxygen, which attacks one carbon atom of the double bond; a hydrogen from the allylic position then migrates to oxygen, with concomitant shift of the double bond. This reaction shows the following general features [27]:

(i) the reaction does not proceed with terminal double bonds;
(ii) reactivities vary depending on the structure of the unsaturated compound;
(iii) regio- and stereo-selectivities are observed;

(iv) the allylic hydroperoxides thus formed are good synthetic precursors for various types of compounds, such as allylic alcohols, alkenes, $\alpha,\beta$-unsaturated ketones, and saturated ketones.

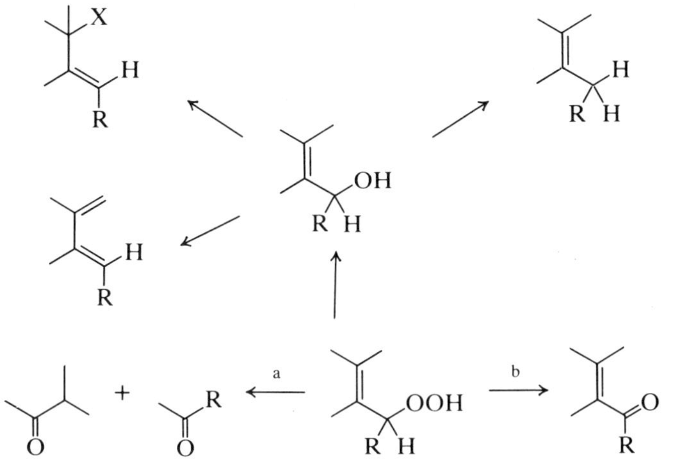

X H R H R H OH R H H R a b + R OOH O O R H O R

The following mechanism has been proposed for the ene reaction [28]:

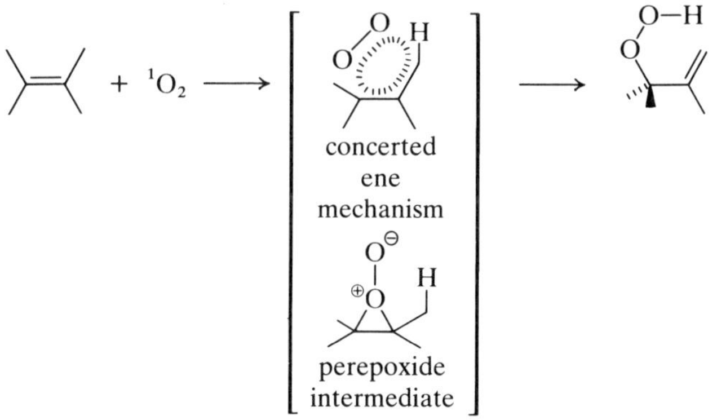

$+ \ {}^1O_2 \longrightarrow$ O O H concerted ene mechanism $O^{\ominus}$ H $^{\oplus}O$ perepoxide intermediate $\longrightarrow$ O–H O

**Experimental 4.3** 2,3-Dimethyl-3-hydroperoxy-1-butene [28]

A solution of 2,3-dimethyl-2-butene (8.4 g, 0.1 mol) and rose bengal (0.5 g, $5 \times 10^{-5}$ mol) in methanol (80 ml) was irradiated internally with a high-pressure mercury lamp (Philips HP-125 W) in a water-cooled (18–20 °C) apparatus through which oxygen uptake was 2.2 l (over 90%). After removal of the solvent at 12 mm Hg (bath below 25 °C), distillation of the liquid residue under reduced pressure gave 2,3-dimethyl-3-hydroperoxy-1-butene (9.5 g, 82%), BP 54–55 °C/12 mmHg.

#### 4.2.2.1 Competitive formation of allylic hydroperoxides and dioxetanes

The ratio of products from the ene reaction of oxygen (hydroperoxide formation) and from 1,2-addition of oxygen (dioxetane formation) has been observed to depend upon the solvent and temperature employed. Vinyl sulphides (**35**) [29], vinyl ethers (**36**) [30] and enamines (**37**) [31] preferentially gave the dioxetanes in protic solvents and also at low temperatures.

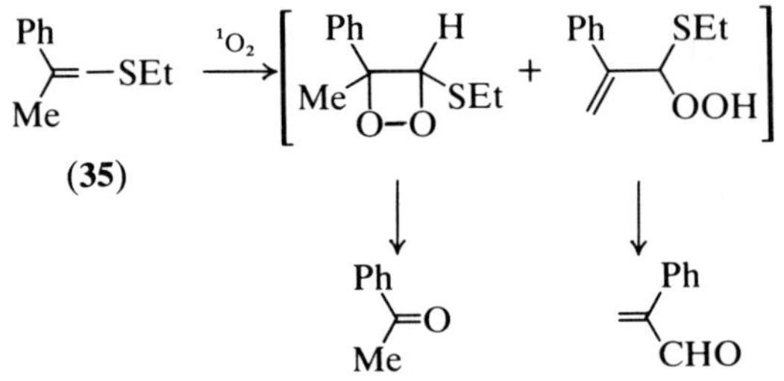

(**35**)

| | °C | % | % |
|---|---|---|---|
| MeOH | 20 | 95 | 5 |
| MeCN | 20 | 27 | 73 |
| $CHCl_3$ | 20 | 21 | 89 |
| $CH_2Cl_2$ | 20 | 21 | 89 |
| | −78 | 88 | 12 |

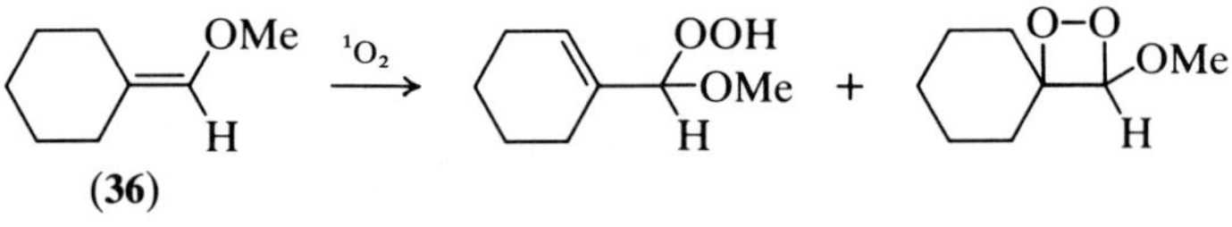

(**36**)

| | °C | % | % |
|---|---|---|---|
| $C_6H_6$ | 20 | 97 | 3 |
| $CH_2Cl_2$ | 20 | 73 | 27 |
| MeOH | 20 | 32 | 58 |
| | −80 | 9 | 54 |

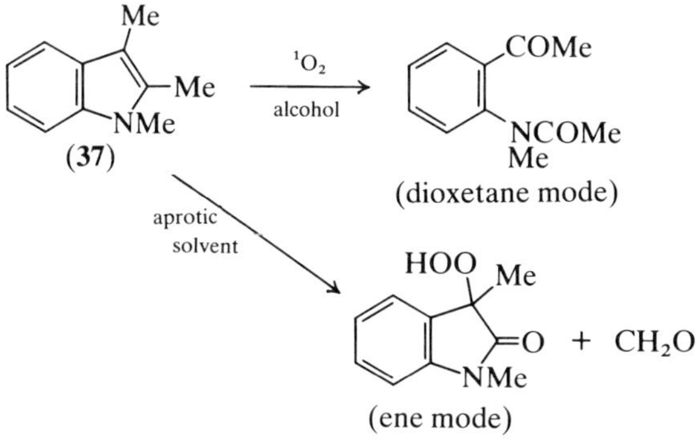

Me
(37) NMe Me
$^1O_2$ alcohol
COMe NCOMe Me
(dioxetane mode)
aprotic solvent
HOO Me =O + $CH_2O$ NMe
(ene mode)

#### 4.2.2.2 Stereoselective formation of allylic hydroperoxides

In contrast with autooxidation, the ene reaction with singlet oxygen proceeds stereospecifically and in a suprafacial manner as follows. Oxygen attacks from one plane of the starting olefinic molecule, and the allylic hydrogen of the same plane is removed [32]. In the following example the 19-methyl group sterically hindered attack of the oxygen molecule from the $\beta$-side, allowing stereospecific $\alpha$-attack of oxygen in the ene reaction and regio- and stereo-selective elimination of the axial hydrogen at the 7-position [33]:

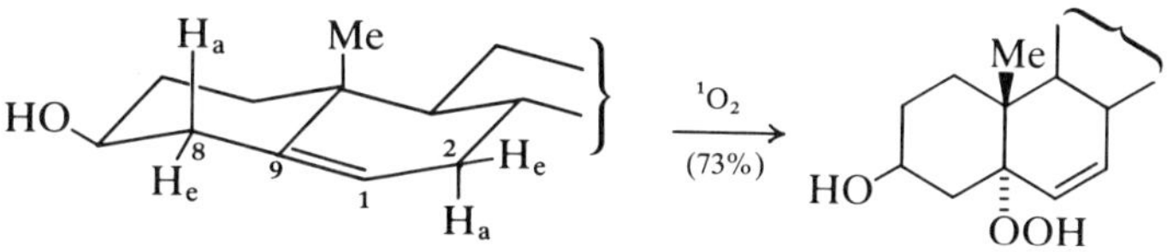

$H_a$ Me HO 8 $H_e$ 9 1 2 $H_e$ $H_a$
$^1O_2$ (73%)
Me HO OOH

#### 4.2.2.3 Regioselectivity in the formation of allylic hydroperoxides

There is no difference in reactivity between methyl and methylene groups, but methine protons are almost inert to this ene reaction owing to steric hindrance [34]. Allylic hydrogen in the *Z*-position in trisubstituted olefins is selectively eliminated [34].

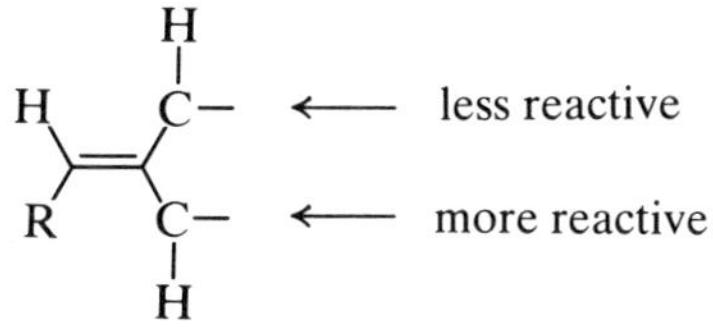

H H C— ⟵ less reactive
R C— ⟵ more reactive H

In polyolefins, the more substituted olefin bond, that is the one of lowest ionic potential, is preferentially oxidized [35]:

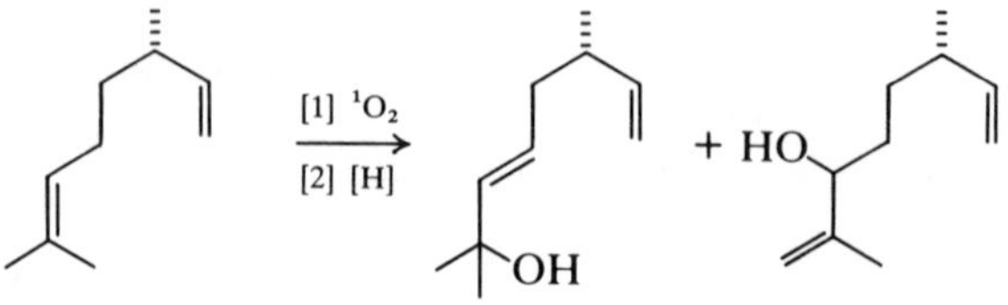

[1] $^1O_2$; [2] [H]

OH (51%) + HO (49%)

### 4.2.3 Preparation of hydroperoxysulphides

Photosensitized oxidation of the $\alpha$-position of heteroatomic compounds gives the corresponding hydroperoxides.

$$R-X-CHR^1R^2 \xrightarrow[X\,=\,\text{hetero atom}]{^1O_2} R-X-C(OOH)R^1R^2$$

Among many heteroatomic compounds, sulphides are the best substrates and give the $\alpha$-hydroperoxides. In 1976 Corey *et al.* [36] found that photosensitized oxidation of the benzyl sulphide (**38**) led to the formation of the aldehyde (**39**) and proposed the persulphoxide (**41**) as a key intermediate.

$$\underset{(\mathbf{38})}{ArCH_2SR} \xrightarrow[5°C]{h\nu/TPPZn/O_2} \underset{(\mathbf{41})}{[ArCH_2\overset{\oplus}{S}(O{-}O^{\ominus})R]} \longrightarrow \longrightarrow [ArCH(OH)S(O)R] \longrightarrow \underset{(\mathbf{39})}{ArCHO} +$$

$$\underset{(\mathbf{40})}{[RSOH]} \xrightarrow[R\,=\,Bu^t]{YCH{=}CHY} RS(O)(Y)CH-CH(H)Y$$

In contrast, Ando and coworkers [37] isolated the $\alpha$-hydroperoxide (**43**) in quantitative yield by photosensitized oxidation of the thiazolidine derivatives (**42**).

(42) → $h\nu$/TPP/$O_2$ (quant.) → (43)

On the basis of the results of a study of substituent and concentration effects, they also proposed the following mechanism [38]:

(43) $R^1$ = COOMe, $R^2$ = COPh

## REFERENCES


1. L. Saltiel, J. D'Agostino, E. D. Megarity, L. Metts, K. R. Neuberger, M. Wrighton and O. C. Zafiriou, in *Organic Photochemistry* (ed. O. L. Chapman), Vol. 3, p. 1. Marcel Dekker, New York, 1973.
2. D. Gegiou, K. A. Muszkat and E. Fischer, *J. Am. Chem. Soc.* **90**, 3907 (1968).
3. F. R. Stermitz, in *Organic Photochemistry* (ed. O. L. Chapman), Vol. 1, p. 247. Marcel Dekker, New York, 1967.
4. E. V. Blackburn and C. J. Timmons, *Q. Rev. Chem. Soc.* **23**, 482 (1969).
5. J. A. Deyrup and M. Betkowski, *J. Org. Chem.* **37**, 3561 (1972).
6. C. W. Mallory and F. B. Mallory, *Org. Photo. Synth.* **1**, 55 (1979).
7. H. Nozaki, Y. Nishikawa, Y. Kamatani and R. Noyori, *Tetrahedron Lett.* **1965**, 2161.
8. P. G. Gassman and E. A. Williams, *Org. Photo. Synth.* **1**, 44 (1979).
9. P. E. Eaton and K. Lin, *J. Am. Chem. Soc.* **86**, 2087 (1964).

10. P. E. Eaton and K. Lin, *J. Am. Chem. Soc.* **87**, 2052 (1965).
11. P. E. Sonnet, *Tetrahedron* **36**, 557 (1980).
12. R. S. H. Liu and A. E. Asato, *Tetrahedron* **40**, 1931 (1984).
13. M. Itoh, in *Chemistry and Biology of Synthetic Retinoids* (ed. M. I. Dawson and W. H. Okamura). CRC Press, Florida, 1989.
14. P. J. Kropp, in *Organic Photochemistry* (ed. A. Padwa), Vol. 4, p. 33. Marcel Dekker, New York, 1979.
15. R. W. Denny and A. Nicken, in *Organic Reactions* (ed. W. G. Dauben), Vol. 20, p. 133. Wiley, New York, 1973.
16. G. O. Schenck, *Angew. Chem.* **64**, 12 (1952).
17. G. O. Schenck and K. Ziegler, *Naturwissenschaften* **32**, 157 (1944).
18. G. O. Schenck and R. Wirtz, *Naturwissenschaften* **40**, 581 (1953).
19. G. O. Schenck and D. E. Dunlap, *Angew. Chem.* **68**, 248 (1956).
20. A. C. Cope, T. A. Liss and G. W. Wood, *J. Am. Chem. Soc.* **79**, 6287 (1957).
21. A. Schönberg, in *Preparative Organic Photochemistry*, pp. 385–389, Springer-Verlag, New York, 1968.
22. P. de Mayo and S. T. Reid, *Chem. Ind.* **1962**, 1576.
23. See Ref. 15, p. 276.
24. C. S. Foote, M. T. Wuesthoff, S. Wexler, I. G. Burstain, R. Denny, G. O. Schenck and K.-H. Schulte-Elte, *Tetrahedron* **23**, 2583 (1967).
25. C. Dufraisse and S. Ecary, *C. R. Acad. Sci. Paris* **223**, 735 (1946).
26. A. A. Gorman and M. A. Rodgers, *Chem. Soc. Rev.* **10**, 205 (1981).
27. L. M. Stephenson, M. J. Grdina and M. Orfanopoulos, *Acc. Chem. Res.* **13**, 419 (1980).
28. A. A. Frimer, in *The Chemistry of Peroxides* (ed. S. Patai), p. 201. Wiley, New York, 1983.
29. W. Ando, K. Watanabe, J. Suzuki and T. Migita, *J. Am. Chem. Soc.* **96**, 6766 (1974).
30. E. W. H. Asveld and R. M. Kellog, *J. Am. Chem. Soc.* **102**, 3644 (1980).
31. I. Saito, S. Matsugo and T. Matsuura, *J. Am. Chem. Soc.* **101**, 7332 (1979).
32. L. M. Stephenson, *Tetrahedron Lett.* **21**, 1005 (1980).
33. G. O. Schenck and O. A. Neumuller, *Liebig's Ann. Chem.* **618**, 194 (1958).
34. M. G. Orfanopoulos, M. B. Grdina and L. M. Stephenson, *J. Am. Chem. Soc.* **101**, 275 (1979).
35. L. A. Paquette and D. C. Liotta, *Tetrahedron Lett.* **1976**, 2681.
36. E. J. Corey and C. Ouannes, *Tetrahedron Lett.* **1976**, 4263.
37. T. Tanaka, K. Hoshino, E. Takeuchi, Y. Tamura and W. Ando, *Tetrahedron Lett.* **25**, 4767 (1984).
38. T. Tanaka, K. Ishibashi and W. Ando, *Tetrahedron Lett.* **26**, **1985**, 4609.

# – 5 –

# Preparation of Nitrogen-Containing Compounds

Photochemical synthesis of nitrogen-containing compounds is of significance for the direct functionalization of unactivated methyl and methylene groups.

## 5.1 PREPARATION OF AMINES

On the basis of the results of thermal cyclization of *N*-halogenated amines (**1**) [1] and (**3**) [2], Wawzopek and Nordstrom [3] proposed that the formation of (**4**) may also proceed in 85% sulphuric acid or trifluoroacetic acid at room temperature by irradiation of (**3**).

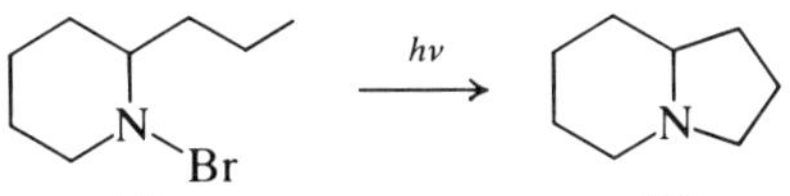

$h\nu$

(**1**) (**2**)

$h\nu$, $H_2SO_4$, 70–80%

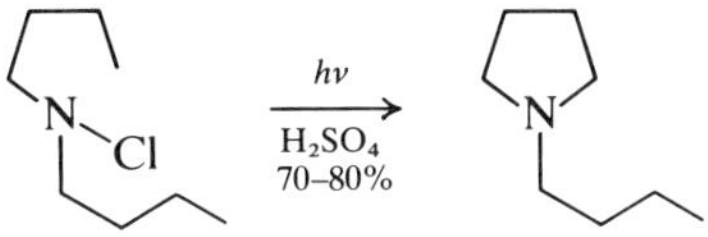

(**3**) (**4**)

Acid treatment of *N*-chloramines first gives a salt (**5**), which then undergoes homolytic cleavage under the influence of light to afford ammonium and chlorine free radicals [4]. The radical (**6**) then abstracts hydrogen intramolecularly from the $\delta$-position in the chain to form a new radical (**7**), which by reaction with (**5**) then gives a $\delta$-chloro derivative (**8**). By final treatment with alkali, ring closure yields an *N*-alkylpyrrolidine (**9**).

$$\underset{(5)}{R(H)N^{\oplus}Cl} \xrightarrow[-Cl^{\cdot}]{h\nu} \underset{(6)}{R(H)N^{\oplus}{}^{\cdot}} \longrightarrow \underset{(7)}{R(H)(H)N^{\oplus}\;\dot{C}H_2} \xrightarrow{(5)} \underset{(8)}{R(H)(H)N^{\oplus}\;Cl} \xrightarrow{OH^-} \underset{(9)}{N{-}R}$$

This cyclization of *N*-halogenated amines is known as the Hofmann–Löffler–Freytag reaction. The effectiveness of the $\delta$-hydrogen abstraction depends strongly on steric factors, as shown by the following two examples. Upon irradiation in sulphuric acid, *N*-chloro-*N*-methylcyclohexylamine (**10**) yielded the bicyclic amine (**11**) in only 11% yield after 30 h [5].

$$\underset{(\mathbf{10})}{Cl{-}N{-}Me} \xrightarrow[90\%\ H_2SO_4]{h\nu} \underset{(\mathbf{11})}{N{-}Me}$$

On the other hand, irradiation of *N*-chlorocamphidine (**12**) in sulphuric acid gavc the cyclocamphidine (**13**) in 67% yield [6].

$$\underset{(\mathbf{12})}{N{-}Cl} \xrightarrow[H_2SO_4]{h\nu} \underset{(\mathbf{13})}{N}$$

An example of the application of the Hofmann–Löffler–Freytag reaction to alkaloid synthesis is the preparation of the Halarrhena alkaloid dihydroconessine (**15**) by the photochemical cyclization of the *N*-chloramine (**14**) [7].

$$(Me)_2N\text{–steroid (H; Me, H–C–NHMe)} \longrightarrow \underset{(\mathbf{14})}{(Me)_2N\text{–steroid (H; Me, H–C–N(Cl)Me)}} \xrightarrow{h\nu}$$

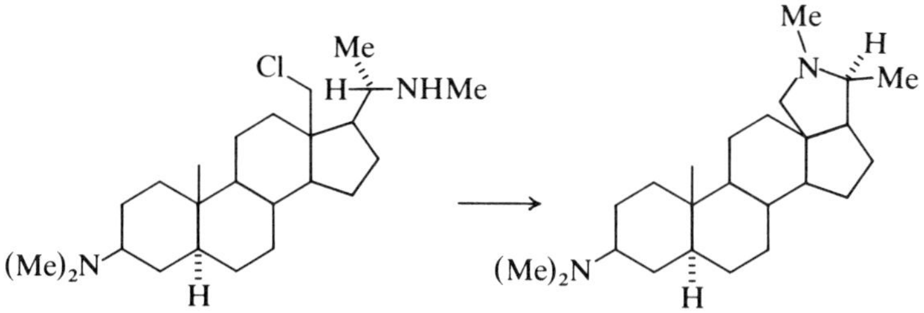

(15)

The basic structure (**17**) of pyrrolidine alkaloids was prepared in 35% yield by photochemical cyclization of *N*,*N*-dibromo-4-heptanamine (**16**) at room temperature [8].

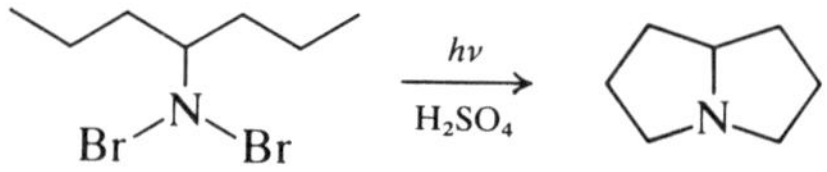

(16) (17)

**Experimental 5.1** Dihydroconessine

3$\beta$-Dimethylamino-20$\alpha$-(*N*-chloro-*N*-methylamino)-5$\alpha$-pregnane (**14**) (90 mg) was dissolved in 90% sulphuric acid (10 ml) at 0 °C. The resulting solution was irradiated in a quartz test tube at 0 °C with a mercury arc lamp under a stream of nitrogen. After 70 min the solution was poured over ice, made alkaline with sodium hydroxide and extracted with ether. The ethereal extracts were evaporated, and the residue was refluxed with ethanol (10 ml) containing potassium hydroxide (1 g) for 30 min. The solution was diluted with water and extracted with ether. The ethereal solution was dried over sodium sulphate and concentrated *in vacuo* to a slightly yellow semisolid (80.3 mg) which was chromatographed on Woelm neutral alumina (7.5 g). Elution with 2 : 1 benzene–ether gave an oil (64.4 mg, 79%) which crystallized from aqueous acetone as flat needles of dihydroconessine (**15**), MP 101.5–102.5 °C.

## 5.2 PREPARATION OF NITROSO COMPOUNDS AND OXIMES

Preparation of nitroso compounds by photochemical transformation of organic nitrites is known as the Barton reaction [9]. In this reaction a hydroxyl group is formed and the nitroso group then replaces hydrogen in the $\gamma$-position. The nitroso compound may then rearrange to the isomeric oxime. This reaction has been widely applied in organic synthesis.

The following two examples are typical of the Barton reaction. n-Octyl nitrite (**18**) and ($\pm$)-menthyl nitrite (**20**) gave 4-nitroso-1-octanol (**19**) [10] and ($\pm$)-10-nitrosomenthol (**21**) [11] respectively.

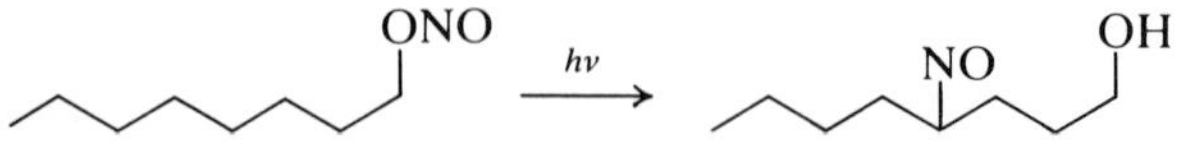

ONO $\xrightarrow{h\nu}$ NO OH

(18) (19)

ONO $\xrightarrow{h\nu}$ OH NO

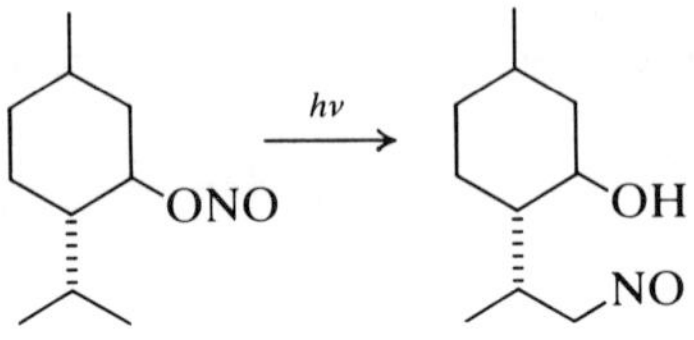

(20) (21)

The Barton reaction involves photochemical cleavage of the N—O bond in the nitrite ester, giving an alkoxy radical and nitric oxide. The $\gamma$-hydrogen is abstracted by the alkoxy radical, and the resulting alkyl and NO radicals combine as shown in the following example:

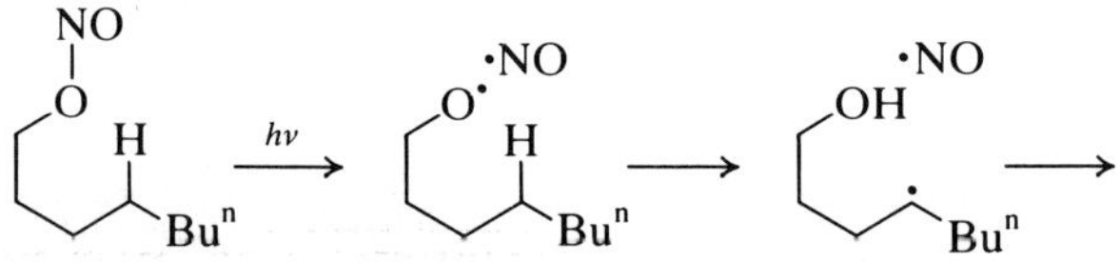

NO O H $Bu^n$ $\xrightarrow{h\nu}$ ·NO O· H $Bu^n$ $\longrightarrow$ ·NO OH $Bu^n$ $\longrightarrow$ (19)

The Barton reaction has contributed greatly to functionalization of unactivated methyl groups in the steroid field, and many biologically important steroids have been prepared. An example is the synthesis of aldosterone 21-acetate (**20**) [12], (**21**) [13] and (**22**) [14].

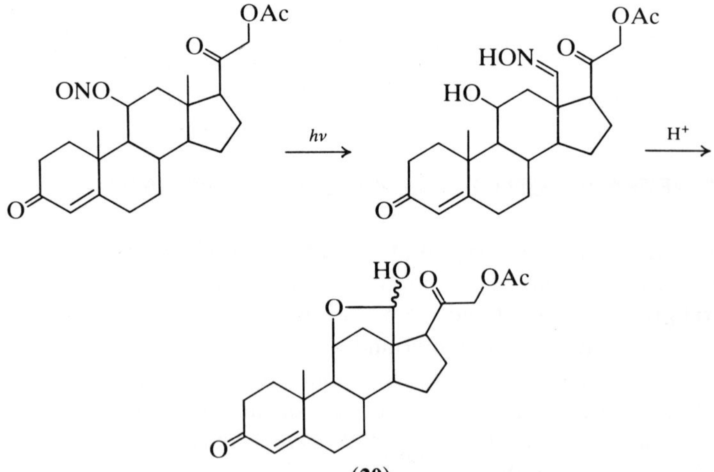

OAc O ONO O $\xrightarrow{h\nu}$ OAc HON O HO O $\xrightarrow{H^+}$

HO O OAc O O

(20)

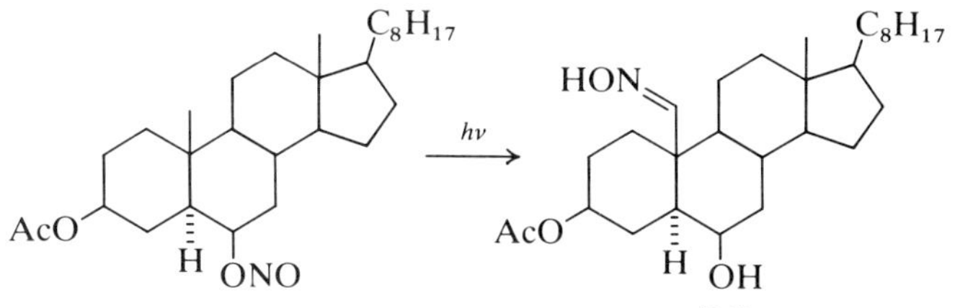

(**21**)

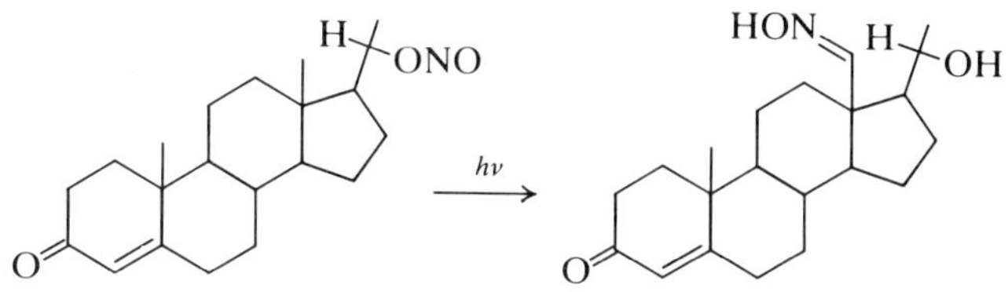

(**22**)

Irradiation of cyclohexane in the presence of chlorine and nitric oxide (1 : 2 v/v) gave 1-chloro-1-nitrosocyclohexane (**23**) [15], while when the $Cl_2$/NO ratio was changed to 1 : 8 v/v nitrosocyclohexane dimer (**24**) was also formed as a by-product [16]. The two products (**23**) and (**24**) were converted into cyclohexanone oxime (**25**). The direct photooximation of saturated hydrocarbons, especially that of cyclohexane, is of great significance, since ε-caprolactam (**26**) is easily obtainable from cyclohexanone oxime (**25**) via the Beckmann rearrangement.

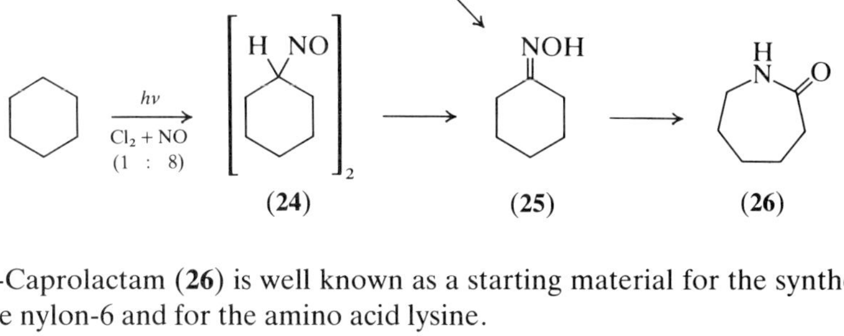

(**23**)

(**24**) (**25**) (**26**)

ε-Caprolactam (**26**) is well known as a starting material for the synthetic fibre nylon-6 and for the amino acid lysine.

## REFERENCES


1. A. W. Hofmann, *Chem. Ber.* **16**, 558 (1883).
2. K. Löffler and C. Freytag, *Chem. Ber.* **42**, 3427 (1909).
3. S. Wawzopek and J. D. Nordstrom, *J. Org. Chem.* **27**, 3726 (1962).
4. S. Wawzopek and T. P. Culbertson, *J. Am. Chem. Soc.* **81**, 3367 (1959).
5. E. J. Corey and W. R. Hertler *J. Am. Chem. Soc.* **82**, 1657 (1960).
6. W. R. Hertler and E. J. Corey, *J. Org. Chem.* **24**, 572 (1959).
7. E. J. Corey and W. R. Hertler, *J. Am. Chem. Soc.* **81**, 5209 (1959).
8. E. Schmitz and D. Murawski, *Chem. Ber.* **93**, 754 (1960).
9. D. H. R. Barton, J. M. Beaton, L. E. Geller, and M. M. Pechet, *J. Am. Chem. Soc.* **82**, 2640 (1960).
10. P. Kabasakalian and E. R. Townley, *J. Am. Chem. Soc.* **84**, 2711 (1962).
11. P. Kabasakalian, E. R. Townley and M. D. Yuduis, *J. Am. Chem. Soc.* **84**, 2716 (1962).
12. D. H. R. Barton and J. M. Beaton, *J. Am. Chem. Soc.* **82**, 2641 (1960).
13. D. H. R. Barton, J. M. Beaton, L. E. Geller and M. M. Pechet, *J. Am. Chem. Soc.* **83**, 4076 (1961).
14. A. L. Nussbaum, F. E. Carlon, E. P. Olivevo, E. Townley, P. Kabasakalian and D. H. R. Barton, *Tetrahedron* **18**, 373 (1962).
15. E. Muller and H. Metzger, *Chem. Ber.* **87**, 1282 (1954).
16. E. Muller and H. Metzger, *Chem. Ber.* **88**, 165 (1955).

# – 6 –

# Preparation of Carbocyclic Compounds

## 6.1 THREE-MEMBERED RINGS

There are three photochemical synthetic routes for cyclopropanes which involve photochemical rearrangement: (i) di-$\pi$-methane systems [1]; (ii) photoextrusion of a small molecule such as nitrogen [2]; and (iii) photoisomerization of 1,3-cyclohexadienes and dihydrofurans.

### 6.1.1 Preparation of three-membered rings by rearrangement of di-$\pi$-methanes

Photochemical rearrangement of di-$\pi$-methanes is a useful synthesis of substituted cyclopropanes. Direct irradiation of *cis*- and *trans*-1,1-diphenyl-3,3-dimethyl-1,4-hexadienes (**1**) and (**2**) gave stereoselectively the *cis*- and *trans*-vinylcyclopropanes (**3**) and (**4**) respectively [3–5].

Ph Ph (**1**) —$h\nu$→ Ph Ph (**3**)   Ph Ph (**5**)

sensitizer

Ph Ph (**2**) —$h\nu$→ Ph Ph (**4**)   Ph Ph (**6**)

From the fact that neither of the isomeric cyclopropanes (**5**) and (**6**) is produced, it is likely that the reaction proceeds via a biradical intermediate (**7**), which is converted into the stable radical (**8**), giving the product (**9**) [6].

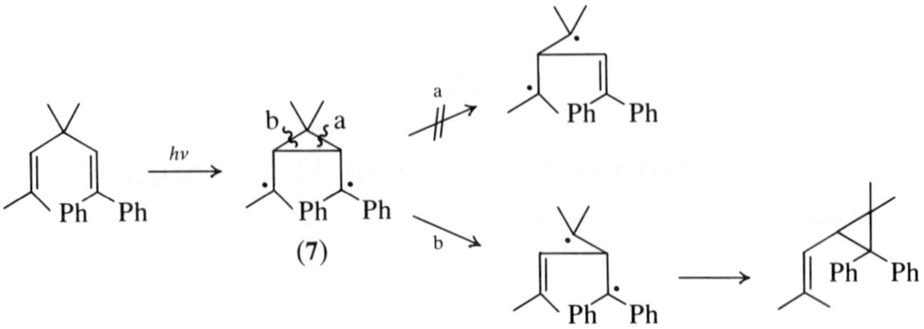

(7) (8) (9)

Multifunctionalized di-$\pi$-methanes gave more complicated products; the $\alpha$-allylcyclopentenones (**10**), (**12**) and (**14**) are good precursors for photochemical synthesis of the cyclopropylcyclopentenones (**11**) and (**13**) and the alcohol adduct (**15**) in good yields [7].

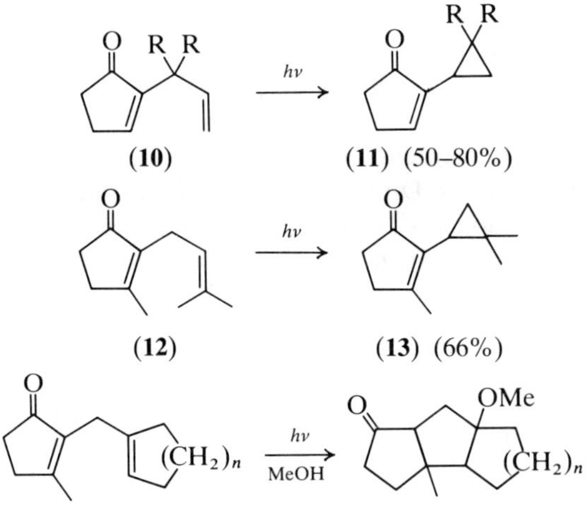

(**10**) (**11**) (50–80%)

(**12**) (**13**) (66%)

(**14**) (**15**) (90%)

Formation of cyclopropanes by photochemical rearrangement of cyclic di-$\pi$-methane systems is also known, and various types of fused cyclopropanes have been synthesized [1]. In particular, considerable effort has been devoted to the photochemistry of benzobicyclic olefins such as benzonorbornadiene, benzobicyclo[2.2.2]octadiene and mono- and di-benzobarralene. For example, acetophenone-sensitized photolysis of benzonorbornadiene (**16**) gave a 95% yield of (**17**) [8, 9], and irradiation of an acetone solution of dibenzobarralene (**18**) produced dibenzosemibullvalene (**19**) in 85% yield [10].

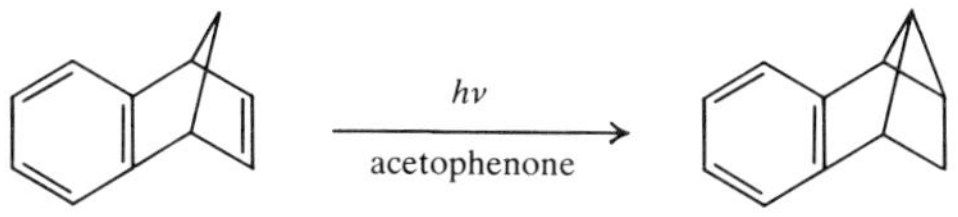

hν
acetophenone

(16) (17) (95%)

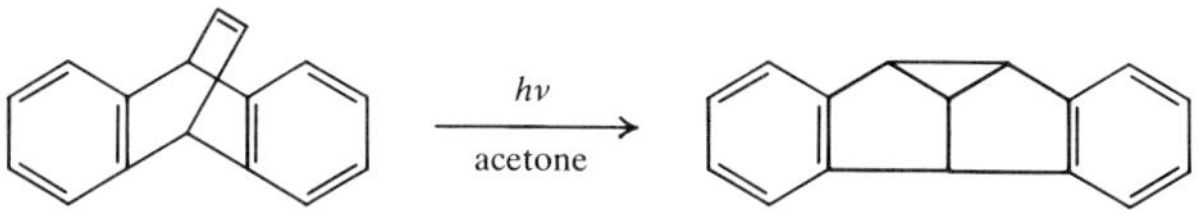

hν
acetone

(18) (19) (85%)

### 6.1.2 Preparation of three-membered rings by photoextrusion

Several syntheses of strained hydrocarbons such as cyclopropanes have included a nitrogen-photoextrusion step [2]. The most frequently employed application is the formation of cyclopropanes by photolysis of pyrazolines. Table 6.1 shows a few interesting examples. The remarkable synthesis of prismane (**20**) by Katz and Acton [11, 12] illustrates the success of this photoextrusion approach.

Perhaps the most common application of this group of reactions in natural-product synthesis has been in the construction of terpenes containing a cyclopropane ring. Table 6.2 shows a few examples of this use of photoextrusion [15].

Two groups have attempted the synthesis of marasmic acid (**21**) by cyclopropanation of the hydrindane (**22**) [17, 20]. Unfortunately, the addition of diazomethane occurs with the wrong stereoselectivity, leading to the formation of the isomarasmic acid framework (**23**).

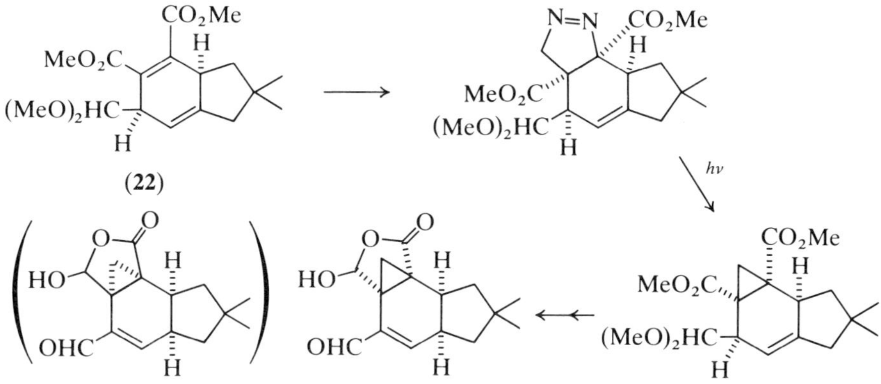

$CO_2Me$ H $MeO_2C$ $(MeO)_2HC$ H

(22)

N=N $CO_2Me$ H $MeO_2C$ $(MeO)_2HC$ H

hν

O O H HO OHC H

(21)

O O H HO OHC H

(23)

$CO_2Me$ H $MeO_2C$ $(MeO)_2HC$ H

TABLE 6.1

Applications of the photoextrusion of nitrogen in strained-ring synthesis

| Reactant | Conditions | Products (Yields) | Ref. |
|---|---|---|---|
|  N=N | 366 nm, $C_6H_6$, 25 °C |  (**20**) 11%; 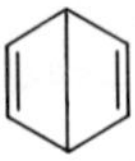 45%;  6%; Benzene 30%; 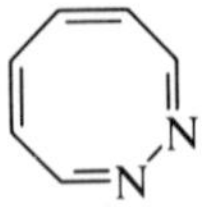 8% | [11, 12] |
| 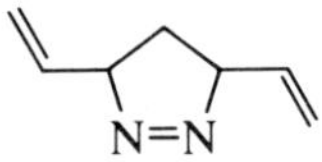 N=N | Pyrex, $CDCl_3$, −50 °C | 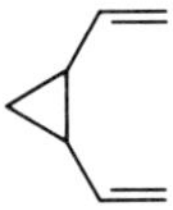 42–56%; 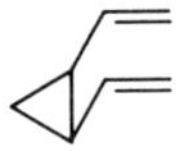 42–56% | [13] |
| 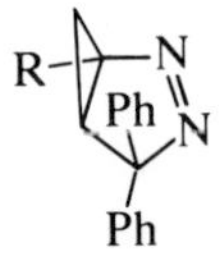 R, N, N, Ph, Ph | Pyrex, pentane, r.t. | 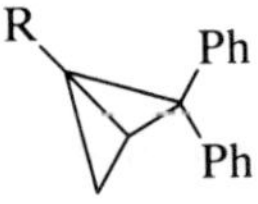 R, Ph, Ph; R = Me 51% | [14] |
| 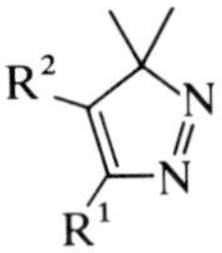 $R^2$, N, N, $R^1$ | 350 nm, $C_6H_6$ |  $R^2$, $R^1$ | [16] |
| $R^1, R^2$ = Ph, Ph | | 92% | |
| $R^1, R^2$ = Me, 2-isopentenyl | | 100% | |

Two sesquiterpenes, (–)-cyclocopacamphene (**25**) and the diastereomeric (+)-cyclosativene (**26**) have been synthesized by the photoextrusion of nitrogen [19]. The excellent yields for the intramolecular cycloaddition and for the photoextrusion of nitrogen make this type of synthetic sequence particularly appealing.

Franck-Newmann [21] investigated the differences in product distribution and yield between thermal and photochemical extrusion of the pyrazoline (**27**) and found that photochemical rather than thermal extrusion of nitrogen is the better method when the two have been directly compared.

TABLE 6.2

Applications of the photoextrusion of nitrogen in natural-product synthesis

| Reactant | Conditions | Product | Ref. |
|---|---|---|---|
| 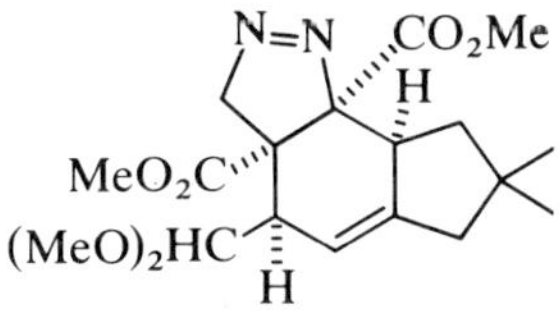 N=N, $CO_2Me$, H, $MeO_2C$, $(MeO)_2HC$, H | Pyrex, ether | 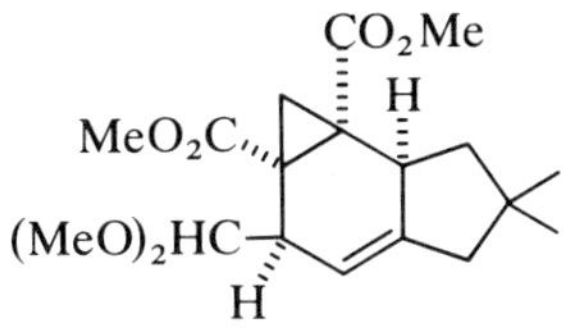 $CO_2Me$, H, $MeO_2C$, $(MeO)_2HC$, H | [17] |
| 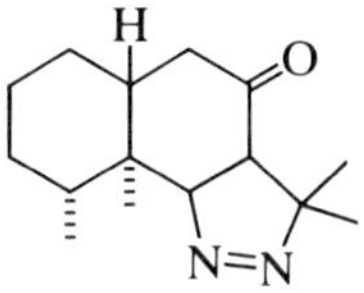 H, O, N=N | Pyrex, benzene | 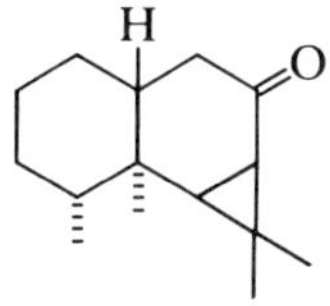 H, O | [18] |
| 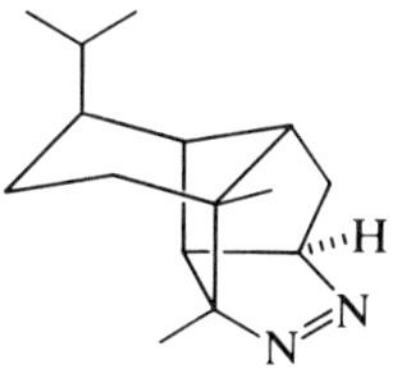 H, N=N | Pyrex, ether, r.t. | 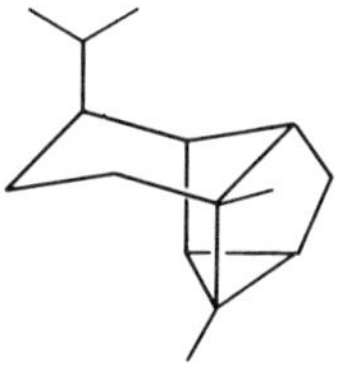 | [19] |

O, H, H → H, N=N (**24**) (81%)

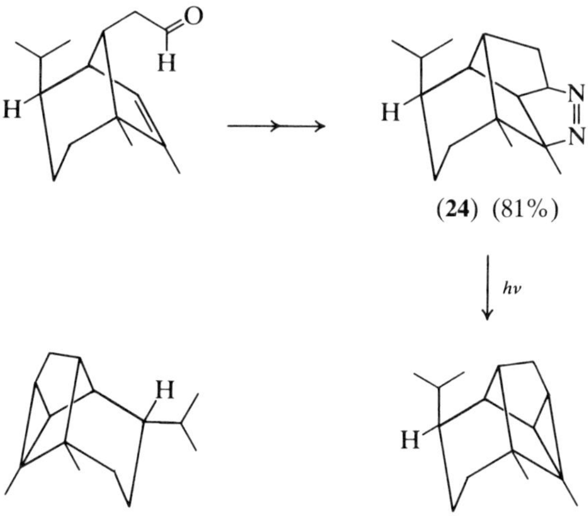

$h\nu$

H (**25**) (92%)

H (**26**)

(27)

| Condition | | % | % |
|---|---|---|---|
| Δ | Refluxing benzene | 0 | 100 |
| $h\nu$ | Direct | 50 | 50 |
| $h\nu$ | $Ph_2CO$-sensitized | 88 | 12 |

The photoextrusion of carbon monoxide from bicyclic ketones is another synthetic method for bicyclic cyclopropanes. For example, (**28**), in which both carbons of the carbonyl are highly substituted and where $\alpha$-hydrogens are not readily accessible, gave an excellent yield of the photo-decarbonylation product (**29**) [22].

OMe OMe O (28) $h\nu > 300$ nm OMe OMe O −CO OMe OMe MeO OMe (29) (88%)

**Experimental 6.1** (–)-Cyclocopacamphene [19]

A solution of the pyrazoline (**24**) (35 mg) in dry ether (35 ml) was irradiated in a Rayonet Reactor, using 3500 Å lamps and a Pyrex filter, for 1 h. Removal of the solvent, followed by distillation (air-bath temperature 80 °C) of the residual oil under reduced pressure (0.3 mmHg) afforded pure (−)-cyclocopacamphene (**25**) (30 mg; 93%).

### 6.1.3 Preparation of cyclopropanes by photoisomerization of 1,3-cyclohexadienes and dihydrofurans

Irradiation of dimethyl-1,3-cyclohexadiene-1,4-dicarboxylate (**30**) gave dimethyl bicyclo[3.1.0]hex-2-ene-1,5-dicarboxylate (**33**) in 60–70% yield via two intermediates (**31**) and (**32**) [23].

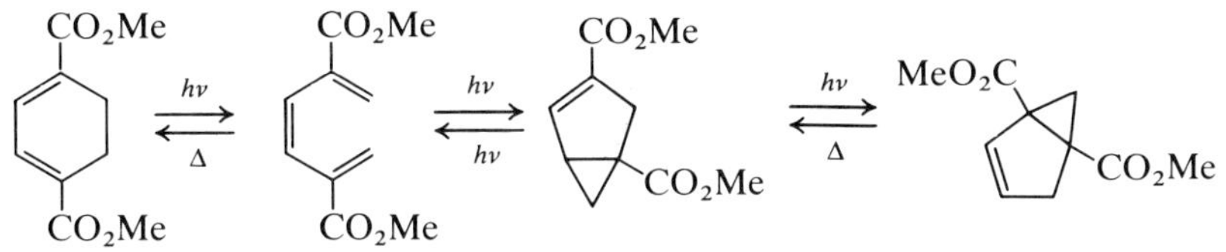

(**30**) (**31**) (**32**) (**33**)

On irradiation of the 2,3-dihydrofuran (**34**), a valence tautomerization took place to give the substituted cyclopropane (**35**) in 35% yield [24].

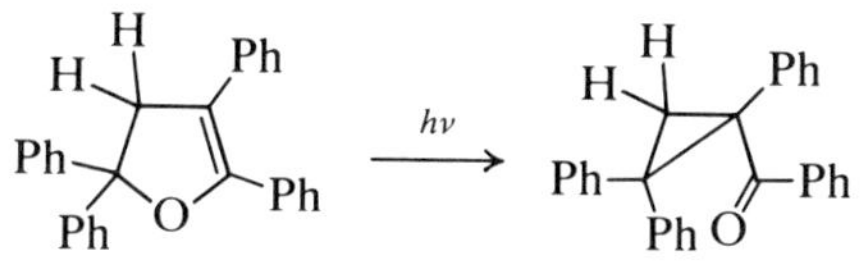

(**34**) (**35**)

Similarly, when the 2,3-dihydrofuran (**36**) was exposed to UV irradiation in ether solution, a 74% yield of 1-acetyl-1-methylcyclopropane (**37**) was obtained, in addition to minor quantities of the geometrical isomers of the enones [25]. The cyclopropanecarbaldehydes (**39**) and (**40**) were prepared from the dihydrofuran (**38**) by the same method [26].

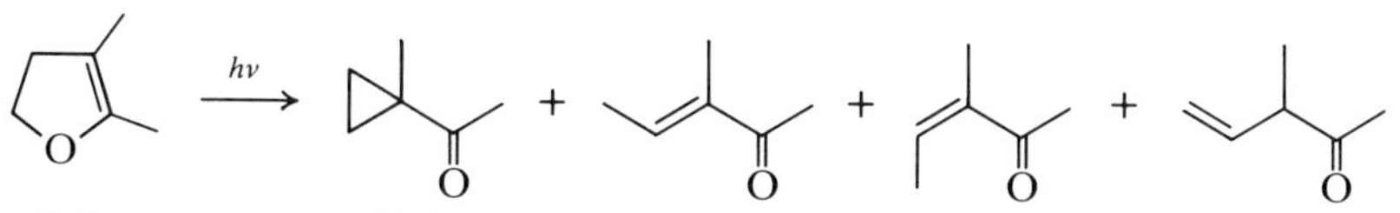

(**36**) (**37**)

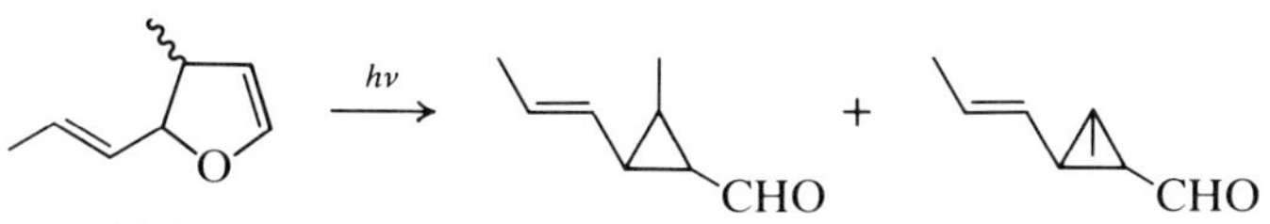

(**38**) (**39**) (**40**)

## 6.2 FOUR-MEMBERED RINGS

### 6.2.1 Preparation of four-membered rings by photocycloaddition

Of several photochemical syntheses of cyclobutanes, photocycloaddition processes are assuming a place of substantial importance in synthetic organic chemistry [27]. For example, photocycloaddition reactions played an important role in the syntheses of cubane (**41**) by Eaton and Cole [28] (see also Section 8.1), caryophyllene (**42**) by Corey *et al.* [29], β-bourbonene (**43**) by White and Gupta [30] and annotinine (**44**) by Wiesner *et al.* [31].

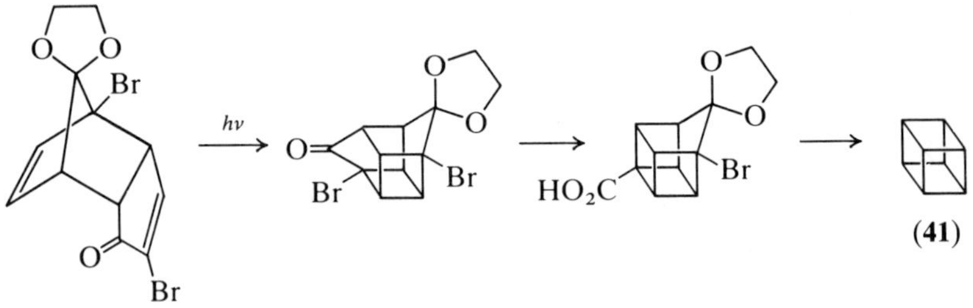

O O Br O Br $\xrightarrow{h\nu}$ O Br Br O O $\longrightarrow$ $HO_2C$ Br O O $\longrightarrow$

(**41**)

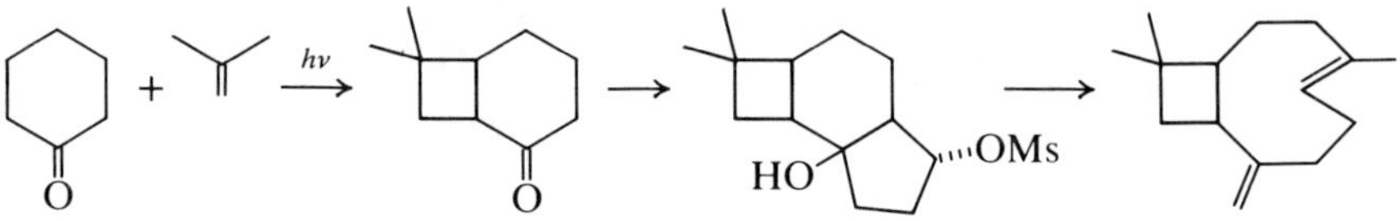

O + $\xrightarrow{h\nu}$ O $\longrightarrow$ HO OMs $\longrightarrow$

(**42**)

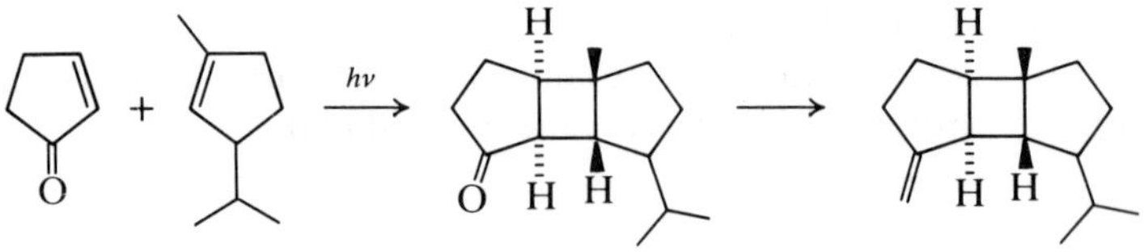

O + $\xrightarrow{h\nu}$ H O H H $\longrightarrow$ H H H

(**43**)

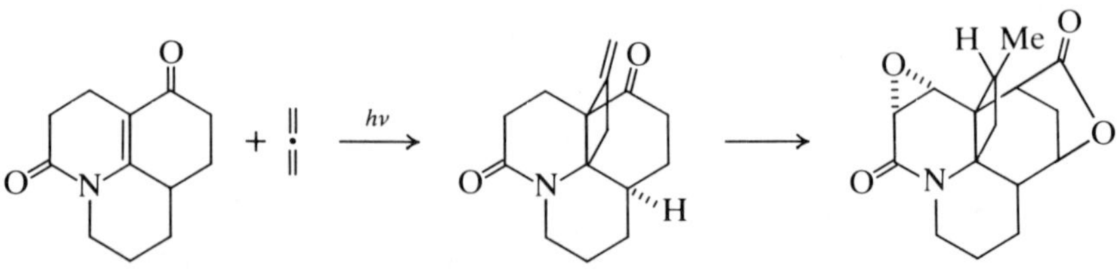

O O N + $\xrightarrow{h\nu}$ O O N H $\longrightarrow$ H Me O O O O N

(**44**)

As is the case with most reactions in organic photochemistry, photocycloaddition covers a variety of types. Certain useful generalizations are emerging that interrelate these reactions and permit a rather general picture of the processes involved to be drawn.

### 6.2.1.1 Combination of two components

The components for cyclobutane formation by [2 + 2] photochemical cycloaddition involve various types of electrophilic olefins which are reacted with an alkene [27]. From the synthetic point of view, useful "enone" components include $\alpha,\beta$-unsaturated ketones, $\alpha,\beta$-unsaturated nitriles, maleic anhydride derivatives and the enolic forms of 1,3-diketones and their analogues.

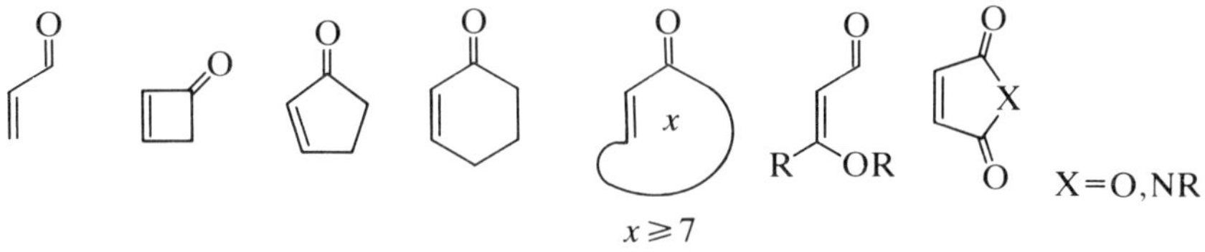

O O O O O O O x R OR X O X=O,NR $x \geqslant 7$

With few exceptions, the double bond of an enone undergoing efficient cycloaddition to an alkene is part of a ring of six or less members. Without this constraint, on excitation the enone double bond undergoes energy-wasting *cis–trans* isomerization. The double bond may be either contained covalently in a ring of four, five or six members, or may be part of the enol tautomer of an intramolecularly hydrogen-bonded $\beta$-dicarbonyl compound. Unless otherwise prevented from undergoing *cis–trans* isomerization, as in the bicyclic enone (**45**) [32], enone rings with seven or more members, such as cyclooctenone (**46**) [33, 34], do not undergo excited-state cycloaddition reactions with alkenes. However, their *trans* double-bond isomers [e.g. (**47**)], formed on irradiation of the *cis* isomer, will react with alkenes in the dark to give *trans* fused products [e.g. (**48**) of reversed orientation [35, 36]. The unstable *trans*-enones also readily undergo [4 + 2] cycloaddition

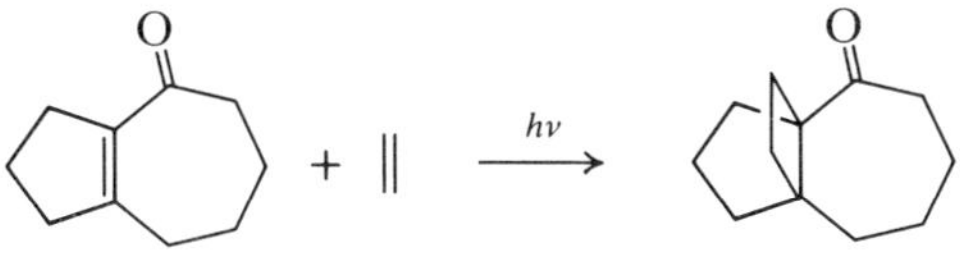

O + || $h\nu$ O

(**45**)

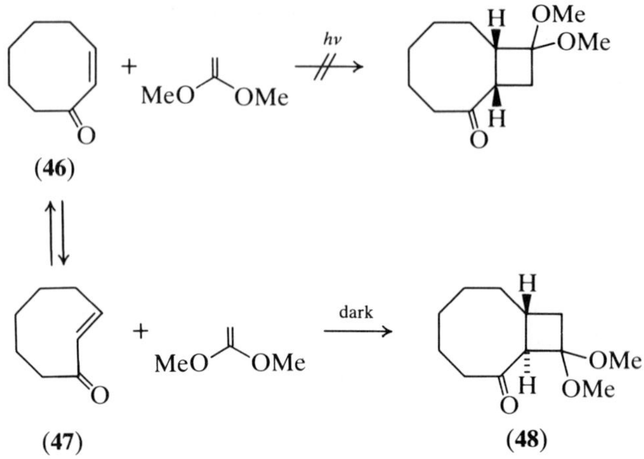

(46)

(47)

(48)

reactions with 1,3-dienes [(**49**)→(**50**)] [37]. A third reaction of the strained *trans* seven- and eight-membered 2-cycloalkenones is their reaction with oxygen and nitrogen nucleophiles, in a Michael sense, to yield products of *cis* addition to the double bond [(**49**)→(**51**)] [38, 39].

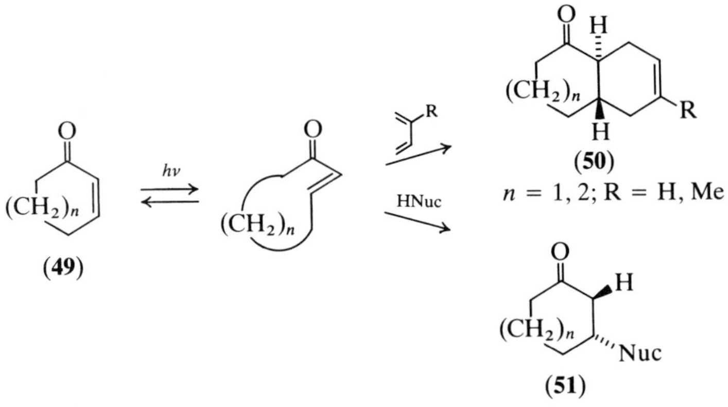

(49)

(50)

$n$ = 1, 2; R = H, Me

(51)

### 6.2.1.2 Cycloaddition of simple enones to C═C double bonds

The first example of a photochemical cycloaddition reaction between the double bond of an $\alpha,\beta$-unsaturated ketone and an alkene was apparently an intramolecular one—the conversion of carvone (**52**) to carvone camphor (**53**) on prolonged exposure to Italian sunlight [40, 41].

In the intervening seven decades, the general reaction has received considerable attention, most notably since 1960, and a number of

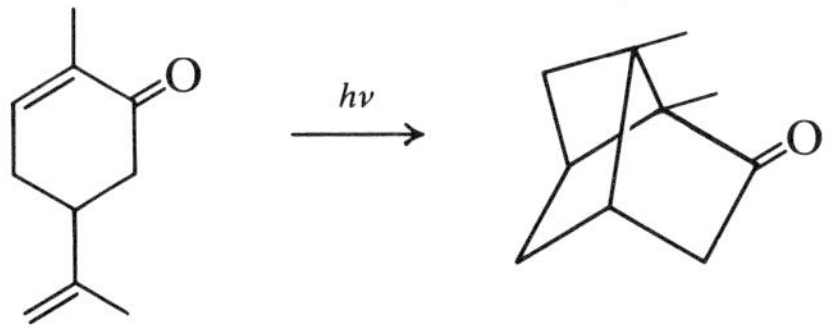

(52) $\xrightarrow{h\nu}$ (53)

informative reviews have been published [27, 42–44]. Only during the last 25 years has the photocycloaddition reaction been utilized for synthetic purposes, but its value would now appear to be well established, in an intermolecular as well as an intramolecular sense.

### *(i) Scope and limitations*

(*a*) *Competing reactions* Cyclobutane formation can be accompanied by a number of competing processes, each of which reduces the efficiency of the desired photocycloaddition reaction [45]. Although a detailed presentation of these processes is beyond the scope of this book, a brief discussion of the

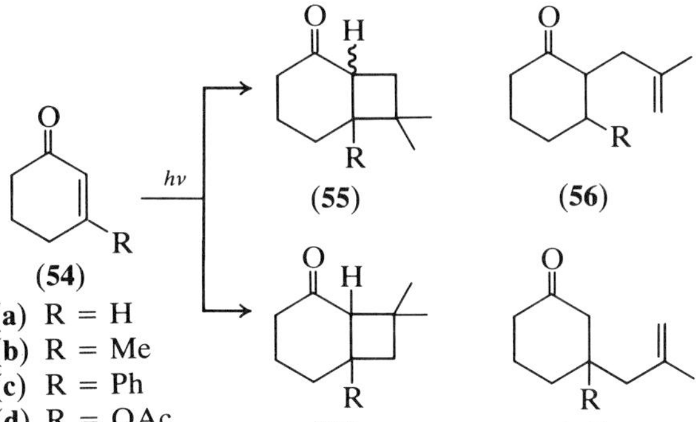

(54) $\xrightarrow{h\nu}$ (55) (56) (57) (58)

**(54)**
**(a)** R = H
**(b)** R = Me
**(c)** R = Ph
**(d)** R = OAc
**(e)** R = CN

Product distributions from isobutene and several cyclohexenones

| (54) | (55) | (56) | (57) | (58) | HT/HH | Ref. |
|---|---|---|---|---|---|---|
| **a** | 33 | 14 | 6 | 8 | 3.4 | [35] |
| **b** | 0 | 27 | 13 | 50 | 0.43 | [46] |
| **c** | 45 | 0 | 0 | 0 | large | [46] |
| **d** | 13 | 17 | 25 | 21 | 0.87 | [46] |
| **e** | 68 | 0 | 10 | 0 | 6.8 | [47] |

more important ones should be of value to the synthetic chemist planning cyclobutane formation. For instance, competitive dimerization of the enone can be a problem, particularly with electron-deficient alkenes. This effect can be partially overcome by increasing the alkene–enone ratio. A second source of inefficiency previously seen is the competitive formation of allylic addition products such as (**56**) and (**58**), presumably via hydrogen abstraction from intermediate 1,4-diradicals.

*Photodecomposition.* There is an additional possibility that the cyclobutane product itself will be photochemically active and suffer degradative reactions. The most common of these degradative reactions is $\alpha$-cleavage, the products of which are derived from radical recombination and, in particular, several disproportionation pathways [48, 49]. A vivid illustration of such competing reactions is the attempted photoaddition of the enone (**60**) to the methylenecyclopentanes (**59**), where in each case the majority of the isolated photoproducts resulted either from secondary decomposition of the desired cyclobutanes [(**61**)$\rightarrow$(**64**) and (**62**)$\rightarrow$(**65**)] or from hydrogen abstraction from the intermediate diradical to give (**63**) [50, 51]. That the amounts of (**64**) and (**65**) could be reduced by the use of longer wavelength light (uranium glass filter, >330 nm) is of interest.

(**59**) (**60**) $h\nu$ (**61**) (**62**) (**63**)

$R^1 = R^2 = H$
$R^1 = Me, R^2 = Pr^i$

(**64**) (**65**)

*Photoenolization.* A further complicating factor is double-bond migration in the starting enone, a process sometimes referred to as *photoenolization* [52, 53]. The reaction between isophorone (**66**) and isopropenyl acetate (**67**) gave the expected product (**68**) in only moderate yield after prolonged irradiation, accompanied by a considerable amount of polymeric material [54]. Irradiation of an ethyl acetate solution of (**66**) (no alkene present) led to a nearly quantitative conversion to (**69**). Thus it has been shown that one of the sources of inefficiency is deconjugation of isophorone to the exocyclic methylene isomer (**69**), which slowly rearranges to (**66**) on standing in the photolysis medium [55]. Interestingly, it was found that polar solvents promoted the equilibration of (**69**) to (**66**), so that carrying out the cycloaddition reaction in methanol resulted in smooth and rapid formation of the cycloadduct (**68**).

(**69**) (**66**) (**67**) (**68**)

Benzene: (**67**)/(**66**) = 20; modest yield
Methanol: (**67**)/(**66**) = 4; 71% yield

*Oxetane formation.* The final competing reaction to be mentioned is oxetane formation (Paterno–Büchi reaction), which is a very interesting process and has been thoroughly studied [56]. It is now clear that steric bulk at either the $\gamma$ or $\alpha'$ positions enhances the amount of oxetane formation, as does the incorporation of an electronegative atom at the $\alpha'$ position. Furthermore, polar solvents tend to increase the cyclobutane : oxetane ratio. Chapman *et al.* [57, 58] have shown that in the addition of 2,3-dimethyl-2-butene (**71**) to 4,4-dimethylcyclohexenone (**70**, R′ = H), nearly

(**70**) (**71**) (**72**) (**73**)

| $R^1$ | $R^2$ | Solvent | Products ratio |
|---|---|---|---|
| H | H | Isooctane | 100 : 0 |
| Me | H | Isooctane | 48 : 52 |
| Me | Me | Isooctane | 70 : 30 |
| Me | F | Isooctane | 10 : 90 |
| Me | F | Acetonitrile | 85 : 15 |

(74) (71)

| R | Solvent | Products ratio |
|---|---|---|
| Me | Cyclohexane | 24 : 76 |
| Me | Acetonitrile | 38 : 62 |
| $(OCH_2)_2$ | Cyclohexane | 47 : 53 |
| $(OCH_2)_2$ | Acetonitrile | 71 : 29 |

equal amounts of cyclobutane (**72**, R′ = H) and oxetane (**73**, R′ = H) were formed. Yoshida *et al.* [59] have also reported similar results with cyclopentenone derivatives. The effect of solvent on the cyclobutane : oxetane ratio is of some interest. Perhaps the most thorough study of the cyclobutane–oxetane balance has been performed by Margaretha [60–62], who found that the addition of 2,3-dimethyl-2-butene (**71**) to a series of cyclopentenones (**74**) in different solvents gave the expected results.

*(b) Regiochemistry* Corey *et al.* [35] have investigated problems involving regio- and stereo-chemistry in the [2 + 2] photocycloaddition of the 2-cyclohexenone system. It has been established that isobutene, 1,1-dimethoxyethylene, and positively substituted olefins in general, react quite rapidly, whereas negatively substituted olefins react very slowly. Competition experiments gave the relative rates of addition to excited 2-cyclohexenone as shown in Table 6.3 [35]. Orientation is not random. With positively substituted olefins (R—CH=$CH_2$) or ($R_2C$=$CH_2$) the 7-substituted or 7,7-disubstituted product is favoured in each case. The products show both *cis* and *trans* ring junctures, with the *trans* isomer predominating in the 7,7-disubstituted products as shown below. Vinylmethyl ether, vinyl acetate and benzylvinyl ether gave in each case three stereoisomers of the 7-substituted bicyclo[4.2.0]octan-2-ones.

Regiochemistry is not a problem with olefins such as cyclopentene, but it is interesting to note that 2-cyclohexenone gave two stereoisomers (*cis* and *trans* ring junctures), while 2-cyclopentenones gave only one product [63].

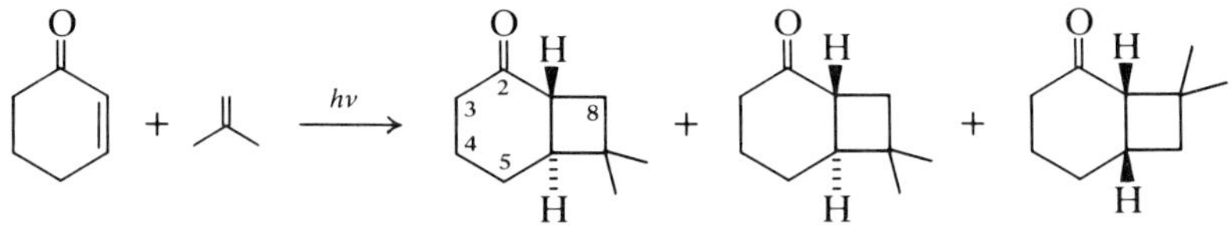

(26%) (6.5%) (6%)

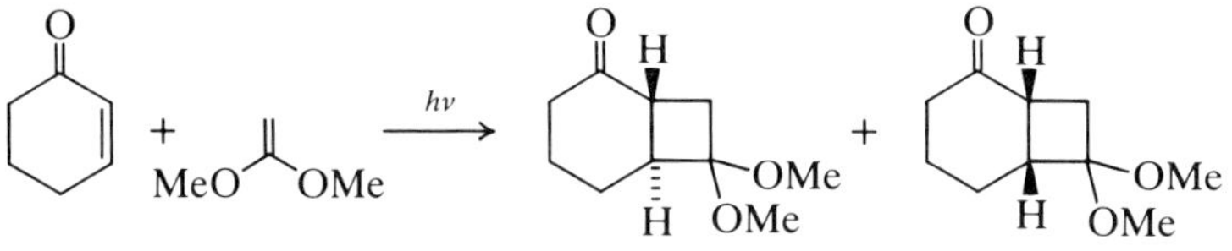

(49%) (21%)

TABLE 6.3

Relative rates of cycloaddition

| Olefin | Relative rate |
|---|---|
| $(MeO)_2C{=}CH_2$ | 4.66 |
| $MeOCH{=}CH_2$ | 1.57 |
| Cyclopentene | 1.00 |
| $(Me)_2C{=}CH_2$ | 0.40 |
| $CH_2{=}C{=}CH_2$ | 0.23 |

Irradiation of 2-cyclohexenone in the presence of either *cis*- or *trans*-2-butene gave the same products (**76**) [35]. No isomerization of either *cis*- or *trans*-butene was observed during the course of the reaction [35].

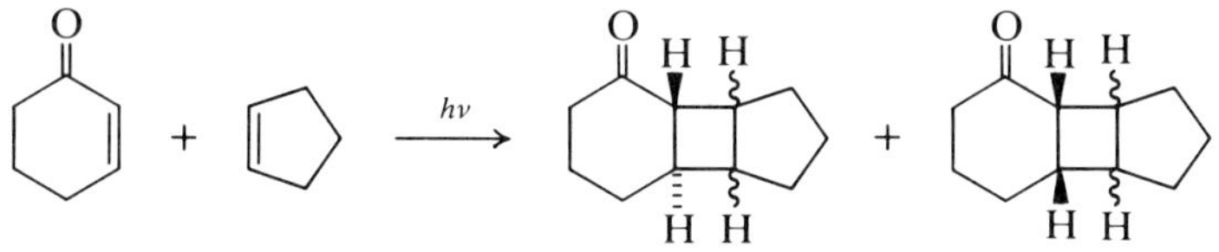

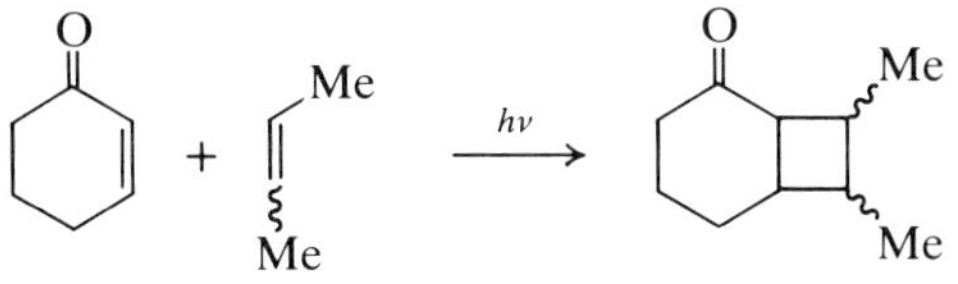

(**76**) (three isomers)

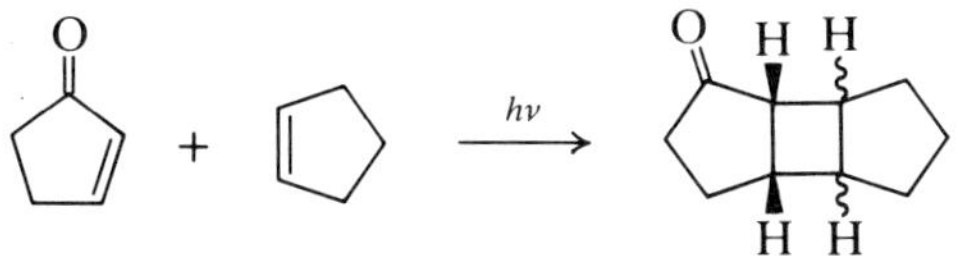

Corey *et al*. [35] have suggested that regiochemistry in the photocycloaddition of cyclic enones to olefins can be accounted for on the basis of a complex between the excited ketone and the olefins in which the olefin serves as the donor and the excited ketone as the acceptor. They have argued that, for an excited ketone, the complex will be of the form (**77**) for positively substituted olefins. The next step in the tentative mechanism is collapse to a biradical intermediate (**78**) [35]. This intermediate is invoked to explain the observed stereochemistry and not regiochemistry. The formation of the biradical involves the $\beta$-position of the $\alpha,\beta$-unsaturated ketone and the more nucleophilic carbon atoms [35].

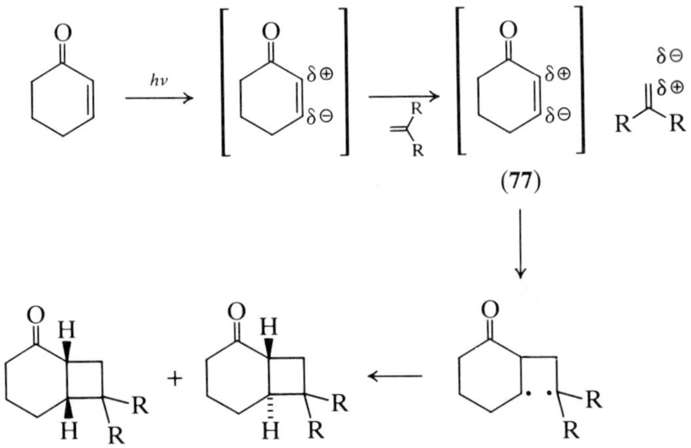

(**77**)

(**78**)

Closure of a highly energetic biradical to a *trans* product is observed without exception. Minor products must come from initial bonding to the $\beta$-position of the enone. The unsaturated by-products obtained from isobutene are easily explained on the basis of a biradical intermediate.

*(c) Stereochemistry* Stereochemistry is as important a consequence of the photocycloaddition reaction as is regiochemistry, and is potentially more complicated. It is true that further transformations of the initial photoproducts (ring fragmentation, equilibration, etc.) often simplify the stereochemical results, but certain factors must be taken into account in planning a synthesis.

If either the enone or alkene possesses stereochemical centres remote from the reacting double bonds, the major photoproduct will usually be the result of the least-hindered approach. The selectivity is often high, as illustrated in the following relatively simple examples. There are instances

(80)

| $R^1$ | $R^2$ | Products ratio | |
|---|---|---|---|
| Ac | H | 1.8 | 1 |
| $SiMe_3$ | H | 1 | 5 |
| $SiMe_3$ | $SiMe_3$ | 1 | 10 |
| Ac | OCOPh | 1.8 | 1 |
| $SiMe_3$ | OCOPh | 1 | 5 |

where the preferred direction of approach is dependent on substituent groups, as seen in the addition of furanone (**79**) to the several cyclopentanone enol derivatives (**80**) [64].

Different stereochemical results at the four carbons of the cyclobutane ring can result in eight stereoisomers for each regioisomer formed during an enone–alkene photoaddition. Acyclic alkenes are usually compromised stereochemically during photoaddition. For instance, the reaction of 2-cyclohexenone with *cis*- or *trans*-2-butene gave rise to the same three photoadducts, each *cis* fused, in approximately the same amounts, suggesting the involvement of rotationally equilibrated common intermediate 1,4-diradicals (see above) [35].

Cyclic alkenes may give rise to either *cis*- or *trans*-fused products, with the expected constraints arising from ring size. Cyclopentenes and smaller cycloalkenes give only *cis* fused products. Cyclohexenes usually give rise to *cis* ring fusions, although occasionally *trans* products have been noted [e.g. (**81**) + (**82**) → (**83**) + (**84**)] [65]. Cycloheptenes and larger cycloalkenes give mixtures of *cis* and *trans* fused cyclobutane products [66].

In a similar fashion, the stereochemistry of the enone–cyclobutane ring fusion may be either *cis* or *trans*. Five-membered ring enones invariably

(81) (82) (83) (84)

yield *cis* fused systems, while six-membered ring enones can give rise to either *cis* or *trans* products. Some trends can be discerned, although clear-cut stereochemical predictions remain difficult at present. For instance, electron-poor alkenes yield a preponderance of *cis* fused adducts, while electron-rich alkenes afford mainly *trans* fused adducts. Alkyl substitution at $\beta$-carbon tends to increase the amount of *cis* fused materials, as does a decrease in the enone triplet energy. These trends can be seen in the following cyclohexenone photocycloaddition reactions. Inspection of the stereochemical results of other six-membered ring enones reveals related dependence on the polarity of alkene and the substitution of enone.

$h\nu$ [35] R = Me, OMe

**(54a)**

$h\nu$ [46] R = Me, OMe

**(54c)**

### *(ii) Solvent and temperature effects*

The effects of solvent and temperature on the course of enone–alkene photocycloaddition are very complicated. There appears to be no unified picture that consistently predicts either the direction or magnitude of the effects irrespective of their strengths. However, changes in solvent and temperature are factors that should not be ignored if one is looking for enhanced selectivity in a given photocycloaddition reaction; this is exemplified by the problem of oxetane formation discussed above.

There have been relatively few studies describing regiochemical solvent effects in other enone–alkene photocycloadditions, although the solvent-dependent product ratios of (**85**) and (**86**) are often quoted [67]. The HH(head-to-head, **85**) : HT(head-to-tail, **86**) ratio varies from 98 : 2 in cyclohexane to 45 : 55 in methanol.

The reactions between several cyclohexanones and terminal alkynes gave increasing proportions of HH adducts with decreasing solvent polarity. It appears that an increase in solvent polarity will result in a decrease in the relative amount of the regioisomer that is favoured in a medium of low

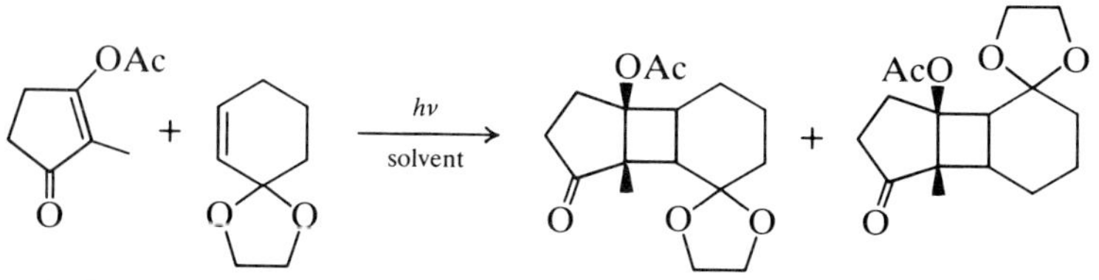

(**81**) (**85**) (HH) (**86**) (HT)

| Solvent | **85 : 86** |
|---|---|
| Cyclohexane | 98 : 2 |
| Methanol | 45 : 55 |

polarity. Variations in stereochemistry due to solvent effects have also been noted, but no systematic study is available. Recently, enone photodimerization in a micellar has been shown to give enhanced photochemical efficiency and regioselectivity [68].

Additional data are available on the influence of temperature on enone–alkene cycloaddition. An increase in quantum efficiency is often found to accompany a decrease in reaction temperature, although deviations have been noted.

The influence of a decrease in temperature on regiochemistry can be significant, as has been noted [(**87**) + (**88**) → (**89**) + (**90**)] [69, 70]. Other related examples include the addition of cyclopentenone (**87**) to trichloroethylene (**91**) [71, 72]. In most cases a decrease in reaction temperature leads to an increase in the proportion of the regioisomer favoured by a consideration of exciplex dipolar interactions.

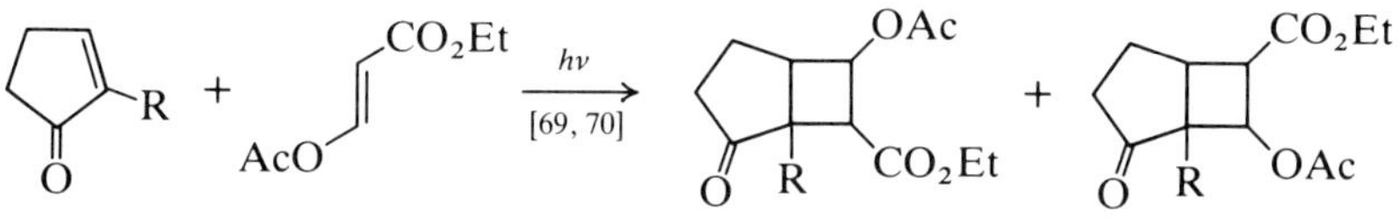

(**87**) (**88**) (**89**) (**90**)

| R | °C | **89** | **90** |
|---|---|---|---|
| H | 25 | 65 | 35 |
| H | −40 | 78 | 22 |
| Me | 25 | 24 | 76 |
| Me | −40 | 38 | 62 |

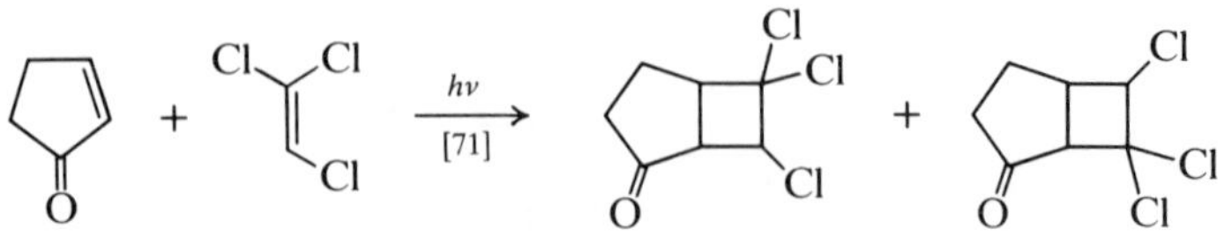

(87) + (91) $\xrightarrow[\text{[71]}]{h\nu}$ (92) + (93)

| °C | **92** | **93** |
|---|---|---|
| 25 | 42 | 58 |
| −75 | 50 | 50 |

**Experimental 6.2** Photoaddition of cyclohexenone to 1,1-dimethoxyethylene [35]

A solution of cyclohexenone (4.09 g, 0.043 mol), 1,1-dimethoxyethylene (53.49 g, 0.61 mol), and 270 ml of pentane was irradiated with a Type L, 450 W Hanovia mercury arc through a Corex-filter for 5 h. The reaction was monitored by measuring IR absorption. Removal of the solvent was followed by distillation on a 46 cm spinning-band column. The forerun, 22.41 g of recovered 1,1-dimethoxyethylene (BP 88–92 °C) was followed by the photoadduct collected in three fractions: (i) 1.24 g, BP 47–73 °C (0.25 mmHg); (ii) 2.80 g, BP 64 °C (0.02 mmHg)–57 °C (0.01 mmHg); and (iii) 3.13 g, BP 61–97 °C (0.01 mmHg). Thus, 7.17 g (91%) of distillate and 0.61 g (7.8%) of brown oily residue were obtained. Analysis of the distillates by vapour-phase chromatography [Model 300 gas chromatography fitted with column (8 feet, 5% Dow Corning F5 1265 on Diatoport S)] indicated product yields of components at the following elution times: 6.9% at 6.0 min, 21.3% *cis*-cyclobutane (12.7 min) and 48.6% *trans*-cyclobutane (15.7 min). Distillation fractions 2 and 3 partially solidified. The solid was twice alternately recrystallized from pentane and evaporatively distilled (65 °C, 0.13 mmHg) to give colourless irregular prisms, MP 51–52.5 °C, of *trans*-cyclobutane.

#### 6.2.1.3 Cycloaddition of $\beta$-dicarbonyl compounds to C═C double bonds (the de Mayo reaction)

One of the earliest examples of the addition of enones to alkenes was the reaction between 2,4-pentanediones, such as acetylacetone (**94**), and several alkenes to form 1,5-diketones, such as (**96**) [73, 74]. This reaction is often referred to as the de Mayo reaction and proceeds through a hydrogen-bonded enol tautomer of (**94**), which in the excited state reacts with alkenes in the same way as do the $\alpha,\beta$-unsaturated ketones discussed in Section 6.2.1.2. The primary photoproducts, acyl cyclobutanols such as (**95**), undergo spontaneous retro-aldol cyclobutane fragmentation.

The majority of synthetic interest in this reaction stems from the ready formation of cyclohexenones on exposure of the 1,5-diketone photo-

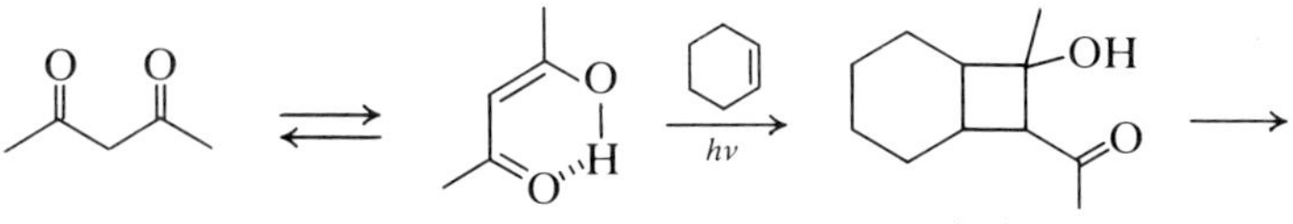

(94) $\xrightarrow{h\nu}$ (95) →

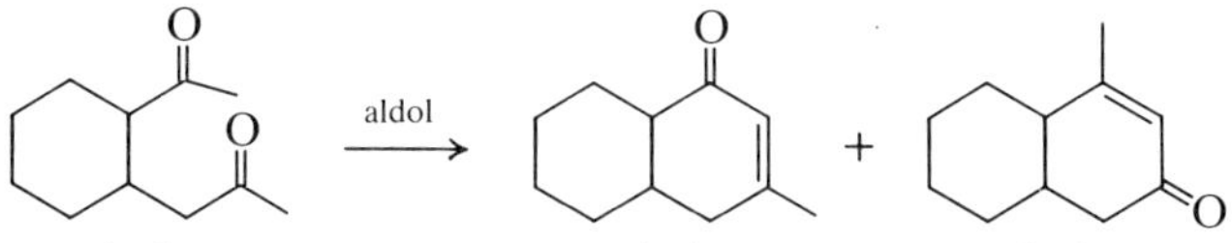

(96) $\xrightarrow{\text{aldol}}$ (97) + (98)

products to aldol condensation conditions, although this reaction is often complicated by competing aldol pathways (**97**) and (**98**).

**Experimental 6.3** Acetonyl-2-acetylcyclohexane [74]

A solution of the diketone (**94**) (15.02 g) in cyclohexene (135 ml) was irradiated under nitrogen with an 80 W water-cooled immersion lamp for 45 h. After removal of the solvent, the product was fractionated on a 40 inch spinning-band column to give the adduct (**96**), BP 89 °C (1.7 mm Hg), with a yield of 20.2 g (78%).

### *(i) Scope and limitations*

As the enone components, several types of $\beta$-diketones have been employed:

Acetylacetone

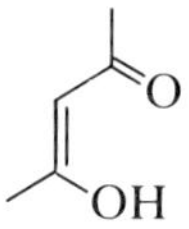

Cyclic 1,3-diketone and its enolacetate and enol ether

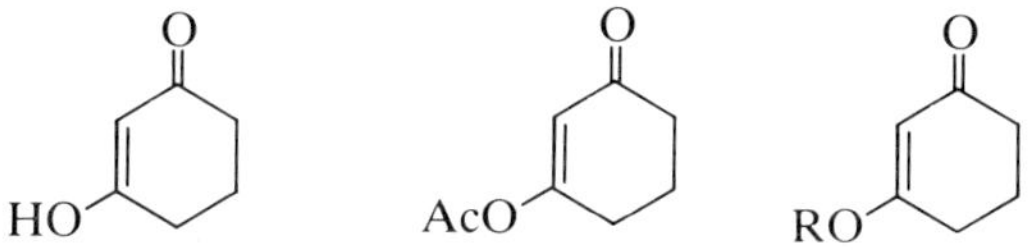

HO AcO RO

Cyclic $\beta$-ketoester and $\beta$-ketoamide

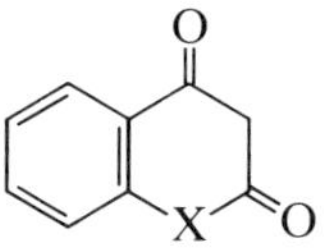

X=O, NMe

The reaction is also possible with cyclic 1,3-diketones, the enone forms of which also behave as $\beta$-hydroxyenones. Thus dimedone (**99**) and cyclohexene react smoothly to give the cyclooctanedione (**101**) after spontaneous retro-aldolization of the primary photoproduct (**100**) [75–77]. Cyclic $\beta$-ketoesters and $\beta$-ketoamides react in a similar fashion to give (**102**) → (**103**) [78]. Enol derivatives (ethers and esters) of cyclic $\beta$-diketones react with alkenes as expected to give stable cyclobutane photoadducts [75, 77], as seen in the reaction between dimedone enol acetate (**104**) and cyclopentene, which gave the stereoisomeric adducts (**105**) and (**106**) [79]. Not surprisingly, similar enol derivatives of acyclic $\beta$-diketones do not give cycloadducts with alkenes, presumably because they are free to undergo energy-wasting *cis–trans* isomerization.

(**99**) (**100**) (**101**)

(**102**) (**103**)

(**a**) X = O (40%)
(**b**) X = NMe (35%)

(**104**) (**105**) (2 : 1) (**106**)

The key structural feature of the acyclic $\beta$-diketones that allows for successful photocycloaddition reactions but is absent in their enol derivatives would appear to be the intramolecular hydrogen bond of the enol tautomer. This hydrogen bond apparently confers sufficient configurational stability on the enol ring to prevent energy-wasting *cis–trans* isomerization.

*(a) Unsymmetrical acyclic $\beta$-diketones* One of the most interesting aspects of the de Mayo reaction concerns unsymmetrical acyclic $\beta$-diketones, where

the presence of two hydrogen-bonded enol tautomers raises the possibility of two photochemically active enol tautomers. Although it is possible to determine the relative amounts of two tautomers in the ground state by conventional NMR techniques [80], such information is of little value in predicting which enol tautomer will give rise to the products actually obtained. For instance, a study of several 2-acetylcycloalkanones (**107**) has shown that, depending on the structure, the cyclobutane product (**108**) can be derived from the enol of highest (**107**, $n = 1$) or lowest (**107**, $n = 2$) ground-state concentration or from a mixture of each [(**109**) → (**110**) + (**111**)] [81]. In the latter case the photoproduct (**110**) gave rise directly to the diketone (**112**), thus providing a convenient route to the steroid skeleton (**113**).

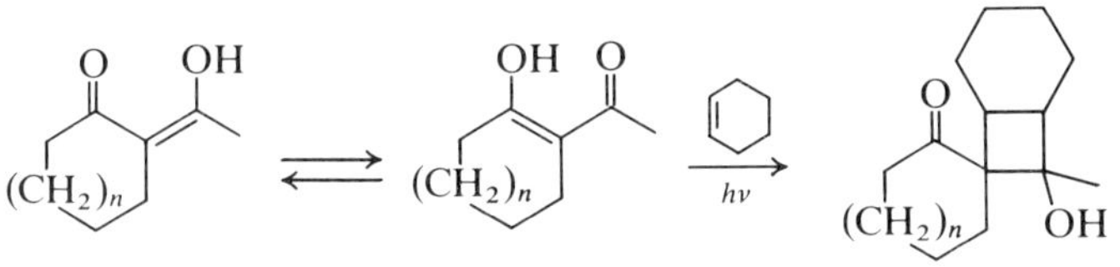

(**107**) $\xrightarrow{h\nu}$ (**108**) $n = 1$ (40%), $n = 2$ (61%)

(**109**) $\xrightarrow{h\nu\ [83\%]}$ (**110**) (1 : 1) (**111**)

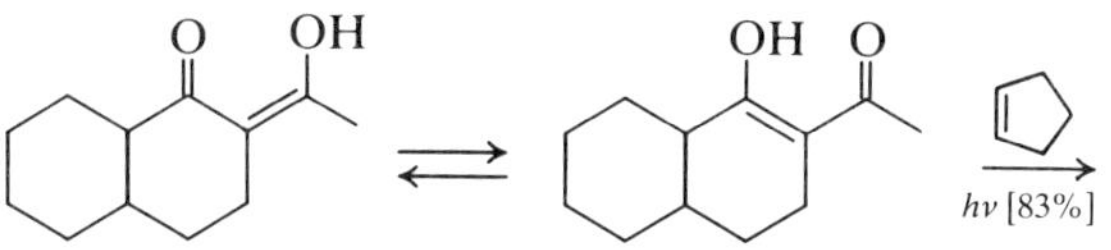

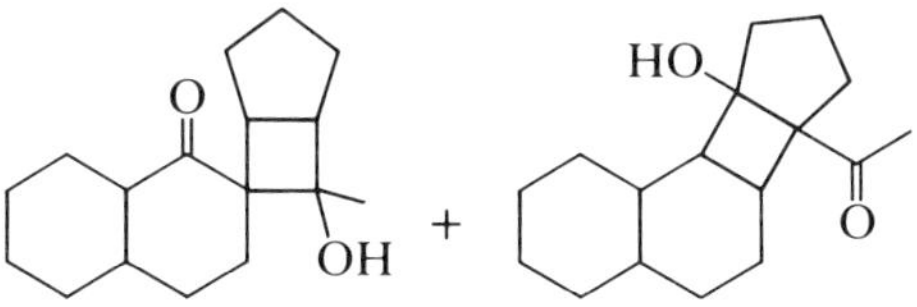

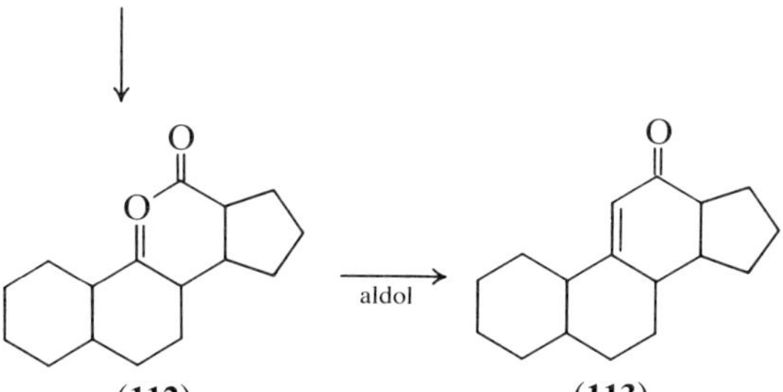

(**110**) → (**112**) $\xrightarrow{\text{aldol}}$ (**113**)

*(b) Acyclic α-formylketones* Acyclic α-formylketones are an interesting class of β-dicarbonyl compounds. Like their diketone analogues, they are highly enolized, with appreciable amounts of each of the two extreme enol tautomers present in the ground state [82–84]. Interestingly, in all cases so far studied photoadditions of alkenes to α-formylketones have been shown to occur exclusively with the enol tautomer that is enolized toward the aldehyde, representing an enol-specific reaction [85, 86]. These reactions are significantly more rapid than those of the corresponding diketones. Thus in a low-conversion competition experiment between acetylacetone (**94**) and the 4-substituted formylacetone (**114**) for excess cyclohexene, there was 3.3 times as much product (**115**) derived from the active enol of (**114**) as (**95**) from the active enol of (**94**) [87].

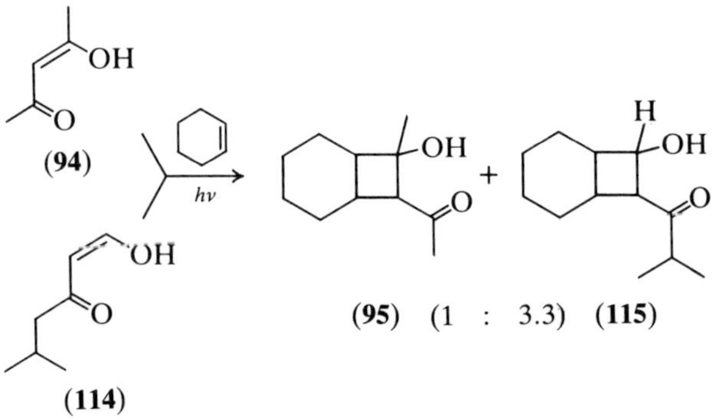

OH O (**94**) $h\nu$ OH O (**114**) OH O + H OH O

(**95**) (1 : 3.3) (**115**)

Acyclic β-ketoesters such as ethyl acetoacetate do not give cyclobutanes on irradiation with alkenes, although oxetanes and free-radical addition products are observed [88]. It seems likely that the decreased acidity of the β-ketoesters relative to β-diketones lowers the enol concentration and weakens the configurationally restricting intramolecular hydrogen bond. Perturbation of the ground-state enol concentration has also been noted on irradiation [89]. There are sufficient examples of successful additions to covalently restricted unsaturated esters (e.g. (**102a**), (**116**) [90, 91] and (**117**) [92]) to demonstrate that there is no problem with the α,β-unsaturated ester chromophore itself.

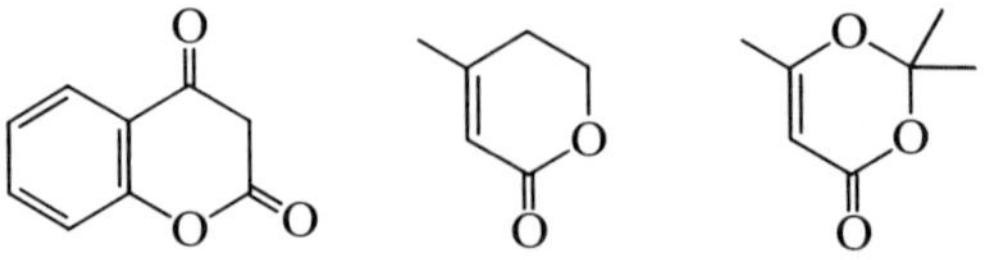

(**102a**) (**116**) (**117**)

### *(ii) Regiochemistry*

The regiochemical picture for the addition of alkenes to $\beta$-dicarbonyl compounds is in general accord with that seen previously for cyclic enones. Thus simple $\alpha$-formylketones give a preponderance of HT adducts with electron-rich alkenes such as isobutene, although the magnitude of the HT preference is not as great as might be expected if the $\alpha$-formylketone is taken as a 4-oxenone (e.g. (**118**) or (**119**)) [93]. However, the reaction still retains much of its value in the preparation of cyclohexenone derivatives. Typical regiochemical ratios for 4-substituted formylacetone derivatives and several alkenes are listed below.

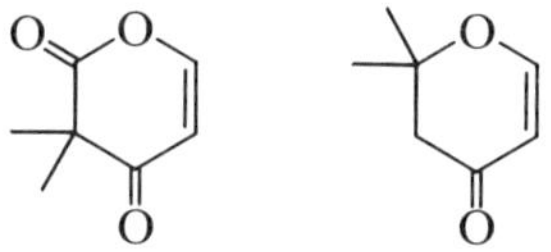

(**118**) (**119**)

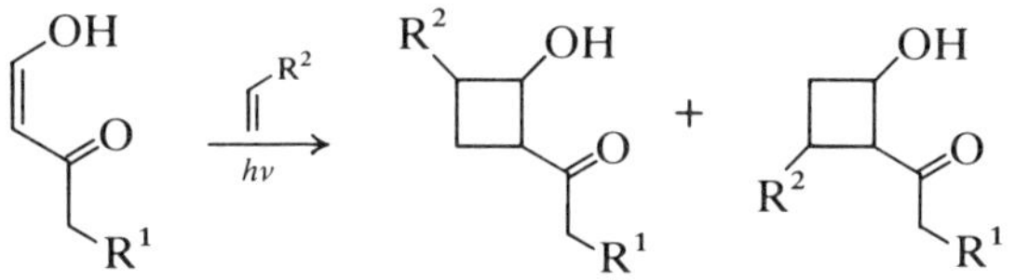

| Alkene | HT | HH |
|---|---|---|
| 1-Butene | 1.3 | 1 |
| Isobutene | 3 | 1 |
| 2,5-Dimethyl-2,4-hexadiene | 5 | 1 |
| 2-Methyl-2-butene | 2.5 | 1 |
| 1-Methylcyclohexene | 3 | 1 |

### *(iii) Solvent effects*

The de Mayo reaction is solvent-dependent, with the rate of reaction being lower in solvents which are hydrogen-bond acceptors such as ether, ethyl acetate and acetonitrile. Presumably such solvents lead to weakening of the intramolecular hydrogen bond in analogy with the ground-state situation.

Photoaddition with acyclic $\beta$-carbonyl derivatives is usually best carried out in non-polar solvents, although ether and methylene chloride have also been used successfully. Polar substituents on the alkene can also retard the reaction for the reasons mentioned previously.

### *(iv) Synthetic usefulness of the de Mayo reaction*

Of particular interest in this photoannellation reaction with $\alpha$-formylketones has been the use of formylacetone (**120**) to provide a specific four-carbon annellation sequence. The particular advantages of this technique derive from the complete enol-specificity of the photoaddition step and from the unidirectional nature of the cyclodehydration step, in contrast with the situation with $\beta$-diketones. This two-step sequence therefore provides a convenient source of cyclohexanones, complementing the traditional Robinson annellation reaction with methylvinyl ketone [94, 95] and the Diels–Alder reaction with derivatives of butadiene [96].

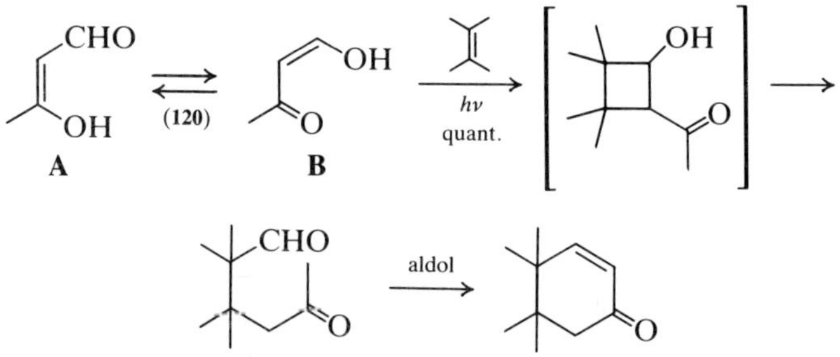

CHO OH A ⇌ (120) OH O B —hν, quant.→ [OH O] →

CHO O —aldol→ O

High enol-specificity has similarly been obtained in the reactions of methyl acetopyruvate (**121**) with C=C double bonds [97]. For instance, the photoaddition of (**121**) to cyclohexene gives a 95% yield of the photoproduct (**123**), derived exclusively from the enol tautomer (**121B**). Similar enol-specificity is also observed with other alkenes [98].

There are several examples of the photoreactions of $\beta,\beta'$-tricarbonyl compounds with alkenes. Dehydroacetic acid (**125**) reacts with cyclohexene

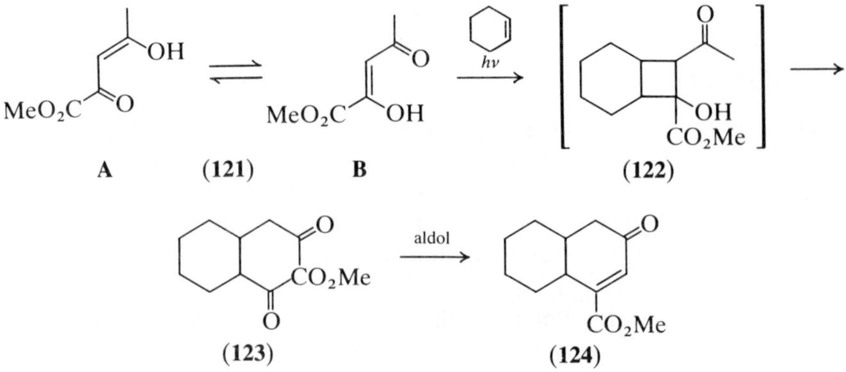

OH $MeO_2C$ O A ⇌ (121) O $MeO_2C$ OH B —hν→ [O OH $CO_2Me$] (122) →

O $CO_2Me$ O (123) —aldol→ O $CO_2Me$ (124)

to give several photoproducts, the two major ones being (**126**) and (**127**) [99]. More useful from the synthetic point of view has been the photochemical cycloaddition reaction of methyl diformylacetate (**128**), which reacts smoothly with alkenes to give the novel enol-hemiacetal structure present in many of the iridoid glucosides. The reaction is interesting because of the several spontaneous tautomeric reorganizations required for the formation of the product (**131**) from the initial photoproduct (**129**).

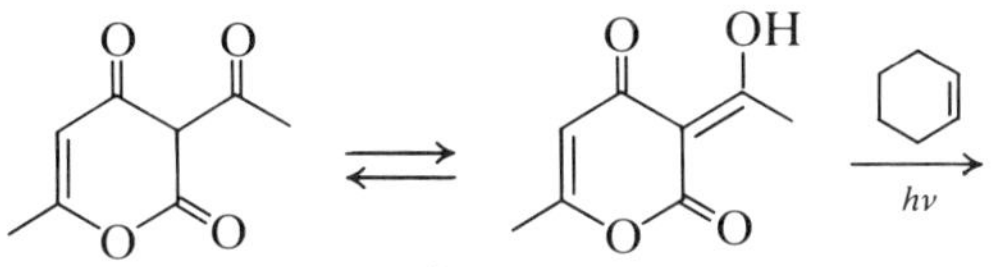

**A** (**125**) **B**

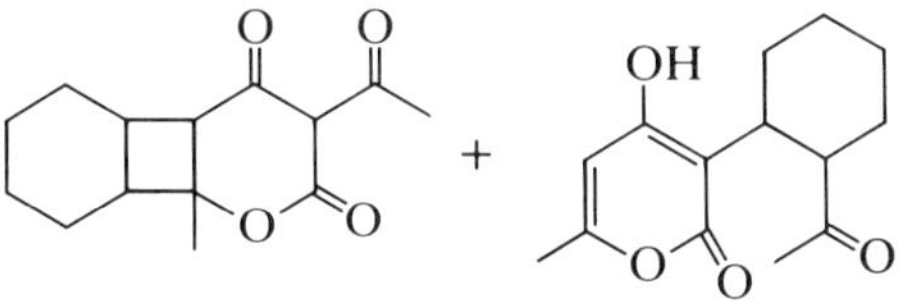

(**126**) (**127**)

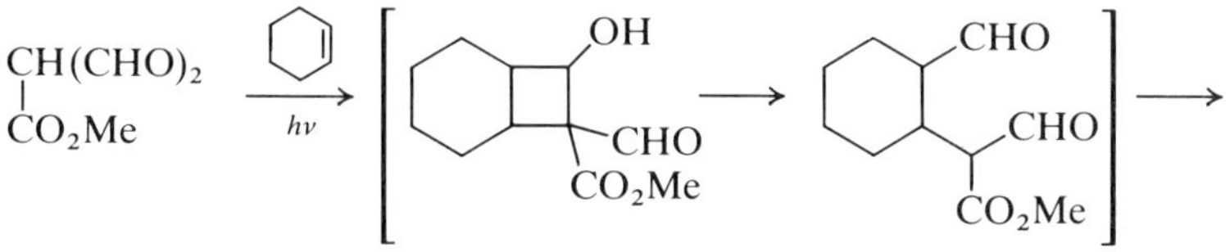

$CH(CHO)_2$ / $CO_2Me$ (**128**) → (**129**) → (**130**) 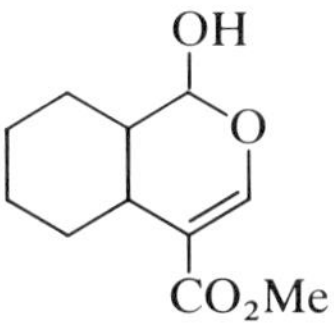 → (**131**)

This photochemical reaction has been utilized in the total synthesis of such important iridoids as loganin (**132**) [100–102], secologanin (**133**) [103], swerside (**134**) [104] and sarracenin (**135**) [105].

It has been possible to convert the photoproducts of the reaction of (**128**) with simple alkenes to $\alpha$-methylene-$\delta$-lactones by a simple reduction–oxidation sequence [106]. Thus reduction of the cyclohexene photoproduct

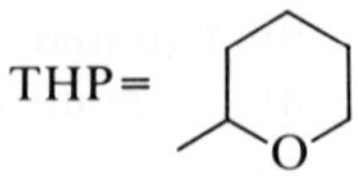

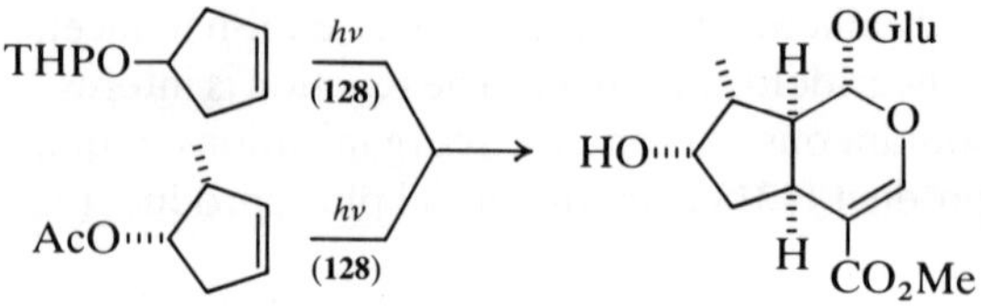

THP=

THPO— $\xrightarrow[(128)]{h\nu}$

AcO···· $\xrightarrow[(128)]{h\nu}$ → HO···· H OGlu O H $CO_2Me$ (132)

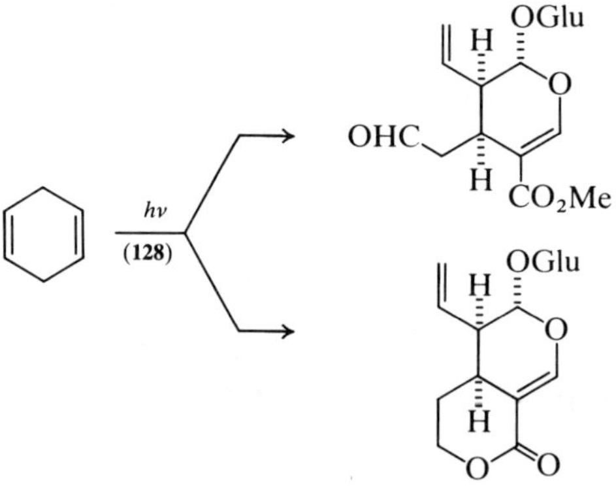

OHC H OGlu O H $CO_2Me$ (133)

$\xrightarrow[(128)]{h\nu}$

H OGlu O H O O (134)

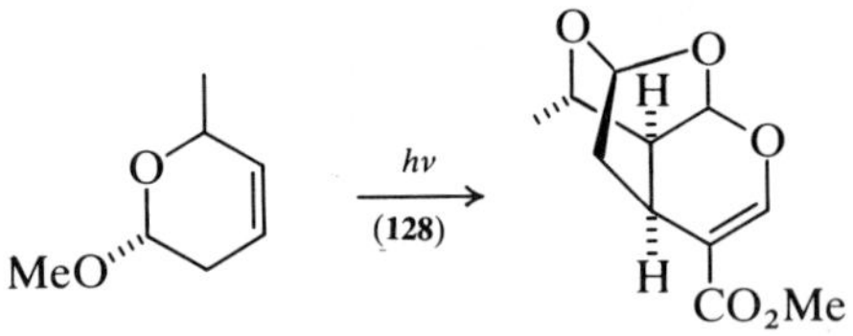

MeO O $\xrightarrow[(128)]{h\nu}$ O O H O H $CO_2Me$ (135)

(131) with lithium aluminium hydride gives the diol (136), which on direct exposure to manganese dioxide gives the methylene lactone (137) in 70% yield [106].

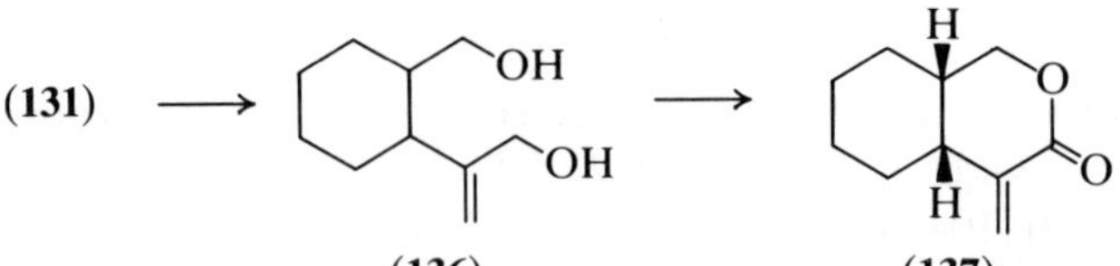

(131) → OH OH (136) → H O O H (137)

### 6.2.1.4 Other examples of cyclobutane formation from other components

#### *(i) Alkyne photoadditions*

Photocycloadditions of enones with alkynes occur with varying efficiency, and are often accompanied by competitive enone dimerization. In the reaction between cyclohexenone and 3-methyl-3-butenyne exclusive cycloaddition to the alkene was observed [107]. The reaction with alkynes is noteworthy because of the highly strained nature of the product cyclobutenes. Acetylene itself appears to act as a (reluctant) photopartner [108], although only modest yields of photoproducts have been reported [109]. The cyclobutene photoproducts are capable of further photochemical reactions, as shown by the rearrangement of (**138**) to a photostationary equilibrium with (**139**) [110]. This rearrangement is not usually observed with cyclohexenone photoproducts.

(**87a**) (**138**) (**139**)

One of the more interesting aspects of alkyne photoadditions involves cycloaddition to terminal alkynes. Thus cyclopentenone (**87a**) and 1-hexyne react to give two primary photoproducts, (**140**) and (**141**), in a 4 : 1 ratio, of which the isomer (**141**) rearranges to a mixture of (**141**) and (**142**) (2.3 : 1) [72]. The related rearrangement of (**140**) is degeneration. In general, preferential HH regiochemistry is observed with terminal alkynes, and is surprising in view of the expected electronic similarity between alkynes and alkenes.

(**87a**) R = $C_4H_9$ (**140**) (**141**) (**142**)

**Experimental 6.4** Photoaddition of trimethylsilylpropyne to cyclopentenone [109]

Anhydrous nitrogen gas was bubbled into a solution of cyclopentenone (**87a**) (3.4 g) and trimethylsilylpropyne (85 g) for 20 min. The solution was irradiated with a mercury lamp (Hanovia 450 W) through a Pyrex filter for 16 h and the reaction was monitored using gas chromatography [Perkin-Elmer F20: Chromosorb W, 4 m, 5% SE 30 (A 700)]. After removal of the remaining acetylene by distillation at ordinary pressure, the residue was distilled under reduced pressure to give an adduct (**138**) (5.6 g, 70%), BP 55–57 °C (0.1 mmHg) which was found to be a mixture of two regioisomers (83 : 17) by analysis of the gas chromatograph and the NMR spectrum. The last fraction of the distillation consisted of the cyclopentenone dimer (0.45 g).

### *(ii) Allene photoadditions*

The cycloaddition of allene (**143**) to enones is accompanied by unusual regiochemical regularity [36]. The stereochemistry of the addition is of interest, as will be seen shortly, and the photoproducts have proved to be particularly valuable synthetic intermediates [35].

$CH_2=C=CH_2$ ; $h\nu$ [62%]

(**87a**) (**143**) (19 : 1)

$CH_2=C=CH_2$ ; $h\nu$ [55%]

(**54a**) (**143**)

**Experimental 6.5** Photoaddition of cyclohexenone to allene [35]

A solution of cyclohexenone (**54a**) (11.0 g, 0.1145 mol), allene (**143**) (100 g, 2.5 mol), and pure pentane (530 ml) was irradiated with a Type L, 450 W Hanovia mercury arc through a Corex filter for 4.5 h. During irradiation the whole vessel was immersed in a dry-ice bath sufficiently deep to cool it even above the level of the internal liquid. If the solution is not cold enough when the allene is introduced, rapid ebullition can occur resulting in loss of sample. At the end of the reaction a long hypodermic needle was inserted into the reaction mixture and a stream of nitrogen was passed through to blow off excess of allene. The cooling bath was slowly lowered and the allene was carefully distilled. Removal of the solvent and distillation gave 8.6 g (55%) of the adduct, BP 43–45 °C (0.65 mmHg).

### *(iii) Photoaddition of maleic acid derivatives*

*(a) To allenes* The photochemical cycloaddition of maleic anhydride to unsaturated compounds has been found to be widely applicable. Maleic

anhydride may be replaced by alkyl maleates, maleimides or alkyl-substituted maleic anhydrides. The alkene component may be open-chain or cyclic [111].

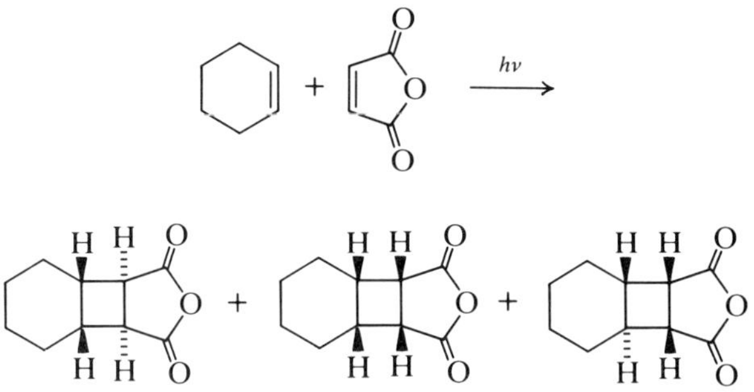

Cyclohexene appears to form charge-transfer complexes with maleic anhydride. Irradiation gives rise to a variety of products consisting of three 1 : 1 cycloadducts [112].

*(b) To aromatic rings* Addition of maleic anhydride to benzene and naphthalene is also known, but the yield is often low. However, five-membered heteroaromatic rings such as furan and thiophene gave respective adducts in good yield [113, 114].

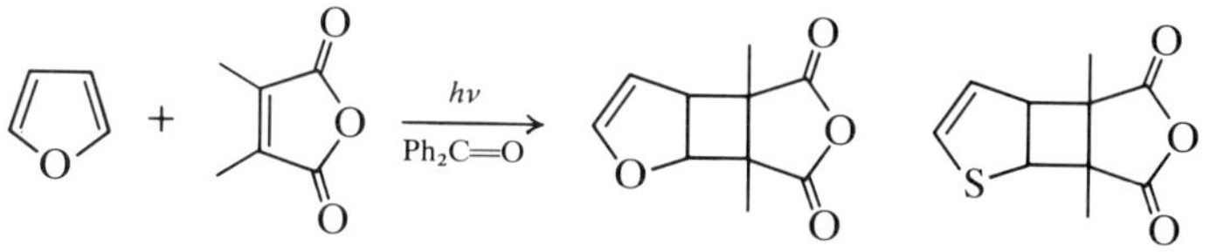

### *(iv) Photocycloaddition of heteroaromatics containing α,β-unsaturated carbonyl systems to alkenes*

Heteroaromatic compounds such as 2-pyridones, 2-pyrones, 2-quinolones, coumarins, thiocoumarins and 1-isoquinolones are excellent enone systems for photocycloaddition reactions with alkenes. In contrast with the photocycloaddition of carbocyclic enone systems [35], the photocycloaddition reaction proceeds regioselectively to afford HT adducts in the case of 2-quinolones [115, 116], coumarins [117] and thiocoumarins [118], and HH adducts in the case of 1-isoquinolones [119]. All products obtained have *cis* stereochemistry without exception.

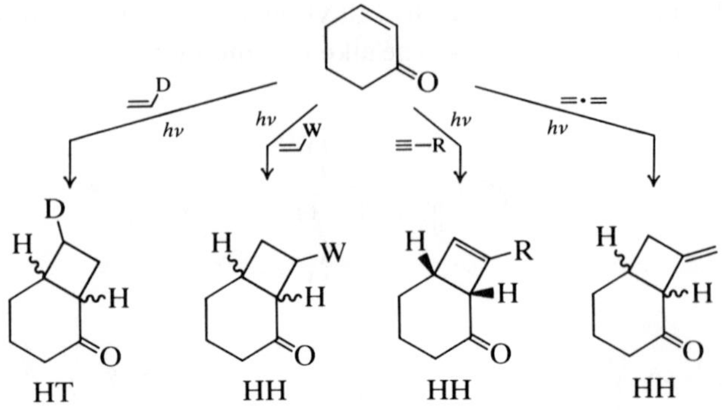

D
$h\nu$
$h\nu$
W
$h\nu$
≡—R
=•=
$h\nu$

HT HH HH HH

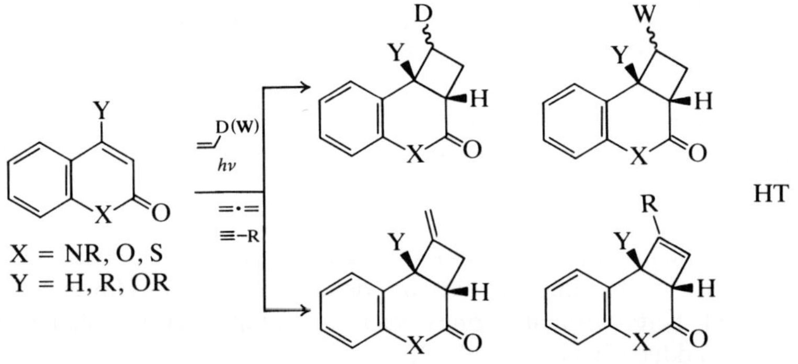

D(W)
$h\nu$
=•=
≡—R

X = NR, O, S
Y = H, R, OR

HT

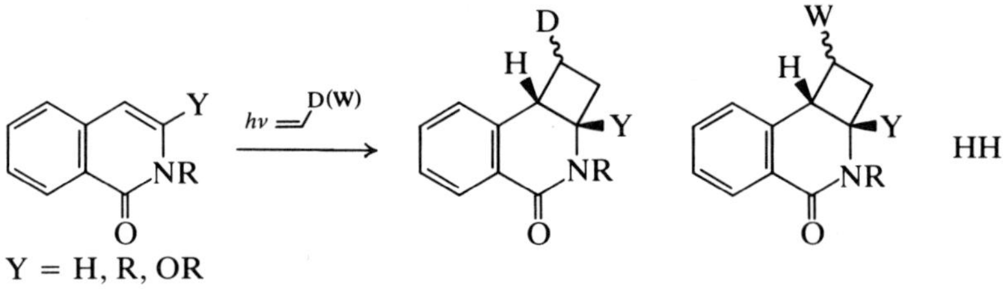

$h\nu$ D(W)

HH

Y = H, R, OR

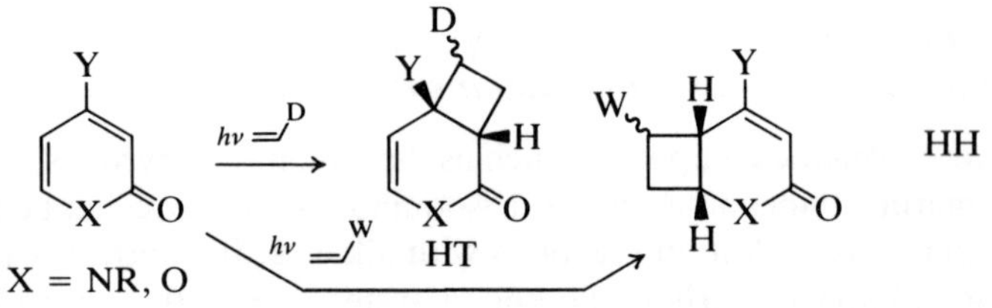

$h\nu$ D

$h\nu$ W

HT

HH

X = NR, O

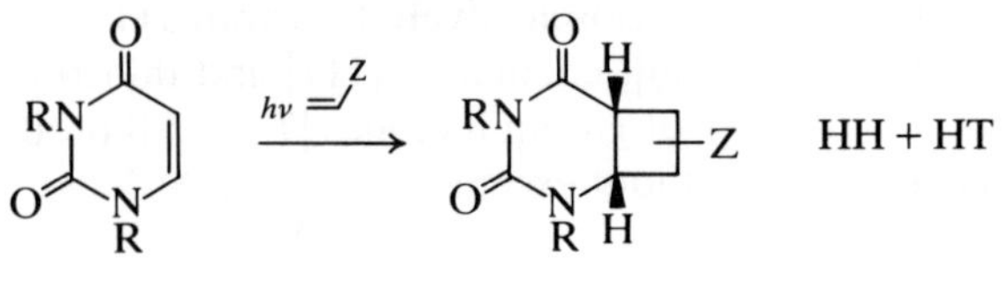

$h\nu$ Z

HH + HT

### 6.2.1.5 Synthetic applications of photocycloaddition

The cyclobutane photoproducts from enone–alkene cycloadditions are particularly versatile synthetic intermediates, which can be used for the preparation of a variety of further products. Space limitations do not permit complete listing of these reactions. However, several recent developments, including those reactions that appear to be of particular synthetic use, will be mentioned. Fortunately, there already exists a thorough review of the application of many of these transformations to the synthesis of natural products [120].

In the conversion of a particular cyclobutane photoproduct to the desired synthetic target, the cyclobutane ring can either be left intact or can be broken. The former situation occurs, for example, in the synthesis of caryophyllene (**42**) [29], the bourbonenes (**43**) [30, 121, 122], grandisol (**144**) [90, 123–128], fragrantol (**145**) [123] and annotinine (**44**) [31]. This section, however, will focus on several of the many ways in which the cyclobutane ring can be further converted to useful products.

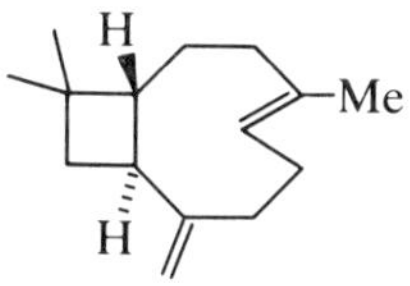

(**42**) 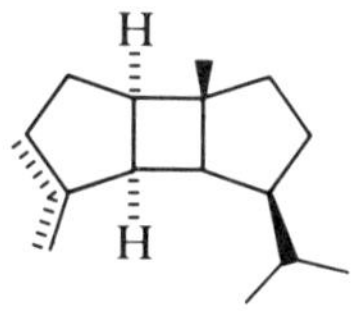 (**43**) 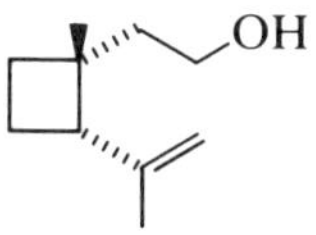 (**144**)

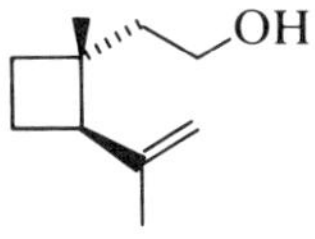

(**145**) 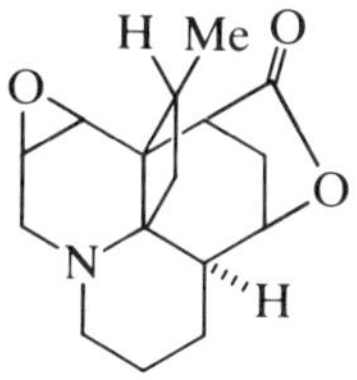 (**44**)

#### *(i) Representative transformation reactions of cyclobutanes*

(*a*) *Base-catalysed elimination* Photoadducts may be fragmented by base-catalysed elimination. For example, the aldehydes (**147**) and (**149**) are formed from the vinylene carbonate adducts (**146**) [129] and (**148**) [130]. It is interesting that the adduct (**150**), which bears no acidic $\alpha$-hydrogen, undergoes a different reaction to give the cyclopropanes (**151**) and (**152**) [129].

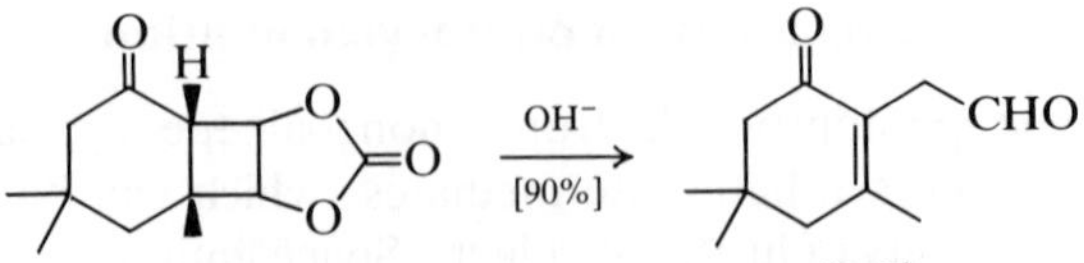

OH⁻ [90%]

(146) (147)

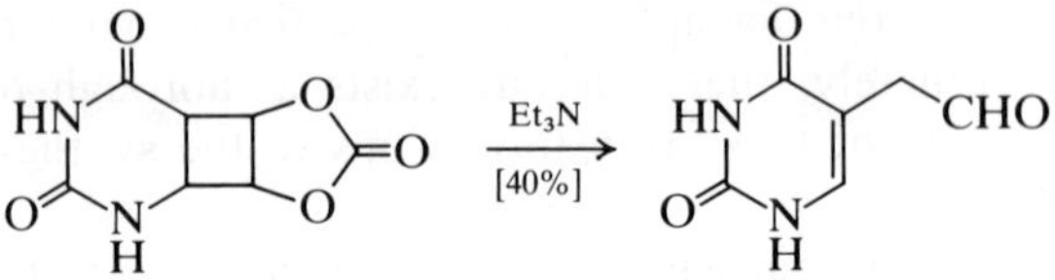

$Et_3N$ [40%]

(148) (149)

OH⁻ [50%] OH⁻ (quant.)

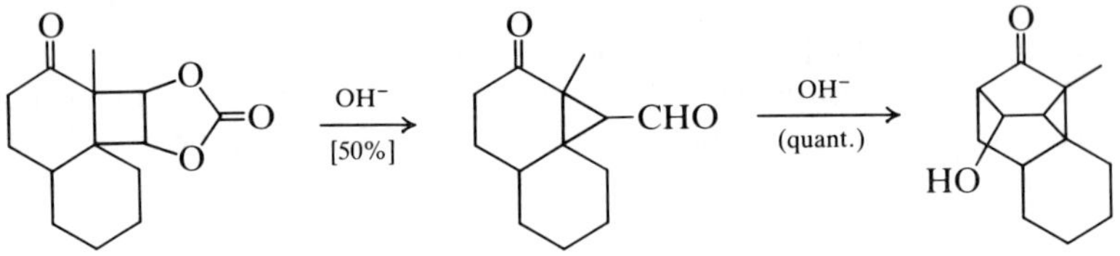

(150) (151) (152)

(*b*) *Retro-aldol reaction* Primary-photoproduct acyl cyclobutanols such as **(95)** undergo spontaneous retro-aldol cyclobutane fragmentation and are seldom observed.

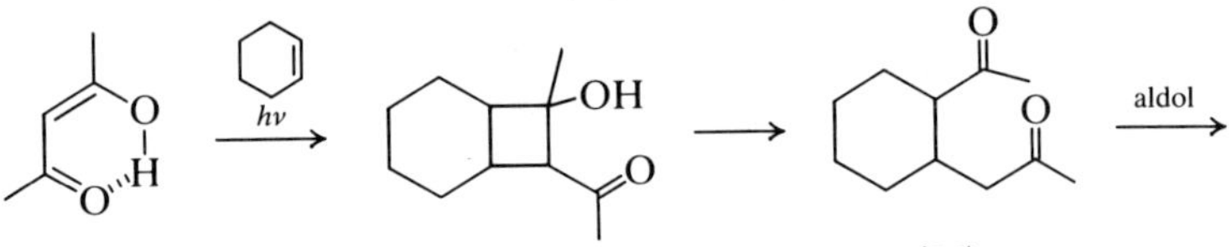

$h\nu$ aldol

(94) (95) (96)

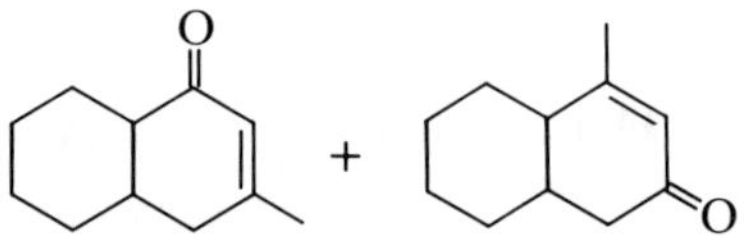

(97) (98)

The principal synthetic interest in this reaction lies in the ready formation of cyclohexenones on exposure of the 1,5-diketone photoproducts to aldol-condensation conditions, although this reaction is often complicated by the competing aldol pathways **(97)** and **(98)** [73, 74].

The reaction is also possible with cyclic 1,3-diketones, the enone forms of which also behave as $\beta$-hydroxyenones. Thus dimedone **(99)** and cyclohexene react smoothly to afford the cyclooctanedione (**101**) after spontaneous retro-aldolization of the primary photoproduct (**100**) [75–77].

**Experimental 6.6** Irradiation of acetylacetone with cyclohexene and retro-aldol reaction and acid-catalysed cyclization of the adduct [74]

A solution of the diketone (**94**) (15.02 g) in cyclohexene (135 ml) was irradiated with an 80-W lamp under nitrogen in a water-cooled immersion apparatus for 45 h. After removal of the solvent, the product was fractionated on a 40-inch spinning band column to give the adduct (**95**) (20.2 g, 78%), BP 89 °C at 1.7 mmHg). The oxime, prepared under the condition using hydroxylamine hydrochloride–pyridine at room temperature, crystallized from methanol had a melting point of 165.5–165.7 °C. A solution of the adduct (**95**) (4.5 g) in ethanol (45 ml) was heated at 80 °C for 80 min with concentrated hydrochloric acid (2 ml). After removal of the solvent, the product was extracted with chloroform, the chloroform solution washed and dried, and the solvent evaporated. Distillation (BP 173–175 °C at 1.5 mmHg) gave a mixture (5 : 3) of two unsaturated ketones (**97**) and (**98**) (4.04 g, 95%). The ketones were separated by chromatography over a 5-feet silicone SE 30 column at 200 °C, and the separated fractions were then distilled at 180 °C (bath temperature) at 35 mmHg to give two ketones (**97**) and (**98**).

(*c*) *Grob fragmentation* Cyclobutane fragmentation can take the form of a fragmentation–elimination reaction, or Grob fragmentation, as seen in the three-step conversion of the photoproduct (**153**) to the cycloheptenone (**155**) via the treatment of (**154**) with base [77, 131], and the related conversion of the intramolecular adduct (**156**) to the bicyclooctenone (**158**) [132].

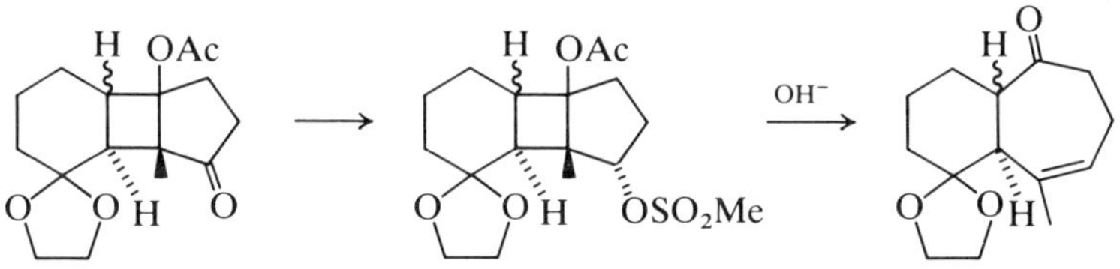

H OAc O O H O (**153**) → H OAc O O H $OSO_2Me$ (**154**) $\xrightarrow{OH^-}$ O H O O H (**155**)

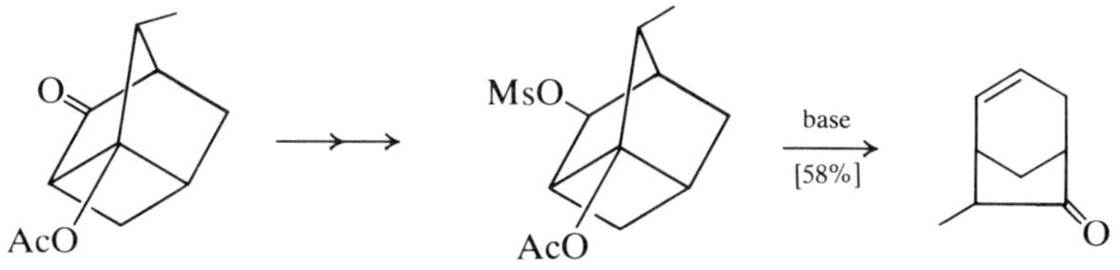

O AcO (**156**) ⟶⟶ MsO AcO (**157**) $\xrightarrow[\text{[58\%]}]{\text{base}}$ O (**158**)

(*d*) *Photochemical cleavage* Another transformation permits the smooth conversion of furanone adducts to cyclohexenones—the same cyclohexenones that are obtained from the photochemical reaction between α-formylketones and alkenes—followed by aldol cyclization as discussed in Section 6.2.1.3. This transformation can be accomplished in two general ways as shown below for the adduct (**160**) from 2,3-dimethyl-2-butene and the furanone (**159**) [133]. These two sequences have recently been employed in the syntheses of the sesquiterpenes occidentalsol (**161**) and acorenone (**162**) [134].

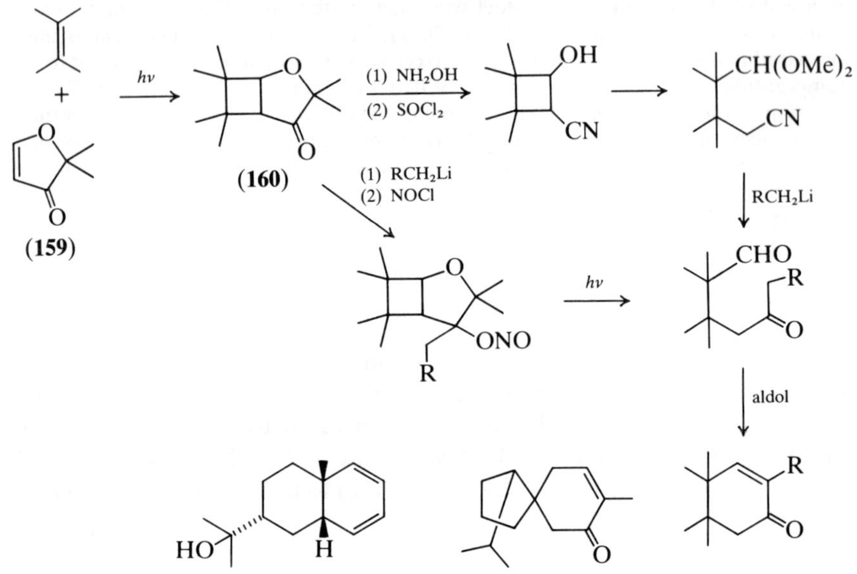
hν
(1) $NH_2OH$
(2) $SOCl_2$
OH
CN
$CH(OMe)_2$
CN
(1) $RCH_2Li$
(2) NOCl
$RCH_2Li$
CHO
R
O
ONO
R
hν
aldol
HO
H
O

(159) (160) (161) (162)

(*e*) *Reductive cleavage* Reductive cleavage of the central σ-bond has been accomplished in several ways to provide a convenient entry to various cyclohexane derivatives. For instance, the photoadduct (**164**) is smoothly reduced to the indane (**165**) (after alkylation of the ester enolate) [135, 136], while application of a similar sequence to the photoadduct (**167**) derived from isophorone (**166**) gave the lactone (**169**), a direct precursor of octalone (**170**) [135–137]. Alternatively, metal–ammonia reduction of the phosphate ester (**171**) derived from (**164**) gave the exocyclic methylene compound (**172**).

(*f*) *Thermal cleavage* Different results are obtained on thermolysis of cyclobutene photoadducts, the products being the consequence of thermal bicyclo[2.2.0]hexane reorganization, followed by a transannular ene reaction. Several examples illustrate the versatility of this method for the construction of complex organic molecules. Thermolysis of (**164**) at 180 °C gives (**173**) as the only rearrangement product, while at 225 °C a 1 : 2 equilibrium mixture of (**173**) and (**174**) is obtained, the latter being derived from (**173**) by a Cope rearrangement [136]. In a similar fashion, thermolysis of the bicyclohexane (**175**), formed by reduction of the photoproduct (**178**) from methylcyclobutene (**177**) and piperitone (**178**), gives rise to several elemene alcohol isomers (**176**) in addition to significant amounts of other products [138]. The presence of a ketone carbonyl (usually from the enone photopartner) leads to further reaction of the initial thermolysis product. Thus the piperitone adduct (**179**) gives *trans*-fused cadinnane (**181**) by transannular ene cyclization of the cyclodecadienone (**180**) [139]. Similar results have been also noted by others [137, 140].

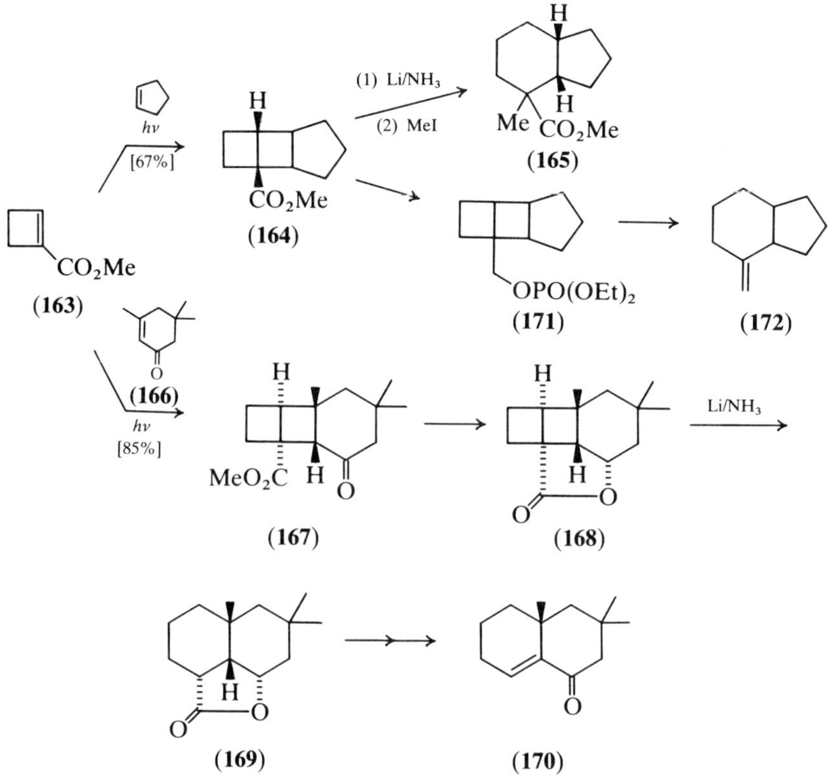

(*g*) *Ring expansion (the Wagner–Meerwein rearrangement)* Relief of the considerable strain that is present in cyclobutane photoadducts can also be accomplished by the Wagner–Meerwein rearrangement of one of the

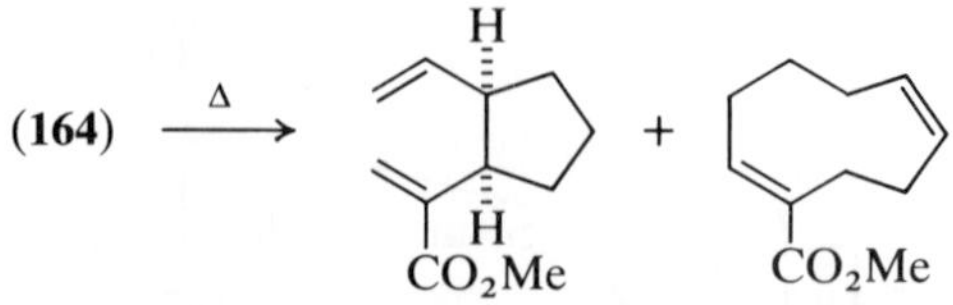

(164) $\xrightarrow{\Delta}$ (173) + (174)

| °C | 173 : 174 | |
|---|---|---|
| 180 | 100 | 0 |
| 225 | 1 | 2 |

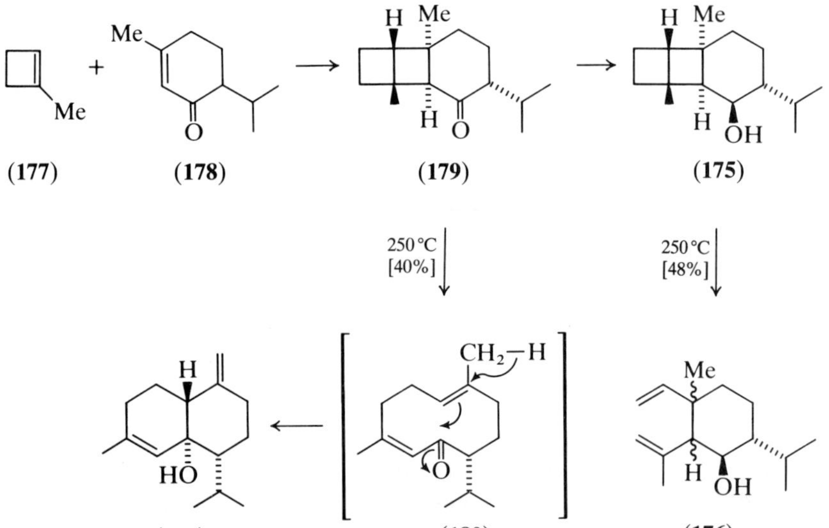

(177) + (178) → (179) → (175)

(179) → 250 °C [40%] → (180) → (181)

(175) → 250 °C [48%] → (176)

cyclobutane σ-bonds. This ring expansion is initiated by the cationic character of an adjacent carbon, usually the enone carbonyl carbon, and results in no net change in the total ring multiplicity of the system. The two alkyl shifts that have the necessary driving force lead to two structurally different systems and are schematically shown below for the photoproduct between an alkene and a six-membered ring enone [141–144]. The bicyclo[4.2.0]octane can give rise to either bridged (a) or fused (b) carbocations, each of which can yield bridged [3.2.1] or fused [3.3.0] products, depending on the nature of X.

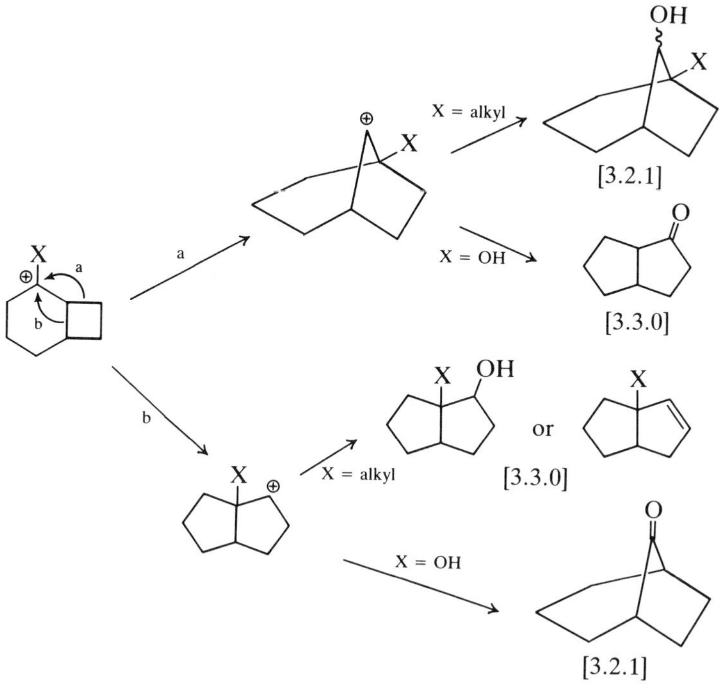

X ⊕ a b

a

b

⊕ X

X = alkyl

OH X [3.2.1]

X = OH

O [3.3.0]

X ⊕

X = alkyl

X OH or X [3.3.0]

X = OH

O [3.2.1]

(*h*) *Radical cleavage* Radical cleavage of [2 + 2] photoproducts, prepared from either excited enones or excited enols and alkenes has provided a new useful photoannellation reaction for various types of compounds as shown in Sections 6.2.1.2 and 6.2.1.3.

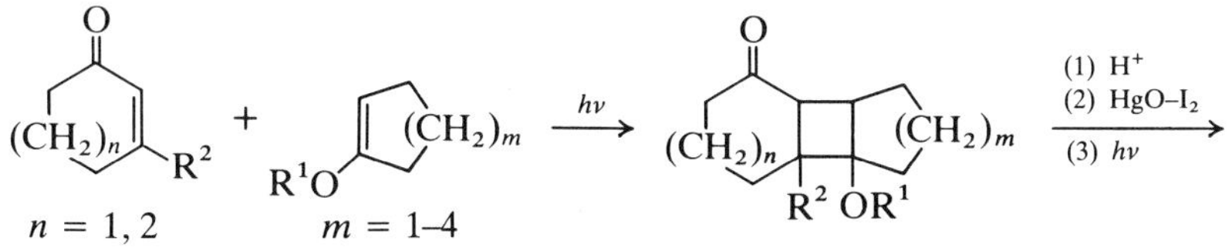

O $(CH_2)_n$ $R^2$ + $R^1O$ $(CH_2)_m$ $\xrightarrow{h\nu}$ O $(CH_2)_n$ $(CH_2)_m$ $R^2$ $OR^1$ $\xrightarrow[(3)\ h\nu]{(1)\ H^+\ (2)\ HgO\text{–}I_2}$

$n = 1, 2$ $m = 1$–$4$

$R = H, Ac, SiMe_3$

(**182**) (**183**) (**184**)

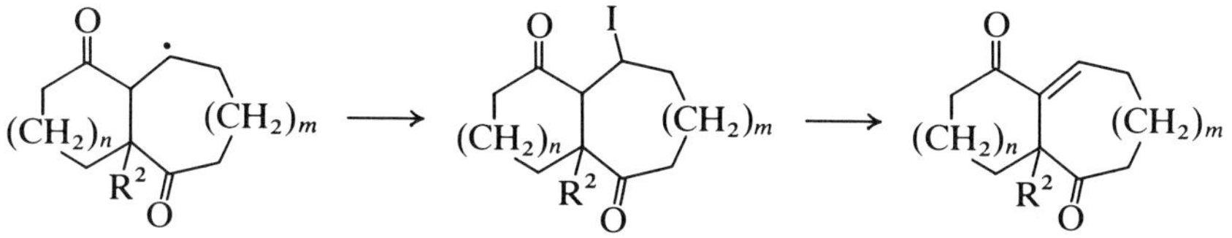

O · $(CH_2)_n$ $(CH_2)_m$ $R^2$ O $\longrightarrow$ O I $(CH_2)_n$ $(CH_2)_m$ $R^2$ O $\longrightarrow$ O $(CH_2)_n$ $(CH_2)_m$ $R^2$ O

(**185**) (**186**) (**187**)

The HT adduct (**184**), prepared from (**182**) and (**183**), was subjected to acid hydrolysis to remove the protective group to give a cyclobutanol, whose hypoiodite was subjected regioselectively to $\alpha$-cleavage to form a mixture of (**186**) and (**187**) via the corresponding alkoxy radical [145–148].

In summary, a variety of C—C bond-fission reactions of a cyclobutane ring formed by photoaddition to alkenes of heteroaromatics containing an $\alpha,\beta$-unsaturated carbonyl system in their ring systems have been described. The reactions are assisted by the strain involved in the four-membered ring and provide unique methods either for introduction of a variety of carbon chains or for annelation to heteroaromatic compounds.

### *(ii) Organic synthesis involving [2 + 2] photocycloaddition*

Since there already exist some excellent reviews on the application of [2 + 2] photocycloaddition to the synthesis of natural products [149, 150], this section will describe recent work on the synthetic application of photocycloaddition.

(*a*) *Intermolecular photocycloaddition of excited enones to C═C double bonds* As described previously (Section 6.2.1.2), intermolecular photoaddition of enones to double bonds is one of the most suitable methods for the synthesis of natural products possessing cyclobutane rings in their structures. This type of photocycloaddition has been established as a potential synthetic route not only for these compounds but also for many target molecules such as sesquiterpenes that are obtainable by modification of the photochemically formed cyclobutane ring.

This section will focus on several examples in which the two starting components, the photoadducts and the final target natural products, are shown in that order.

(*1*) *Total synthesis of (±)-calameon* (**188**) [155]

$CO_2Me$ + → $h\nu$ → $MeO_2C$ H H O → 250°C →

(**163**) (**166**) (**167**)

[ $CO_2Me$ O ] → [97%] → HO $MeO_2C$ → → H OH HO

(**188**)
(±)-calameon

(2) *Synthesis of (±)-sativene* (**189**), *(±)-copacamphene* (**190**), *(±)-*cis-*sativenediol* (**191**) *and (±)-helminthosporal* (**192**) [156]

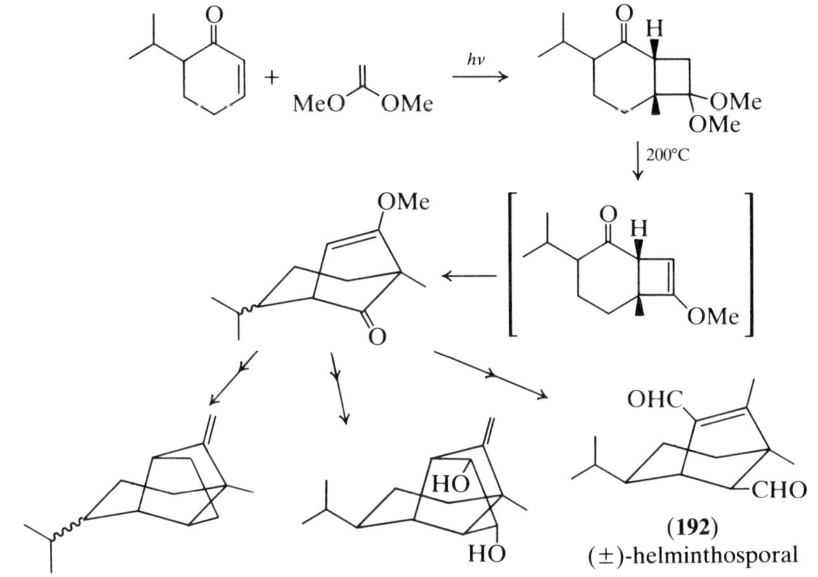

MeO OMe
$h\nu$
H
OMe
OMe
200°C
OMe
H
OMe
OHC
HO
CHO
HO
(**192**)
(±)-helminthosporal

β-Pr[i]: (**189**) (±)-sativene
∝-Pr[i]: (**190**) (±)-copacamphene

(**191**)
(±)-*cis*-sativendiol

(3) *Formal total synthesis of (±)-atractylon* (**193**) *and (±)-isoalantolactone* (**194**) [157, 158]

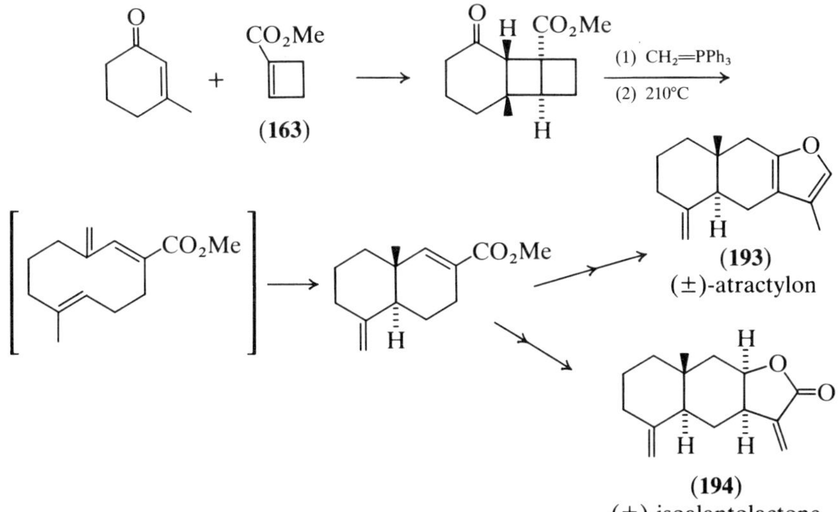

$CO_2Me$
(**163**)
H $CO_2Me$
H
(1) $CH_2{=}PPh_3$
(2) 210°C
$CO_2Me$
$CO_2Me$
H
H
(**193**)
(±)-atractylon
H
H H
(**194**)
(±)-isoalantolactone

(*4*) *Total synthesis of (±)-isabelin* (**195**) [159]

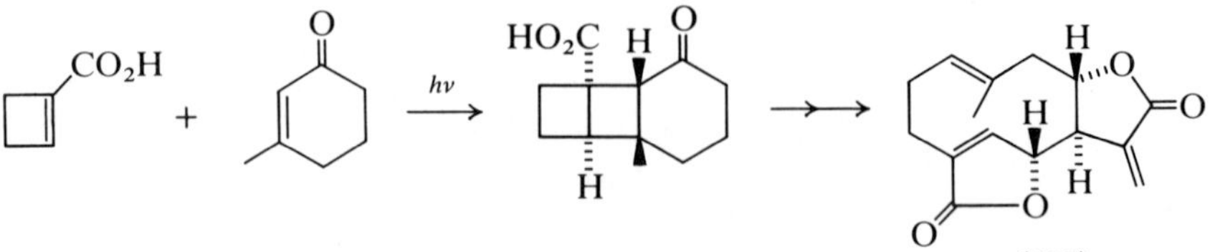

$CO_2H$ + O $\xrightarrow{h\nu}$ $HO_2C$ H O H $\longrightarrow\!\!\!\!\rightarrow$ H O O H H O O O

**(195)**
(±)-isabelin

(*5*) *Asymmetric synthesis of (−)-grandisol* ((−)-**144**) *involving asymmetric [2 + 2] photocycloaddition* In 1986, Meyers and Fleming [160] succeeded in the asymmetric synthesis of (−)-grandisol by the first asymmetric photocycloaddition of a chiral enone to an alkene. Recently, Demuth [161] has also reported an asymmetric synthesis of (+)-grandisol ((+)-**144**) by using asymmetric photocycloaddition of a different chiral enone of (**196**).

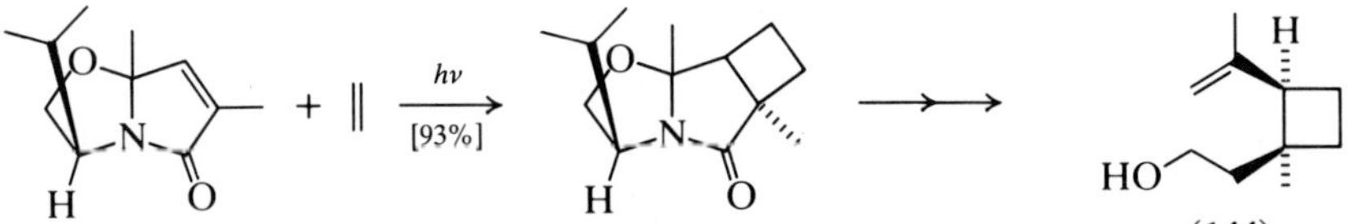

O N H O + ‖ $\xrightarrow[\text{[93\%]}]{h\nu}$ O N H O $\longrightarrow\!\!\!\!\rightarrow$ H HO

(+)-enone

**(144)**
(−)-grandisol

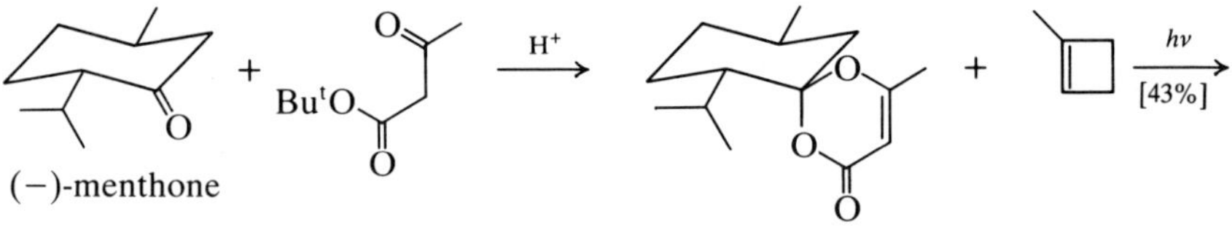

O + $Bu^tO$ O O $\xrightarrow{H^+}$ O O O + $\xrightarrow[\text{[43\%]}]{h\nu}$

(−)-menthone

**(196)**

O O O $\longrightarrow\!\!\!\!\rightarrow$ H HO

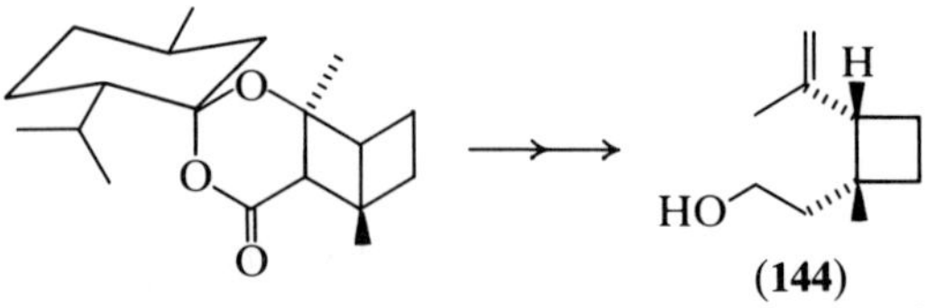

**(144)**
(+)-grandisol

**Experimental 6.7** Photoaddition of (+)-enone to ethylene [160]

The irradiation apparatus was charged with a solution of the enone (5.4 g, 0.0277 mol) in reagent-grade dichloromethane (*ca.* 100 ml) and immersed in a dry-ice/2-propanol bath while the chilled solution was saturated with ethylene. The lamp (G.E. H1000-A36-15, Westinghouse H-36GV-1000) was inserted into the well and turned on. After *ca.* 8 h most of the starting material had reacted. At this time, the lamp was turned off and the apparatus was removed from the cooling bath. The reaction mixture was degassed with a slow stream of nitrogen while it warmed to room temperature, dried over magnesium sulphate, and concentrated with a rotary evaporator at a temperature below 30 °C. The adduct was obtained in 93% yield (5.74 g) containing 7–8% of the *endo*-cyclobutane fused product. The pure product (BP 62–65 °C (3.5 mmHg)) was obtained by chromatography (ether–hexane (1 : 1)) but showed some instability during the separation.

(*b*) *Intramolecular photocycloaddition of excited enones to C=C double bonds* Intramolecular photocycloaddition is expected to proceed in a more stereo- and regio-selective manner than the intermolecular reaction. Wiesner's total synthesis of 1,2-epilycopodine (**197**) is the first synthetic application of intramolecular photocycloaddition of enones to double bonds [162].

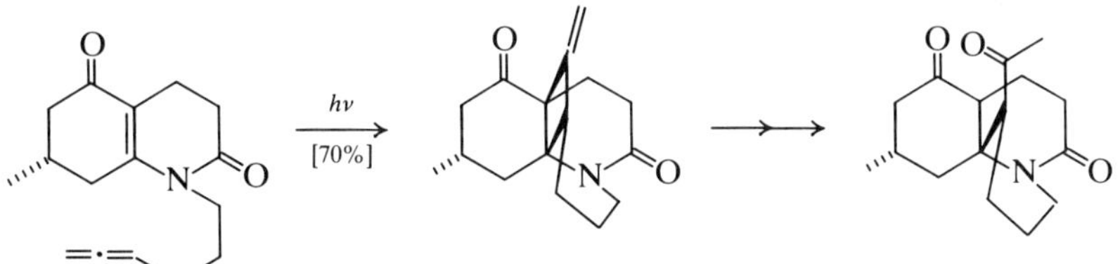

hν [70%]

(**197**)

During the course of the investigation by Wiesner *et al.* of the intramolecular reaction, it was established that, as in the case of radical cyclization, a five-membered ring is preferentially formed whenever two double bonds are connected by either two or three carbon chains [163–165].

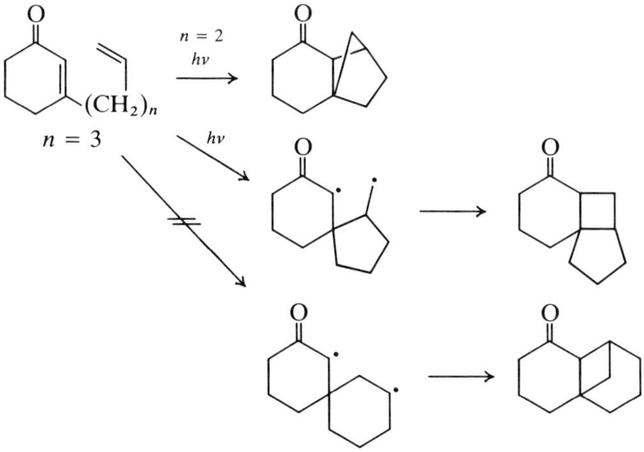

$(CH_2)_n$ $n = 2$ hν $n = 3$ hν

An elegant synthesis of (±)-isocomene (**199**) has been achieved using this preferential formation of five-membered rings [166]. A ring transformation of the 6/4 system in the photocycloadduct (**198**) into a 5/5 system was carried out through a regioselective 1,2-rearrangement of a carbocation. Such a transformation was also involved in the two-step synthesis of the propellane (**200**) by Cargill *et al.* [167].

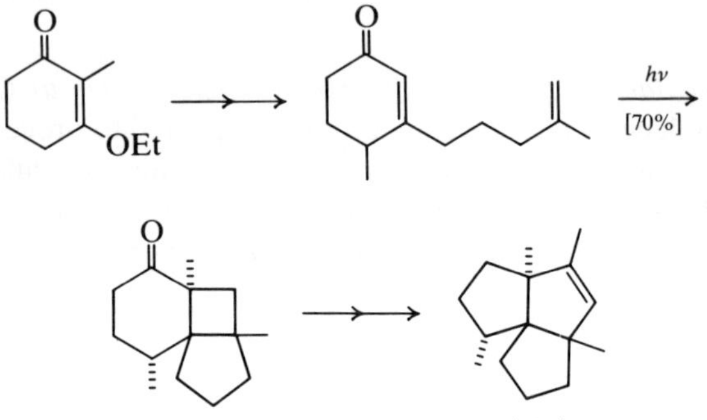

O OEt O $h\nu$ [70%]

O (**198**) (**199**)

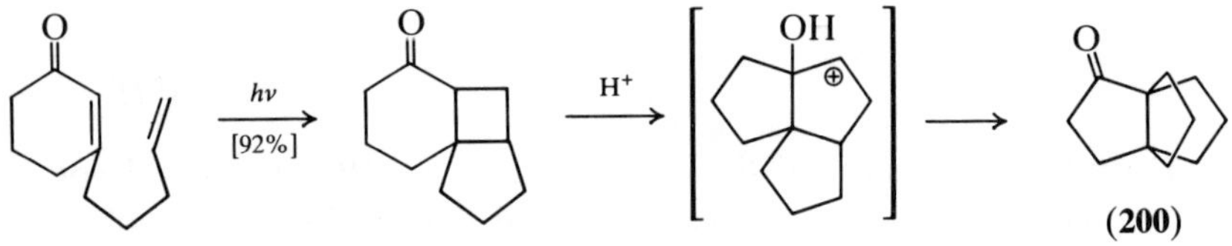

O $h\nu$ [92%] O $H^+$ OH ⊕ O (**200**)

Other examples are as follows.

(*1*) *Total synthesis of (−)-β-panasinsene* (**201**) [168]

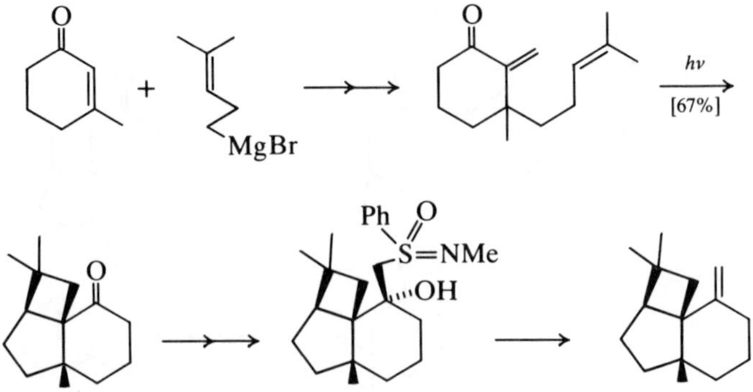

O + MgBr O $h\nu$ [67%]

O Ph O S=NMe OH (**201**)

(2) *Total synthesis of (±)-hirsutene* (**202**) [169, 170]

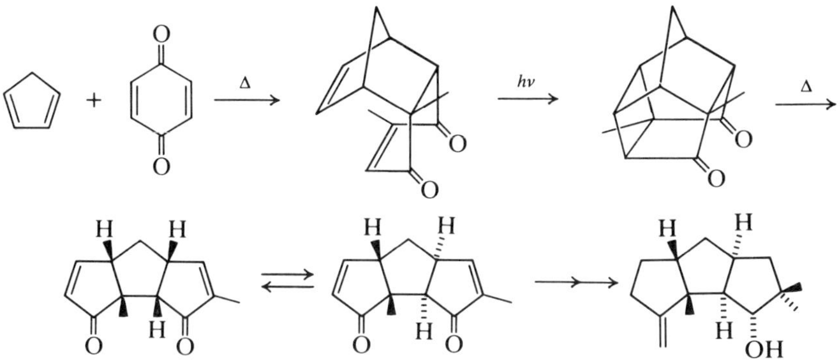

**(202)**
(±)-hirsutene

(3) *Total synthesis of (±)-α-acoradiene* (**203**) [171]

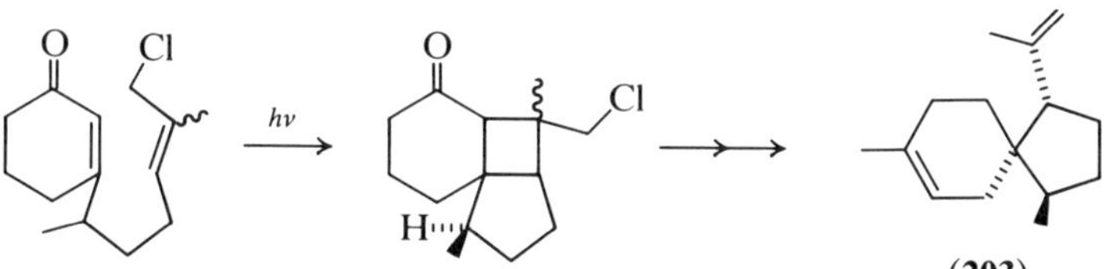

**(203)**
(±)-α-acoradiene

**Experimental 6.8** Intramolecular photoaddition of an allylic chloride [171]

A solution of the allylic chloride (102 mg, 0.424 mmol) in benzene (100 ml) was irradiated with a 125-W high-pressure mercury lamp through a Pyrex filter at room temperature under nitrogen bubbling for 9 h. After removal of the solvent, the residue was chromatographed on silica gel (Merck, silica gel 60) with n-hexane : ether (95 : 5) to give the starting enone (9 mg, 9%) and a mixture of four photoproducts (77 mg, 76%) which was separated by medium-pressure chromatography (prepacked columns, Merck, LiChro-Prep Si 60, 125 mm Hg) to yield the two desired adducts, the main product (oil, 29 mg) and the minor polar cycloadduct (oil, 10 mg).

(c) *Inter- and intra-molecular photocycloaddition of excited 1,2- and 1,3-diketone enols to C═C double bonds (the de Mayo reaction)* As an extension of the de Mayo reaction, which is the photocycloaddition of enolated 1,3-diketones to double bonds (Section 6.2.1.3), enol esters and enol ethers have been employed as enone components since the corresponding adducts can be readily isolated and characterized.

Since Hikino and de Mayo [75] applied this photocycloaddition to the synthesis of $\gamma$-tropolone (**204**) in two steps and in 45% yield, many natural products have been synthesized via the de Mayo reaction.

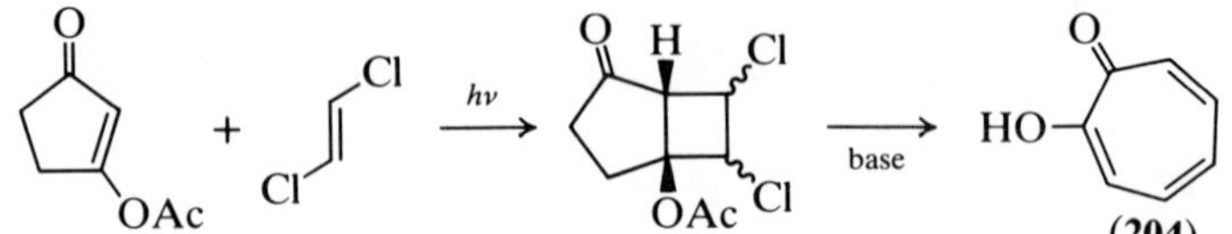

(**204**)

Oppolzer *et al*. have applied the de Mayo reaction to the elegant total synthesis of a number of sesquiterpenes, such as longifolene (**205**) [172]. The following are four examples of natural-product synthesis using the de Mayo reaction published after 1979.

(*1*) *Total synthesis of (±)-hirsutene* (**202**) [173, 174]

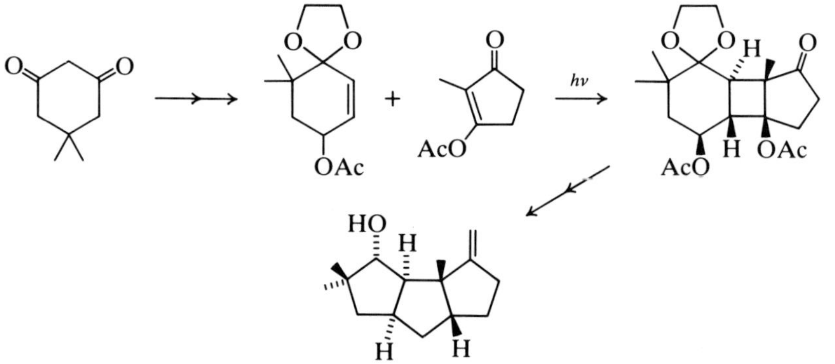

(**202**)

(*2*) *Total synthesis of (±)-longifolene* (**205**) [175]

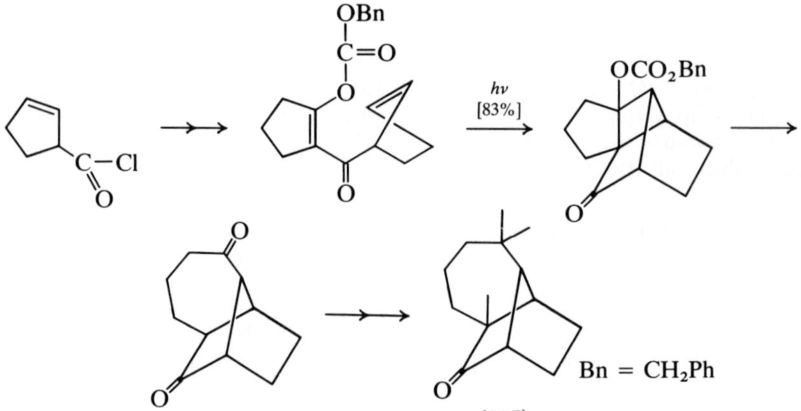

Bn = $CH_2Ph$

(**205**)

(*3*) *Total synthesis of (±)-β-bulnesene* (**206**) [176]

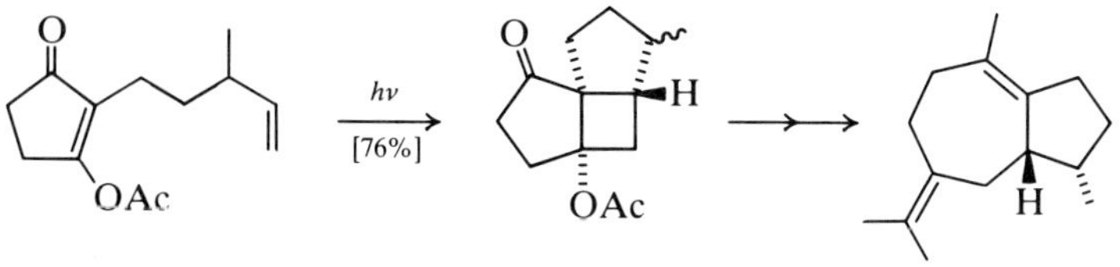

(**206**)
(±)-β-bulnesene

(*4*) *Total synthesis of (−)-sarracenine* (**207**) Recently, Baldwin and Crimmins [105] have achieved an efficient asymmetric synthesis of (–)-sarracenin (**207**) from L-ethyl lactate in just seven steps; the same compound had been synthesized in 15 steps by Whitesell *et al.* in 1978 [177].

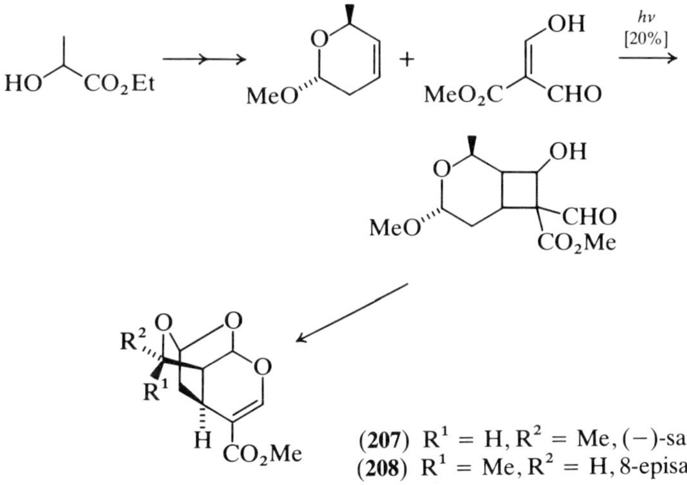

(**207**) $R^1 = H, R^2 = Me$, (−)-sarracenin
(**208**) $R^1 = Me, R^2 = H$, 8-episarracenin

Takeshita and Tanno [97] discovered that dioxopentanoic acid ester (methyl acetopyruvate) (**121**) is a most valuable enolic 1,3-dione, especially for the synthesis of many terpenoids. In most organic solvents this ester is enolized to a greater extent than acetylacetone, which was first used in the de Mayo reaction.

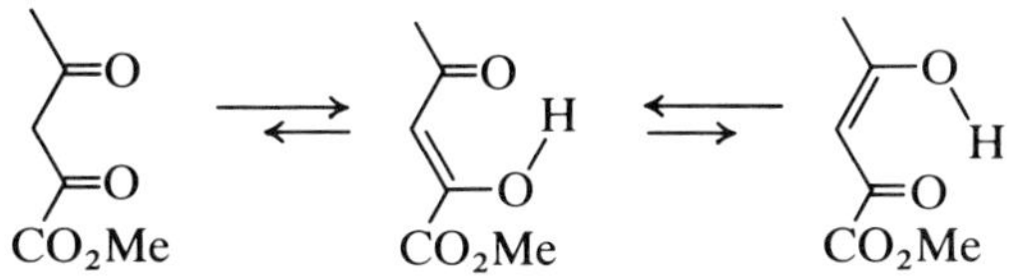

**A** (**121**) **B** **C**

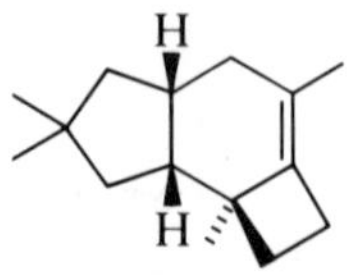

(**209**)
protoillud-6-ene

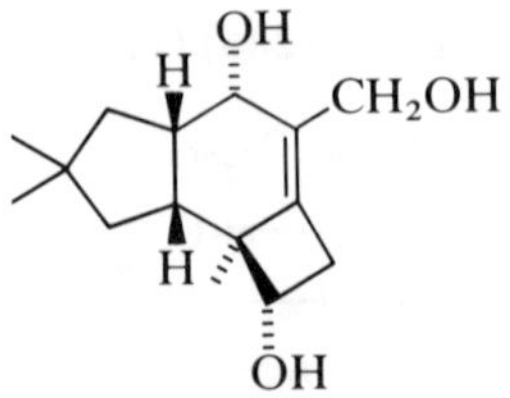

(**210**)
illudol

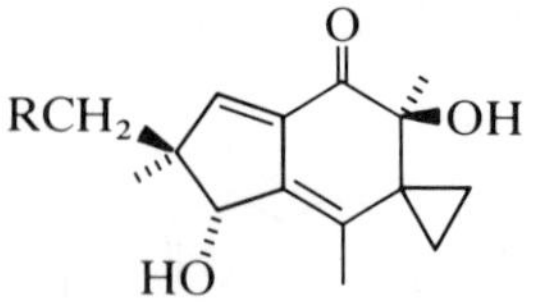

(**211**) R = H, illudin M
(**212**) R = OH, illudin S

The tricyclic sesquiterpenes, protoillud-6-ene (**209**), illudol (**210**), illudin M (**211**) and illudin S (**212**) contain the protoilludane skeleton, which was elegantly synthesized using two-step photoadditions according to the following retrosynthetic route [178]:

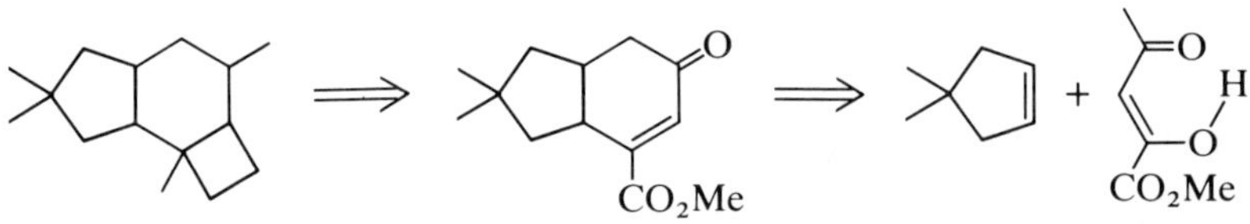

protoilludane

(**121**)

Photocycloaddition of the ester (**121**) to 4,4-dimethylcyclopentene gave a simple adduct, which was further cyclized to the bicyclic cyclohexenes (**213**) and (**214**). In the subsequent photoaddition of (**213**) and (**214**) to ethylene the thermodynamically stable (**213**) was inactive, while the *cis* compound (**214**) gave the adducts (**215**) and (**216**), the latter of which was finally converted into (**217**) and (**218**). Similarly, the enone (**219**) derived from the *cis*-enone (**213**) was used in a photoaddition reaction with 1,1-dimethoxyethylene to give an adduct that was finally converted into protoillud-7-ene (**220**).

Trimethylcyclohexene derivatives such as deoxytrispolone (**222**) have also been synthesized via the de Mayo reaction using the ester (**221**) [179].

Suginome *et al.* [145] have developed a photoannellation reaction involving radical cleavage of cyclobutanol photoadducts that has provided a new route for the conversion of photoadducts into a variety of compounds such as bicyclic rings, benzohomotropolones, furanoquinolones and furanocoumarins. The synthesis of the ifflaiamine analogue (**225**) from the photoadduct (**223**) is shown here as a representative example.

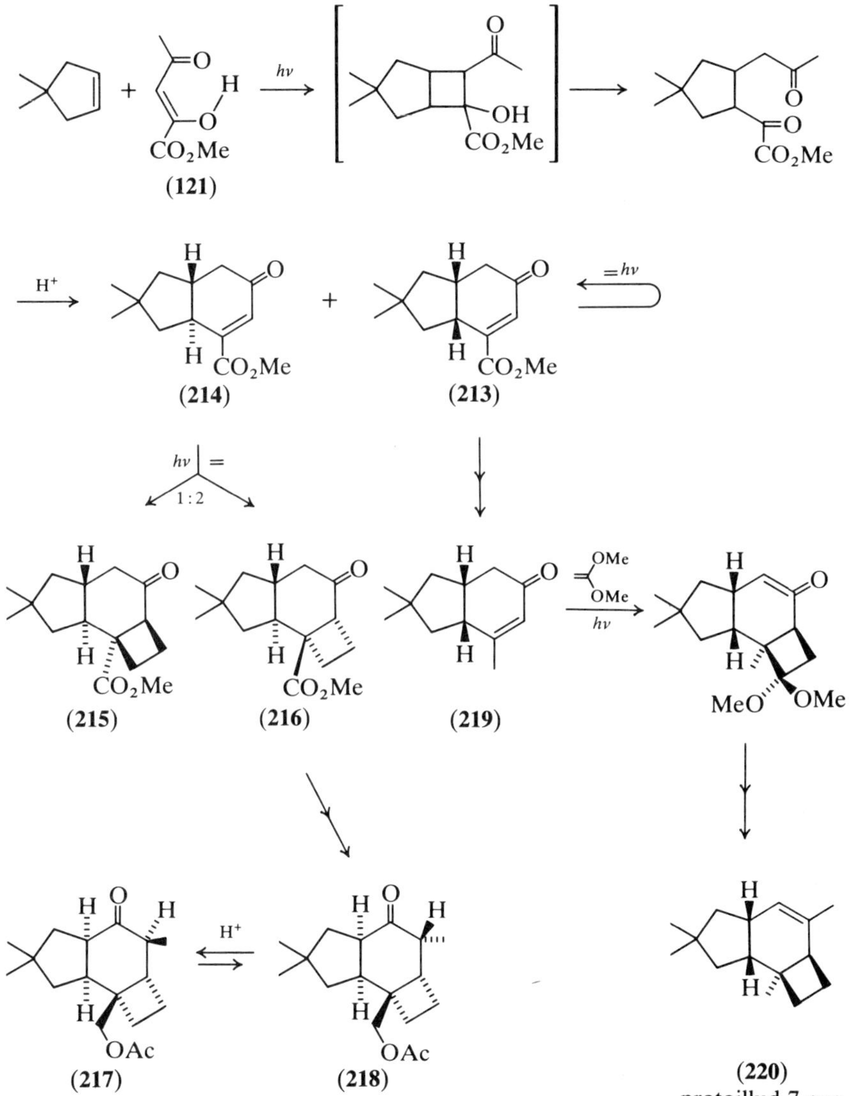

(**121**)

(**214**) (**213**)

(**215**) (**216**) (**219**)

(**217**) (**218**)

(**220**)
protoillud-7-ene

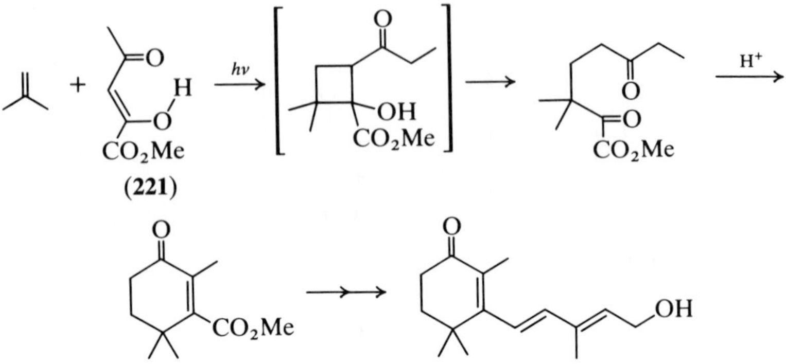

(221)

(222) deoxytrispolone

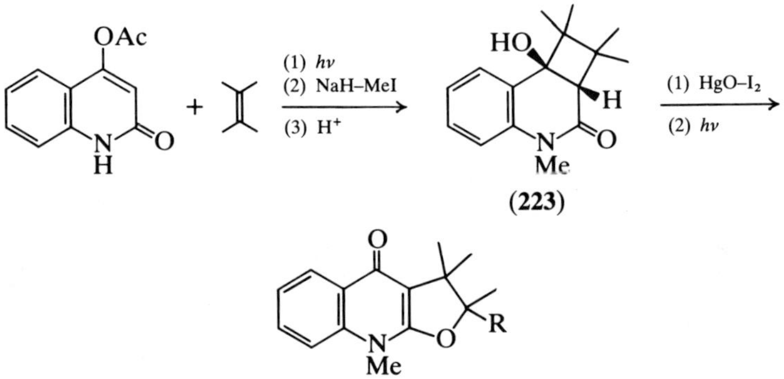

(223)

(224) R = Me
(225) R = H

**Experimental 6.9** Photoaddition of a diketoester to 2-methylpropene [179]

In ethyl acetate (30 ml), 2-methylpropene (*ca.* 20 ml), and the diketoester (**221**) (10.0 g) were internally irradiated by means of a 100-W high-pressure mercury lamp at −50 to −60 °C under nitrogen atmosphere for 37 h. Silica gel chromatography of the mixture from hexane–ethyl acetate (9 : 1) eluted the diketoester, which is obtained by photoaddition followed by a retro-Michael reaction, as a colourless oil (6.7 g, 56%).

### (*iii*) *Organic synthesis involving photocycloaddition of heteroaromatics containing enone systems to C═C double bonds*

Kaneko *et al.* [180–182] have investigated the photocycloaddition of heteroaromatics containing an $\alpha,\beta$-unsaturated carbonyl system to C═C double bonds and have also developed new C—C bond-fission reactions of the photocycloadducts as a versatile method for the introduction of a variety of carbon chains and for annellation to heteroaromatic compounds.

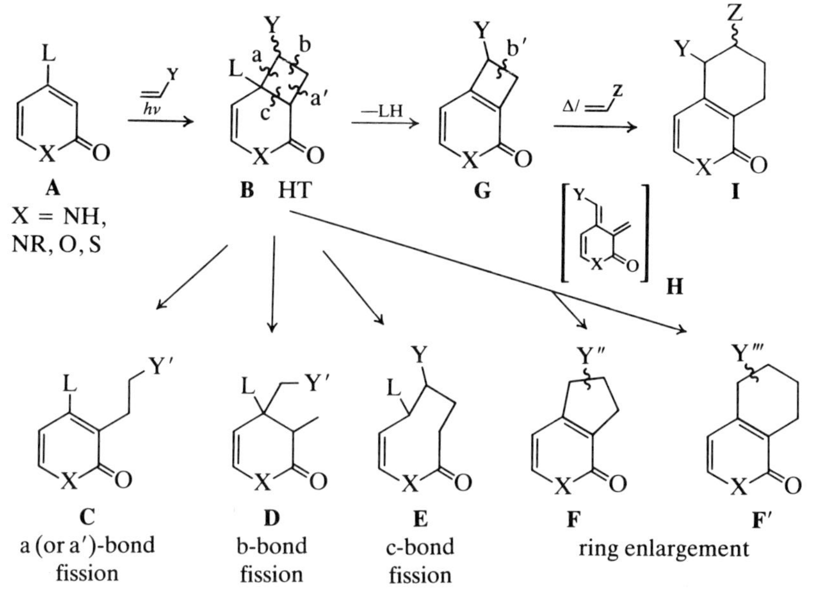

**A** X = NH, NR, O, S — $h\nu$ → **B** HT — −LH → **G** — Δ/ → **I**

[**H**]

**C** a (or a′)-bond fission

**D** b-bond fission

**E** c-bond fission

**F** **F′** ring enlargement

(*a*) *Introduction of carbon chains into heterocycles via a-bond fission* Photocycloaddition of 4-methoxy-2-quinolone (**226**) to vinylene carbonate proceeded smoothly to give the adduct (**227**), which was subjected to ring opening by treatment with base to give the aldehyde (**228**). Cyclization of (**228**) completed the synthesis of dictamnine (**229**) [182].

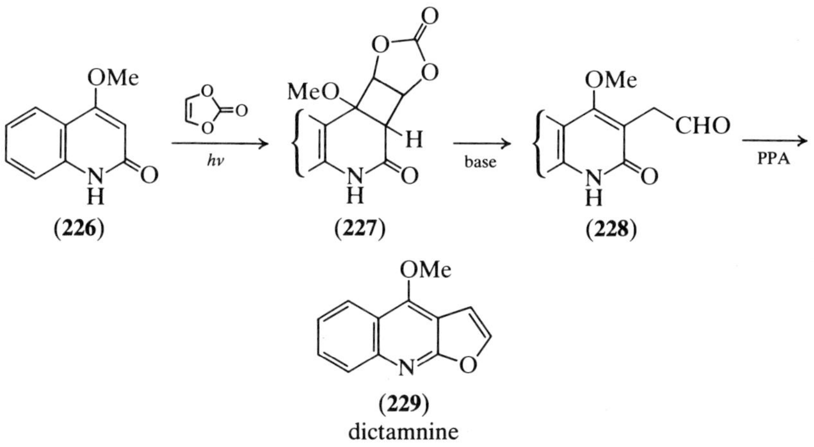

(**226**) — $h\nu$ → (**227**) — base → (**228**) — PPA →

(**229**)
dictamnine

Regioselective photocycloadditions of 2-quinolones and 2-pyridones giving HT adducts have provided a new isoprenylation of the parent heterocycles as shown in the synthesis of khaplofoline (**230**) [183].

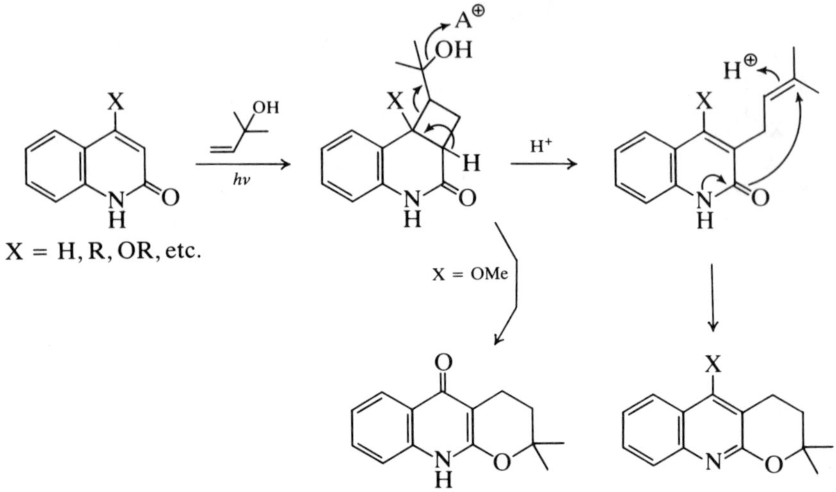

X = H, R, OR, etc.

(**230**)
khaplofoline

(*b*) *Introduction of carbon chains into neighbouring positions of enone systems via b-bond fission* In the reaction sequence A→B→G→ H→I (p. 101) above, the reactions G → H → I correspond to the well-known benzocyclobutene methods [184, 185], which are useful synthetic routes for various six-membered ring systems formally analogous to the Diels–Alder reaction [180].

**Experimental 6.10** Photoaddition of 4-methoxy-1-methylquinolin-2(1H)-one to allene [183]

Photolysis was carried out in a Pyrex immersion apparatus equipped with an Ushio 450-W or Toshiba 400P high-pressure mercury lamp (this corresponds to irradiation at 300 nm) cooled internally with running water. Allene was bubbled into a solution of the quinolone (255.8 mg) in 280 ml of methanol for 10 min and the whole vessel was irradiated for 1 h. After removal of the solvent, the residue was chromatographed on silica gel (100–200 mesh, Kanto Chemical Co., Inc.) (25 g). Elution with dichloromethane–hexane–ether (5 : 5 : 1, v/v/v) afforded first the pure head-to-head adduct, then a mixture of the two regioisomers, and finally the head-to-tail adduct. The portion containing the mixture was separated further by preparative TLC (silica gel with dichloromethane–ether (10 : 1, v/v) as a developing solvent). Combined yields of HT (MP 80.5–82 °C; ether–hexane) and HH adducts (oil) were 164.9 mg (53%) and 30.1 mg (10%), respectively.

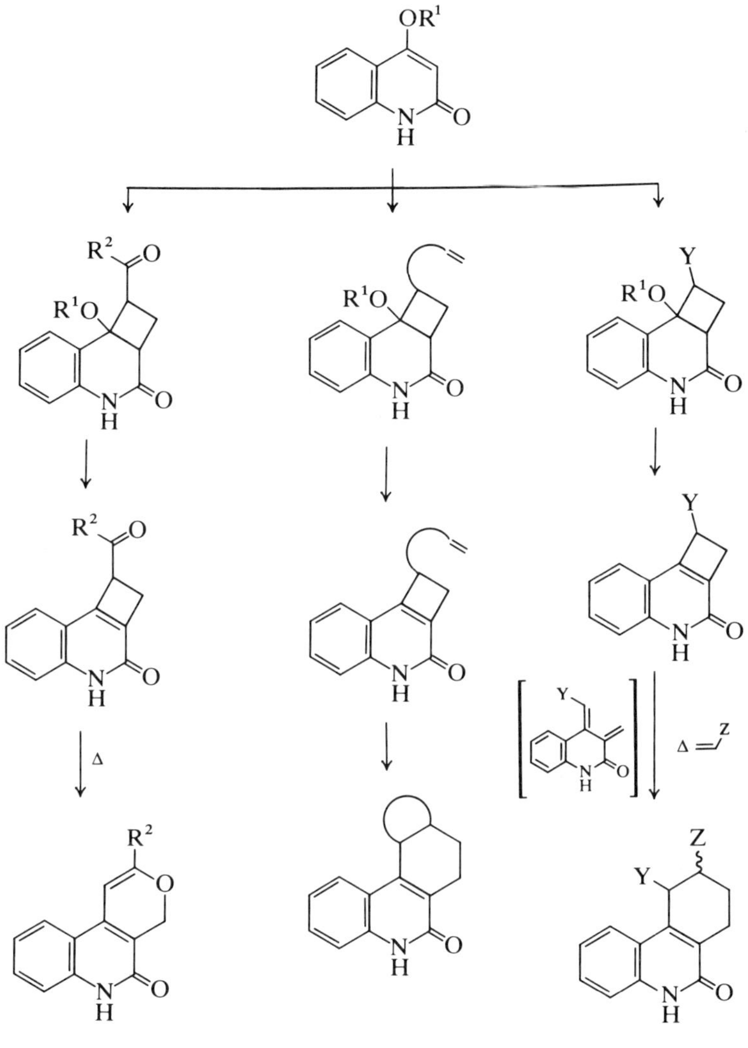

(*c*) *Ring expansion and related reactions via c-bond fission* There are many examples of synthesis of seven- and eight-membered heterocycles via photocycloaddition of five- and six-membered heteroaromatics and C═C double bonds, followed by a retro-aldol reaction under basic conditions [186–190].

Azatropolones have been synthesized by photocycloaddition of dioxopyrrolines to C≡C triple bonds followed by thermal ring opening (addition to double bonds gives dihydroazatropolones) [191–194].

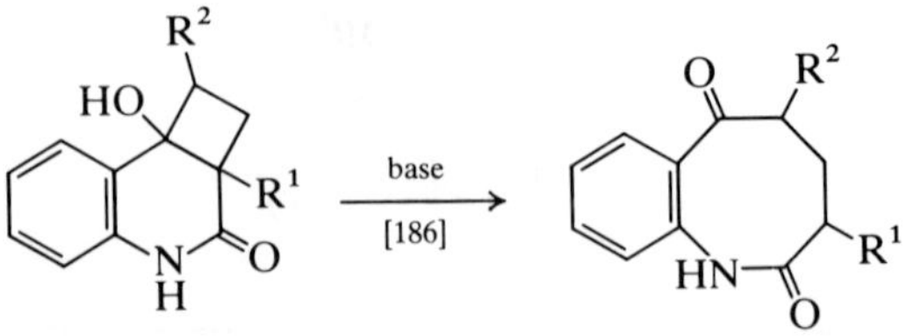

base
[186]

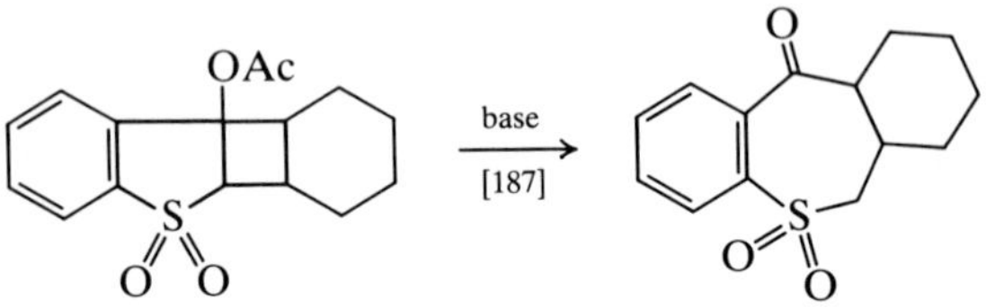

base
[187]

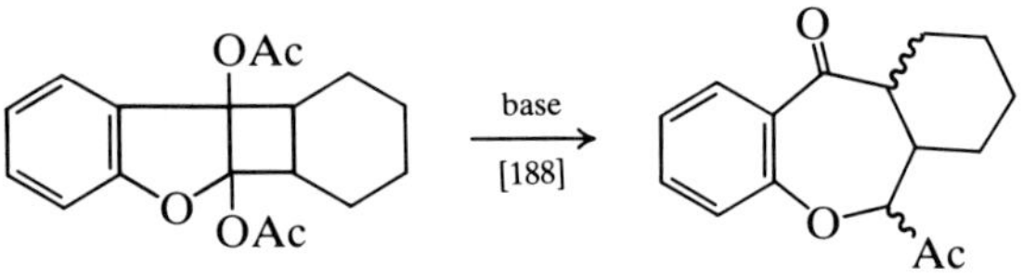

base
[188]

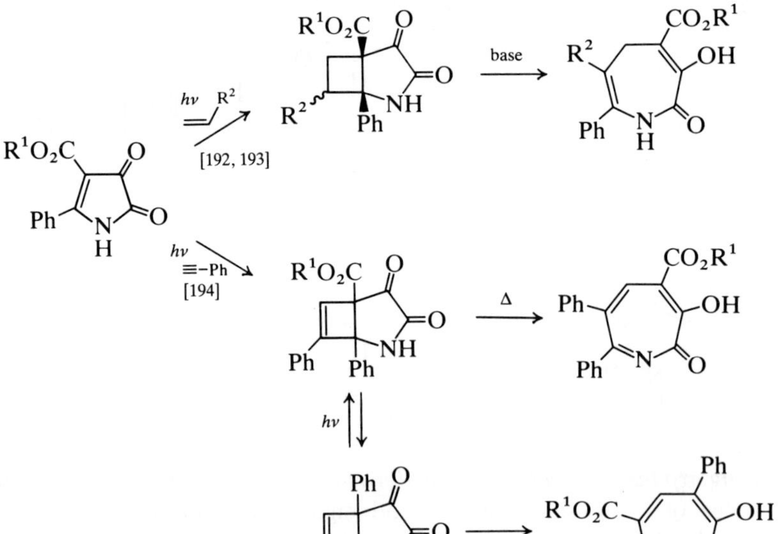

hν R[2]
[192, 193]

base

hν
≡−Ph
[194]

Δ

hν

In addition to the examples shown above, ring expansion of cyclobutanes has also provided a new annellation reaction for the synthesis of cyclopentane rings [195].

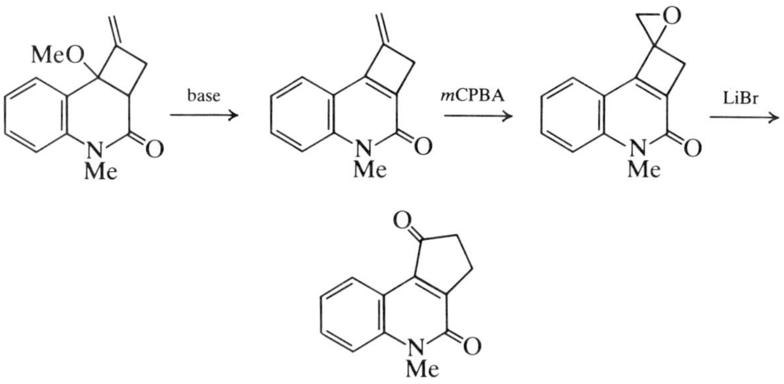

base mCPBA LiBr

### 6.2.2 Preparation of four-membered rings by photocyclization of conjugated dienes

#### 6.2.2.1 Preparation from acyclic 1,3-dienes

Irradiation of conjugated dienes is known to give various types of products, depending on the reaction condition used. In the absence of sensitizers, photolysis of simple conjugated dienes take a variety of different courses. In solution photolysis, remarkably different reaction courses are observed. Thus irradiation of 1,3-butadiene in dilute solution with a high-pressure mercury lamp leads to the formation of cyclobutene (**231**) and bicyclo[1.1.0]butane (**232**) [196–198]. The product ratio obtained depends upon the solvent employed as shown.

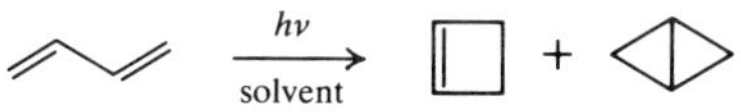

$h\nu$ / solvent

(**231**) (**232**)

| Solvent | **231** : **232** |
|---|---|
| Isooctane | 16 : 1 |
| Ether | 7 : 1 |
| Ether/$CuCl_2$ | 6 : 1 |

At high diene concentrations direct photolysis gives rise to some dimerization. A prime example is the photocyclization of 2,3-dimethyl-1,3-butadiene (**233**) to give 1,2-dimethylcyclobutene (**234**) [197].

Generally, cyclizations of dienes proceed as orbitally symmetric electrocyclic processes as predicted by the Woodward–Hoffmann rules involving

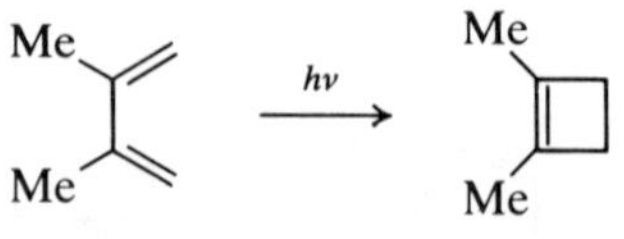

Me Me (233) $\xrightarrow{h\nu}$ Me Me (234)

HOMOs in thermal and LUMOs in photochemical conditions respectively. Although the precise nature of the structure-controlling factors in photocyclization of 1,3-butadiene derivatives is not known, a few typical cyclizations are worthy of mention. Photocyclization of 1,3-pentadiene (**235**) to give the corresponding cyclobutene (**236**) occurs only with the *trans* isomer of (**235**) [197]. In the case of 2,4-dimethyl-1,3-pentadiene (**237**), however, photocylization is not observed. Steric factors would render an *s-cis* conformation for (**237**) unfavourable and accordingly would prevent photocyclization [197].

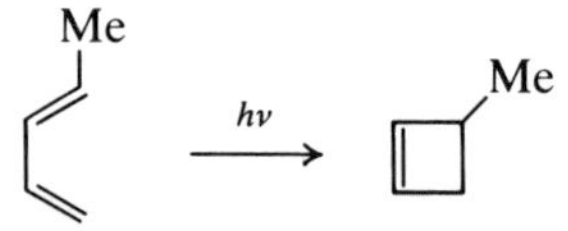

Me (235) $\xrightarrow{h\nu}$ Me (236)

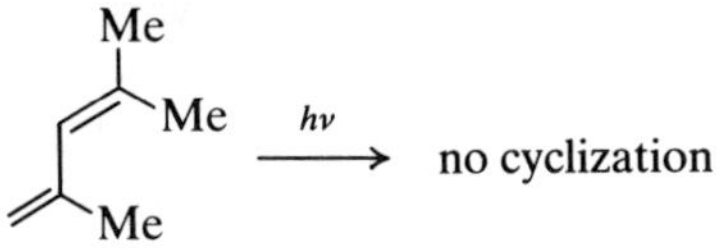

Me Me Me (237) $\xrightarrow{h\nu}$ no cyclization

### 6.2.2.2 Preparation from cyclic 1,3-dienes

As might be anticipated from the behaviour of 1,3-butadiene and its derivatives, cyclic 1,3-dienes often exhibit a marked tendency toward photocyclization to cyclobutene derivatives. The first example of such a process was the formation of photoisopyrocalciferol (**239**) and photopyrocalciferol (**241**) from isopyrocalciferol (**238**) and pyrocalciferol (**240**) respectively [199, 200].

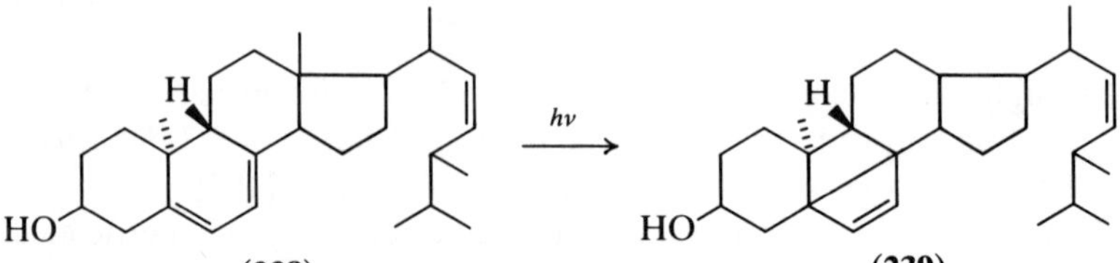

H HO (238) $\xrightarrow{h\nu}$ H HO (239)

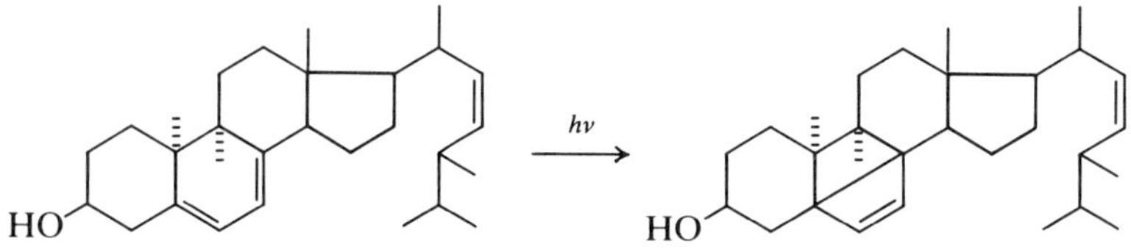

(240) (241)

With those larger-ring dienes whose photochemical behaviour has been examined so far, the principal photochemical transformation is photocyclization. Thus 1,3-cycloheptadiene (**242**) has been photocyclized to 6-bicyclo[3.2.0]heptene (**243**) in solution [201, 202].

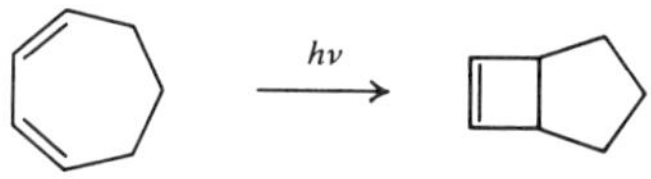

(242) (243)

### 6.2.2.3 Preparation from heterocyclic 1,3-dienes

Numerous heterocyclic molecules possessing a conjugated diene system in their structures undergo photochemical transformations resembling those of carbocyclic dienes. Thus the anhydride of muconic acid (**244**) photocyclizes to the cyclobutene derivative (**245**) [203].

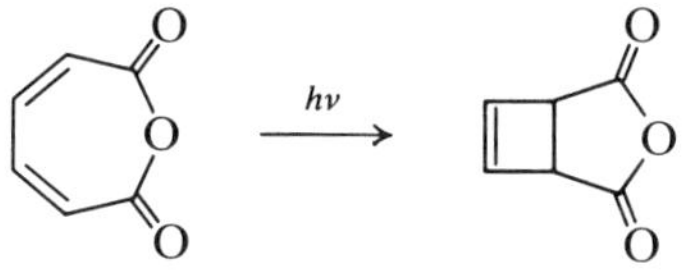

(244) (245)

Among many photochemical reactions of 2-pyrone (**246**) of importance is photocyclization and fragmentation to cyclobutadiene [204]. At 8–20 K in an argon matrix irradiation gives a 5-oxoketene, which can be identified by addition of alcohol. At 77 K formation of the lactone (**247**) occurs, with competition from ketene formation. Prolonged irradiation of 2-pyrone gives cyclobutadiene (**248**) via elimination of carbon dioxide from (**247**).

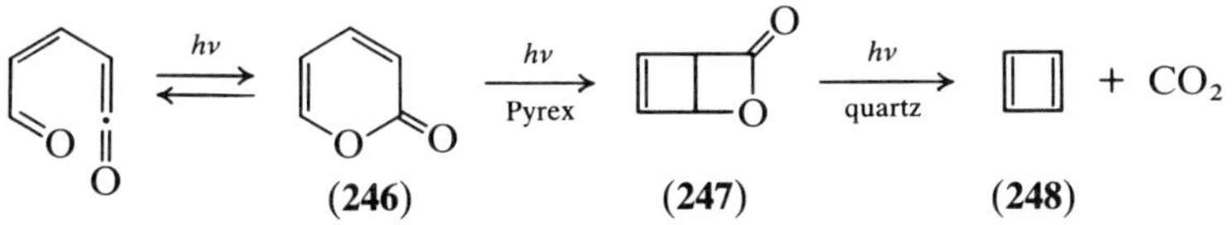

(246) (247) (248)

Six-membered heterocycles (e.g. (**249**)) containing nitrogen can also be photocyclized to give cyclobutenes (**250**) [205].

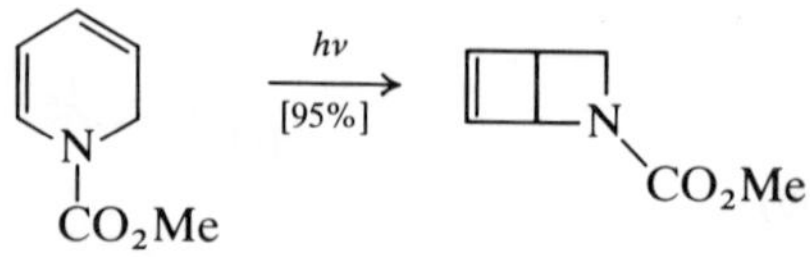

N, $CO_2Me$ (**249**) —$h\nu$ [95%]→ N, $CO_2Me$ (**250**)

Irradiation of 2-pyridone (**251**) gave the cyclobutene (**252**) at low temperature and low concentration and the dimer (**253**) at high temperature and high concentration [204, 206, 207].

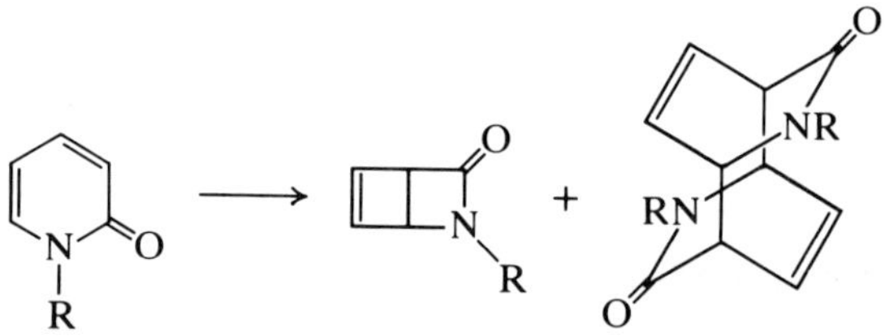

N, O, R (**251**) R = H, Me, COOMe → O, N, R (**252**) + O, NR, RN, O (**253**)

Seven-membered azepine derivatives (**254**) [208, 209] and (**255**) [210, 211] are good precursors for the photochemical synthesis of cyclobutenes.

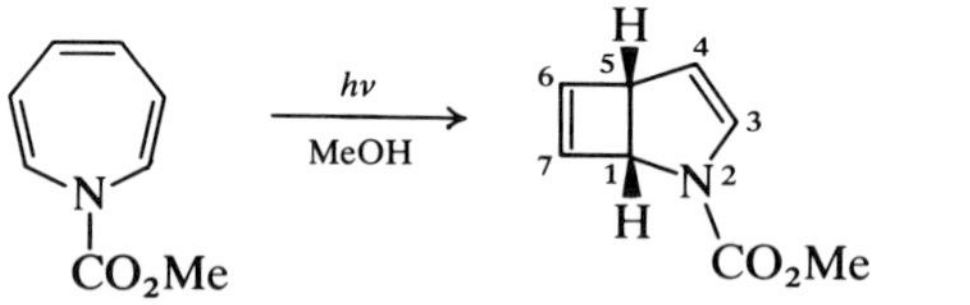

N, $CO_2Me$ —$h\nu$, MeOH→ H, 5, 4, 6, 3, 7, 1, N 2, H, $CO_2Me$

(**254abc**)
(**a**) 2-Me
(**b**) 3-Me
(**c**) 4-Me

(**a**) 3-Me (93%), 1-Me (7%)
(**b**) 4-Me (50%), 7-Me (50%)
(**c**) 5-Me (60%), 6-Me (40%)

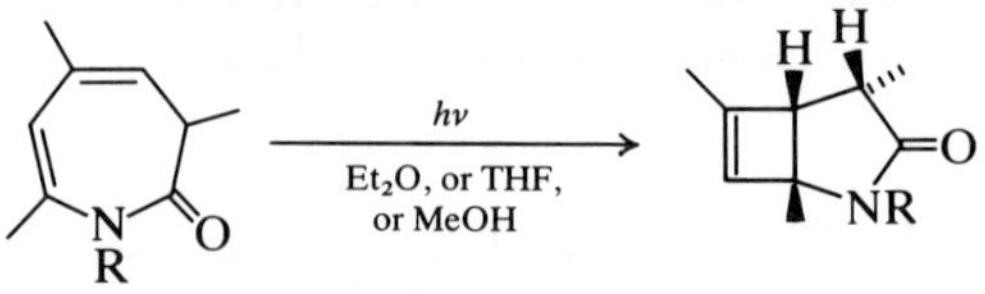

N, O, R —$h\nu$, $Et_2O$, or THF, or MeOH→ H, H, O, NR

(**255**) R = H, Me

(55–78%)

**Experimental 6.11** *N*-Carbomethoxy-2-azabicyclo[2.2.0]hex-5-ene [205]

A 5% solution of 1,2-dihydropyridine (**249**) in methylene chloride was irradiated using a Rayonet photochemical reactor (RP-300 lamps) until the NMR spectrum showed the consumption of all the starting material. Removal of the solvent gave an orange oil. The NMR spectrum showed this to be *ca*. 85% pure. Purer (>95%) (**250**) product could be obtained by passing the crude product through basic alumina with ether.

### 6.2.2.4 Synthetic utility of photocyclization of 2-pyridones

Since the discovery of the formation of photopyridone by the photocyclization of 2-pyridones by Corey and Streith [204], this cyclization has provided a useful synthetic route to $\beta$-lactam antibiotics, since the photocyclized products have not only the $\beta$-lactam structure but also a cyclobutene structure, which can easily be ring opened to give useful compounds [212, 213].

Brennan's group [212]

Kametani's group [213]

Kaneko *et al.* [214–217] have systematically developed this synthetic methodology for $\beta$-lactams and have succeeded in the stereoselective and chiral synthesis of useful synthons of carbapenem $\beta$-lactam (see Section 7.2.2.3).

### 6.2.3 Preparation of four-membered rings by the Norrish type II reaction

Of two characteristic photochemical reactions (Norrish types I and II) of excited carbonyl groups, the Norrish type II reaction has provided a useful synthetic method as exemplified by the elegant syntheses of cage compounds (Section 8.2) and $\beta$-lactams (Section 7.2.2.1).

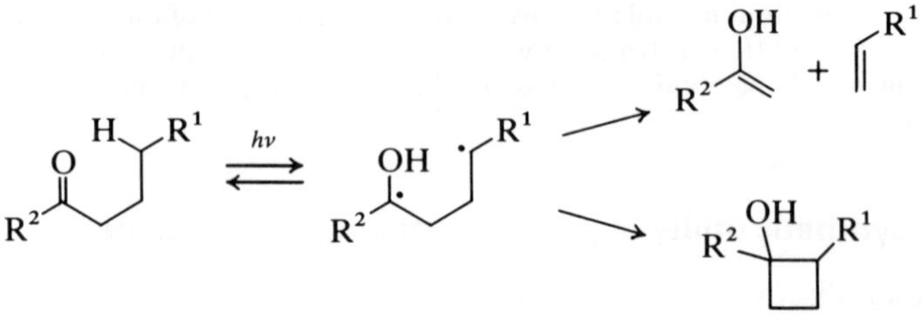

Acyclic ketones undergo preferential cleavage of a biradical to form olefins, while cyclic ketones give cyclobutanols. Photochemical formation of cyclobutanols, (**257**) and (**259**), is strongly affected by the stereochemistry of the starting ketones, (**256**) and (**258**), as shown by the following examples [218–220].

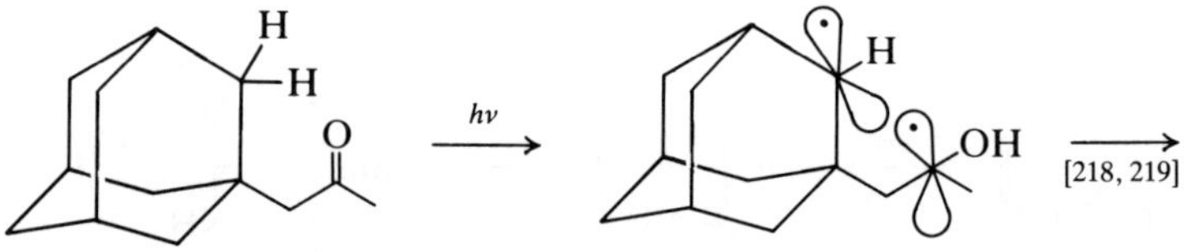

(**256**)

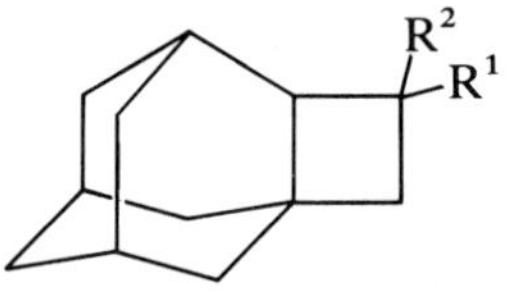

(**257ab**)
(**a**) $R^1$ = OH, $R^2$ = Me (24%)
(**b**) $R^1$ = Me, $R^2$ = OH (61%)

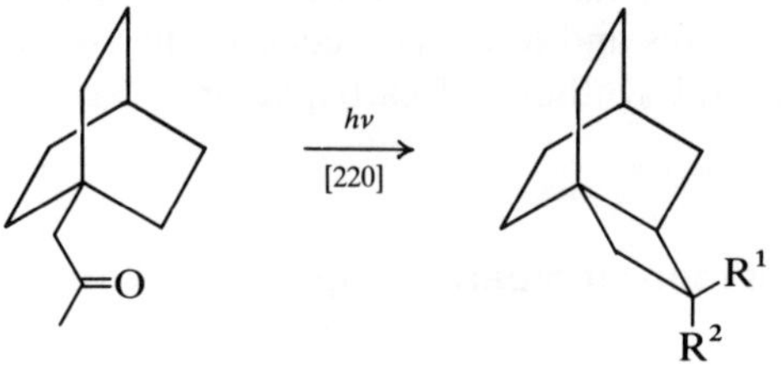

(**258**)

(**259ab**)
(**a**) $R^1$ = Me, $R^2$ = OH (59%)
(**b**) $R^1$ = OH, $R^2$ = Me (24%)

Cyclobutanones have been effectively synthesized by irradiation of $\alpha$-diketones [221].

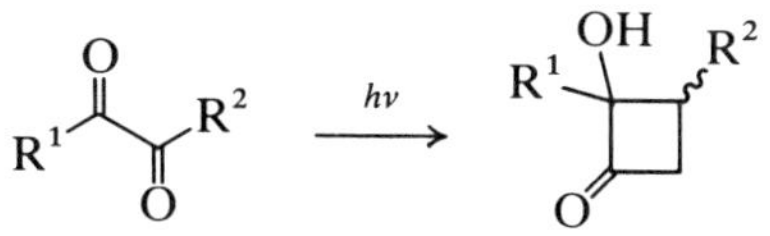

$h\nu$

(**a**) $R^1 = R^2 = Bu^n$ (**a**) $R^1 = Bu^n, R^2 = Et$ (89%)
(**b**) $R^1 = R^2 = Pr^n$ (**b**) $R^1 = Pr^n, R^2 = Me$ (92%)
(**c**) $R^1 = Me, R^2 = Et$ (**c**) $R^1 = Me, R^2 = H$ (49%)

In the photochemical reaction of $\alpha$-diketones cyclization is generally preferred to olefin formation owing to the stabilization of the intermediate biradical by the presence of the second carbonyl group [222].

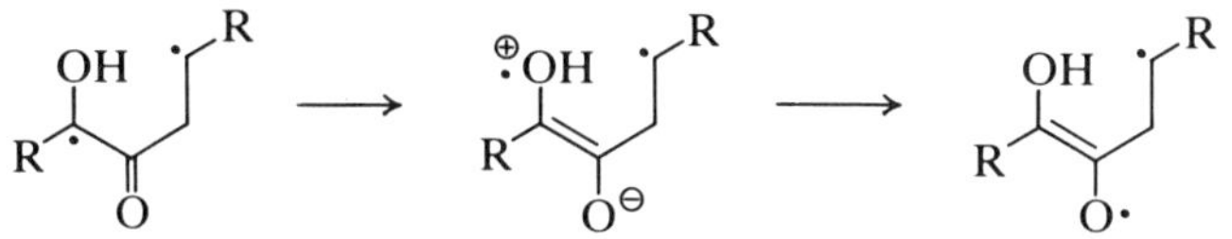

An application of this reaction is the synthesis of $\beta$-lactams, with $\beta$-keto-amides (**260**) and (**261**) playing the role of the $\beta$-diketone (see Section 7.2.2.1) [223, 224].

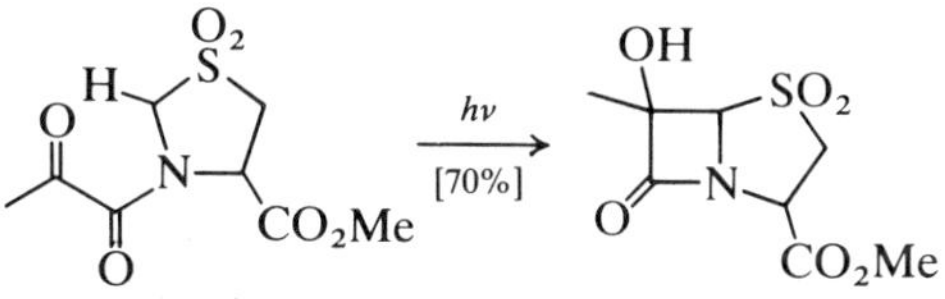

$h\nu$ [70%]

(**260**)

$h\nu$ solvent

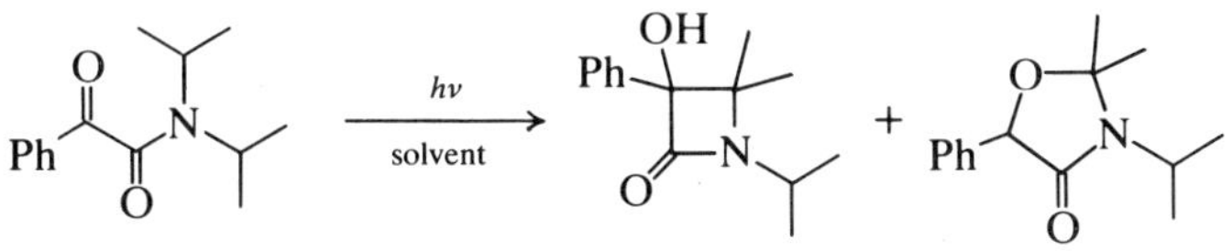

(**261**)

| Solvent | % | |
|---|---|---|
| Neat | 83 | trace |
| $C_6H_6$ | trace | 62 |
| MeOH | 22 | 58 |

**Experimental 6.12** Irradiation of 1-adamantylacetone [220]

A solution of 1-adamantylacetone (**256**) (1.02 g) in 105 ml of t-butyl alcohol was irradiated for 16 h with a medium-pressure mercury arc (450 W, Hanovia type L) through a Corex filter in a water-cooled immersion well through which oxygen-free nitrogen was bubbled. At the end of the irradiation period, the solvent was removed by evaporative sublimation at 250 °C and *ca.* 0.2 mmHg into a receiver cooled in liquid nitrogen. About 5% of the starting material remained and two photoadducts were detected by GC (Aerograph Model A90P on the column 5–18% Carbowax 20M, 5–15 feet 0.25 inch) in 24 and 61% yields. These two products were isolated by preparative GC and characterized. The product (**257a**) with long retention time was a viscous oil and the one with short retention time (**257b**) was a solid, MP 101–104 °C.

## 6.2.4 Preparation of four-membered rings by ring contraction

Photolysis of cyclic $\alpha$-diazoketones usually involves ring contraction with formation of ketenes. In an aqueous medium the reaction may proceed further to give carboxylic acids.

$$(C_xH_y)_n\,C(=N_2)\text{–}C(=O) \xrightarrow[-N_2]{h\nu} (C_xH_y)_n\,C{=}C{=}O \xrightarrow{+H_2O} (C_xH_y)_n\,C(H)(CO_2H)$$

Irradiation of 16-diazo-3$\beta$-hydroxyandrostan-17-one (**262**) leads to a D-ring contraction to give (**263**), thus completing the first synthesis of a *trans*-bicyclo[4.2.0]octane system [225].

$$\textbf{(262)} \xrightarrow[-N_2,\ +H_2O]{h\nu} \textbf{(263)}$$

Cava *et al.* [226, 227] have reported the ring contraction of 2-diazo-1-indanone (**264**) to give benzocyclobutenecarboxylic acid (**265**).

$$\textbf{(264)} \xrightarrow[-N_2,\ +H_2O]{h\nu} \textbf{(265)}$$

**Experimental 6.13** 3$\beta$-Hydroxy-D-norandrostane-16$\xi$-carboxylic acid [225]

The diazoketone (**262**) (4 g) dissolved in distilled tetrahydrofuran (160 ml) and water (40 ml) was irradiated by immersing a Hanau S mercury lamp in a jacketed flask in the solution and maintaining the temperature of the solution at 10 ± 1 °C. After 1 h a sample showed complete disappearance of the 2075 $cm^{-1}$ band from the IR spectrum and that the reaction was therefore complete. The mixture was poured into 600 ml of water and extracted with ether; the organic layer was extracted with a saturated solution of sodium bicarbonate and on acidification with hydrochloric acid the D-nor acid (**263**) precipitated as a foamy solid. After being kept for 6 h at 5 °C, the product was filtered off, washed with water, decolourized with Norit charcoal and recrystallized from methanol, MP 200–205 °C (3 g, 75%). By recrystallization from methanol, the MP was raised to 205–206 °C.

## 6.3 FIVE-MEMBERED RINGS

Not many photochemical synthetic methods for the construction of five-membered carbocyclic systems have been described.

### 6.3.1 Preparation of five-membered rings by photocyclization

A representative example is the formation of bicyclo[3.2.0]heptanes by irradiation of cyclic dienes and dienones. The photochemistry of tropolones has been studied for a considerable time.

O
OMe
(**266**)
$h\nu$
360°C
O
OMe
(**268**)
$h\nu$
O⊖
⊕
MeO
(**267**)

Irradiation of an aqueous solution of $\gamma$-tropolone methyl ether (**266**) gives the cyclopentenone (**268**) in 60% yield as the sole product via the zwitterionic intermediate (**267**) [228]. $\alpha$-Tropolone (**269a**) and its methyl ether (**269b**) underwent the same type of photocyclization to give 4-oxo-2-cyclopenten-1-acetic acid (**270a**) and its methyl ester (**270b**) respectively in

(270ab)

(269ab)
(a) R = H
(b) R = Me

(271) (272) (274) (273)

50% yields in aqueous solution, while in methanol two primary products (**271**) and (**272**) were obtained [229].

Thermally, at 360 °C, the cyclopentenone (**271**) reverted to the starting $\alpha$-tropolone methyl ether (**269b**), and (**272**) was converted into $\beta$-tropolone methyl ether (**273**), while at 200 °C (**271**) isomerized to (**274**) [230]. These two processes are allowed by the Woodward–Hoffmann rules. Similar photocyclizations of substituted $\alpha$-tropolones to cyclopentenone esters have been reported [231].

**Experimental 6.14** Irradiation of $\alpha$-tropolone methyl ether [229]

A solution of $\alpha$-tropolone methyl ether (**269b**) (5.0 g) in distilled water (2.5 l) was flushed with nitrogen, sealed in a Pyrex flask under reduced pressure, and placed on the roof of the laboratory for 14 days (8 days actual sunshine). The solution became turbid, and a red amorphous powder precipitated. Centrifugation of the solution gave a dark-red amorphous powder (1.45 g). The solution, after centrifugation, was extracted with three portions of methylene dichloride. The extract was dried over sodium sulphate and concentrated under reduced pressure. Distillation of the extract residue gave a colourless liquid, 5-methoxybicyclo[3.2.0]hepta-3,6-diene-2-one (**272**), BP 34 °C (0.1 mmHg) (1.19 g).

### 6.3.2 Preparation of five-membered rings by ring contraction

Ring contraction of six-membered ring systems is another synthetic route to five-membered rings. Irradiation of diazonorcamphor (**275**) in methanol gives methyl bicyclo[2.1.1]hexane-5-carboxylate (**276**) in 54% yield [232].

During the course of a study of the differences in reactivity of diazocamphor (**277**) towards photolysis and pyrolysis, Horner and Spietschka [233] found that ring contraction occurs only in the photolytic process, yielding the cyclopentanecarboxylic acid (**278**) when photolysis was carried out in an aqueous medium.

$$(275) \xrightarrow[-N_2\ +MeOH]{h\nu} (276)\ CO_2Me$$

$$(277) \xrightarrow[-N_2\ +H_2O]{h\nu} (278)\ CO_2H$$

Only a few examples of the photochemical formation of five-membered rings by exceptional abstraction of $\delta$-hydrogen via seven-membered intermediates are known [234].

$$\xrightarrow[[91\%]]{h\nu}$$

**Experimental 6.15** Methyl bicyclo[2.1.1]hexane-5-carboxylate [232]

A solution of crude diazonorcamphor (**275**) (73.6 g, 0.54 mol) in anhydrous methanol (4.5 l) was irradiated with a water-cooled Hanovia 500 W immersion quartz lamp using a Corex filter. After 24 h 95% of the diazo compound had reacted. Most of the solvent was removed by distillation through a 30-inch column packed with Heli-Pak. Water (1 l) and crushed ice were added to the residue, and the mixture was extracted with four 1 l portions of pentane. The pentane solution was dried over anhydrous sodium sulphate and the solvent was removed by distillation through the same column. Distillation of the residue gave methyl bicyclo[2.1.1]-hexane-5-carboxylate (**276**) (41 g, 54%) BP 67–71 °C (17–18 mmHg).

## 6.4 SIX-MEMBERED RINGS

### 6.4.1 Preparation of six-membered rings by electrocyclic reactions

The six-membered ring system is the most common among organic compounds. 1,3,5-Hexatriene systems are appropriate conjugated precursors for photocyclization to give 1,3-cyclohexadiene rings.

$h\nu$ $h\nu$

Among many such systems, alkenes conjugated with aromatic rings are good potential starting compounds for the synthesis of cyclohexane derivatives since their photocyclization proceeds very smoothly and in high yield [235, 236].

This type of photocyclization is a useful synthetic route to many different polynuclear aromatic systems and occurs with a wide range of substituted and related molecules including various polycyclic and heterocyclic analogues. In addition, certain systems with a single heteroatom (nitrogen, oxygen or sulphur) in place of the central $\pi$ bond undergo photocyclization, creating new five-membered heterocyclic rings.

The discovery and early development of stilbene photocyclization has been surveyed and several general reviews have appeared. For the synthetic chemistry the review [237] by Mallory and Mallory is excellent as many examples are tabulated therein.

#### 6.4.1.1 Synthesis of phenanthrenes and polyaromatics

Thermodynamically stable *trans*-stilbene (**279**) isomerizes to the *cis* isomer (**280**) on irradiation. Further irradiation of the isomerized *cis*-stilbene (**280**) at 254 nm leads to the formation of dihydrophenanthrene (**281**), which is easily oxidized by molecular oxygen or iodine to give phenanthrene (**282**). Generally, photolysis is carried out in the presence of an oxidant such as molecular oxygen or iodine [236]. (See Table 6.4.)

Although this type of photocyclization is an effective method for the synthesis of substituted phenanthrenes, irradiation under too high a concentration or for too long a time results in low yields of products because of photodimerization of the starting stilbenes and photodecomposition of the products [238].

*trans*-$\alpha$,$\beta$-Dicyanostilbene (**283**), diphenylmaleimide (**284**) and diphenylmaleic anhydride (**285**) are also good starting compounds for the photo-

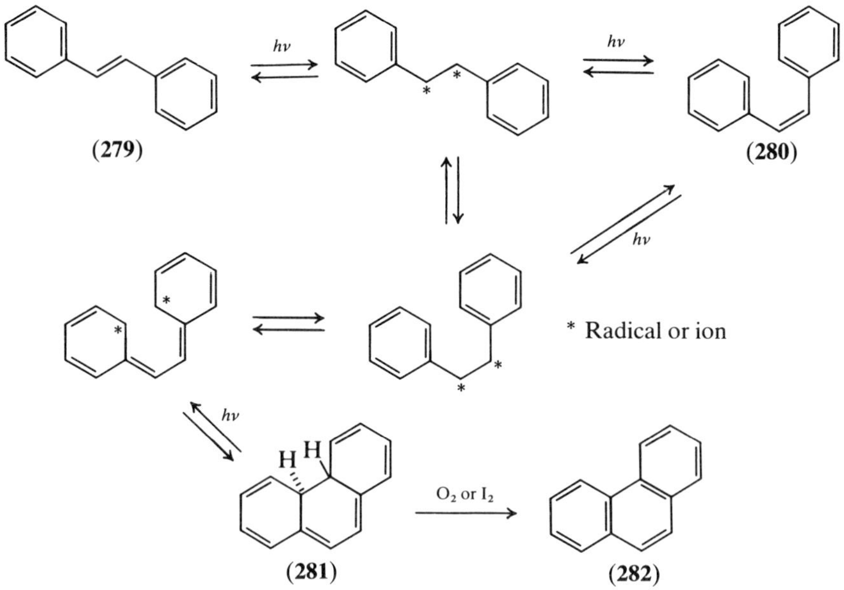

(279) (280)

* Radical or ion

(281) (282)

TABLE 6.4

Photocyclization of stilbenes

| Substituted stilbenes | Substituted phenanthrenes (Yield, %) |
|---|---|
| 4-F | 3-F (98) |
| 4-Br | 3-Br (76) |
| 4-OMe | 3-OMe (42) |
| 4-$CF_3$ | 3-$CF_3$ (67 |
| 2-Cl | 1-Cl (60) |
| 2-Me | 1-Me (57) |

chemical synthesis of phenanthrenes and the corresponding dihydro derivatives (**286**) [239, 240].

Stilbene photocyclization has been applied to phenyl-substituted trienes and is a useful synthetic method for various types of polyaromatic compounds including naphthalenes, chrysenes, coronenes and many others [237, 241].

There are several good reviews of the photocyclization of stilbene and related compounds [235–237].

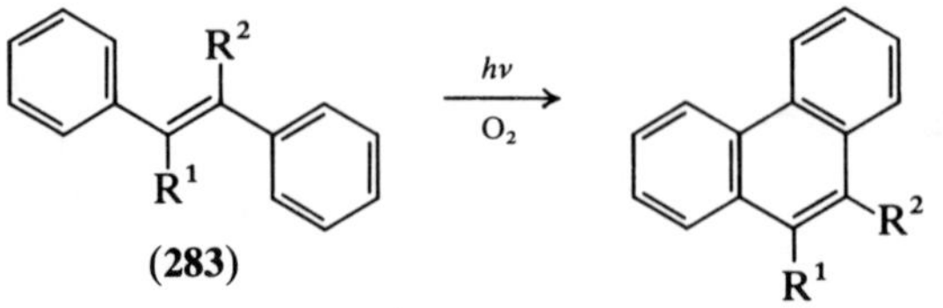

$R^2$ $R^1$ (**283**) $\xrightarrow[O_2]{h\nu}$ $R^2$ $R^1$

(**a**) $R^1 = R^2 = CN$
(**b**) $R^1 = CN, R^2 = H$
(**c**) $R^1 = CONH_2, R^2 = H$
(**d**) $R^1 = Me, R^2 = H$
(**e**) $R^1, R^2 = Me, Ph$

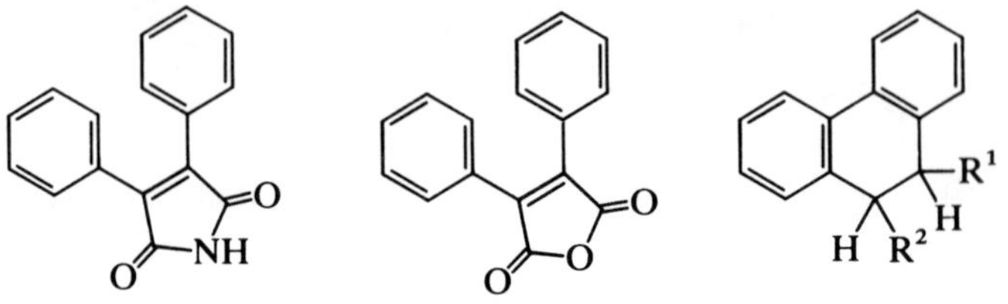

O NH O (**284**)

O O O (**285**)

$R^1$ H H $R^2$ (**286**)

Ph $\xrightarrow{h\nu}$ Ph

$\xrightarrow{h\nu}$

Ph $\overset{h\nu}{\rightleftarrows}$ H Ph $\xrightarrow{I_2}$ Ph H

Ph

benzocoronene $\xleftarrow[I_2]{h\nu}$

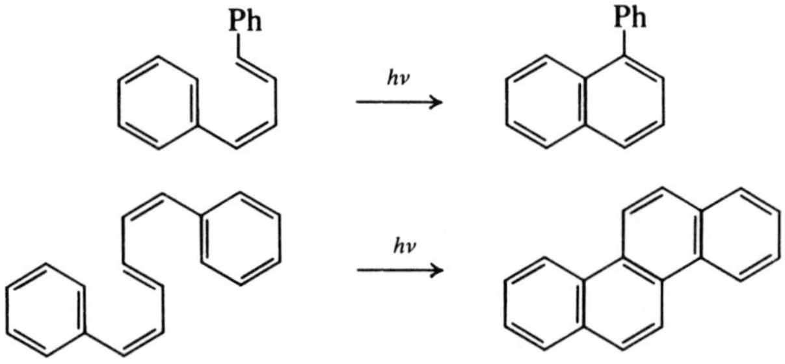

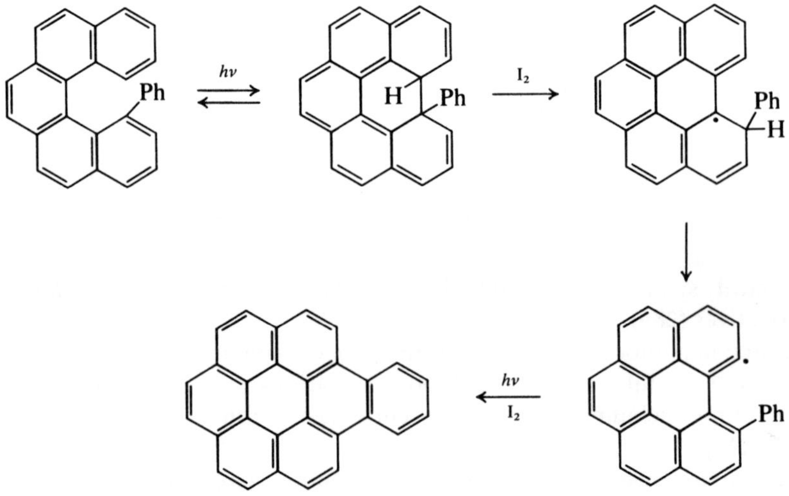

benzocoronene

**Experimental 6.16** *trans*-9,10-Dicyano-9,10-dihydrophenanthrene [239]

A solution of *trans*-$\alpha$,$\alpha'$-dicyanostilbene (5.22 g, 0.023 mol) in purified benzene (275 ml) was degassed at −78 °C under high vacuum. The solution was then irradiated with a Hanovia 400 W medium-pressure lamp through a water-cooled Pyrex jacket for 15¾ h. During irradiation the solution was maintained under vacuum and was stirred magnetically. After filtration, removal of the solvent and recrystallization of the residue from acetone, *trans*-9,10-dicyano-9,10-dihydrophenanthrene (**286**) (3.23 g, 85%) was obtained as prisms, MP 199–204 °C.

### 6.4.1.2 Synthesis of cyclohexane derivatives

By the Woodward–Hoffmann rules, 1,3,5-hexatrienes can undergo photochemical electrocyclization to give cyclohexene derivatives, as shown in the synthesis of vitamin $D_2$ from ergosterol. These reactions illustrate the contrasting photochemical and thermal behaviours of conjugated trienes where the photochemical process is conrotatory and the thermal process disrotatory [242].

$C_9H_{17}$ HO ergosterol $h\nu$ pre-vitamin $D_2$ $\Delta$ pyrocalciferol + iso-pyrocalciferol vitamin $D_2$ lumisterol OH

However, in the photoequilibrium between the conjugated triene (**287**) and the cyclohexadiene (**288**), the ring-opened triene (**287**) is preferred [243, 244].

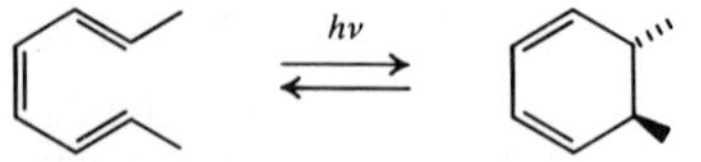

(287) (90–95%) (288) (5–10%)

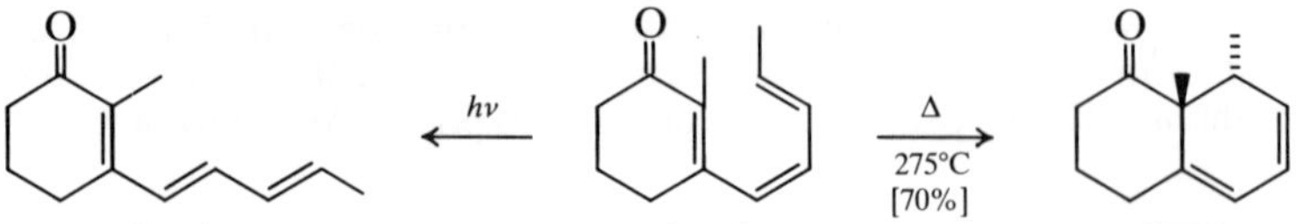

(291) (289) (290)

The trienone (**289**) can be thermally cyclized to the cylohexadiene (**290**), which photochemically isomerizes exclusively to the corresponding *trans*-triene (**291**). Thus in the conjugated triene system the ring-opening reaction is photochemically preferred to the ring closure except in the aforementioned photocyclization of stilbenes and related compounds. Isoe and White have independently discovered a photocyclization of the triene

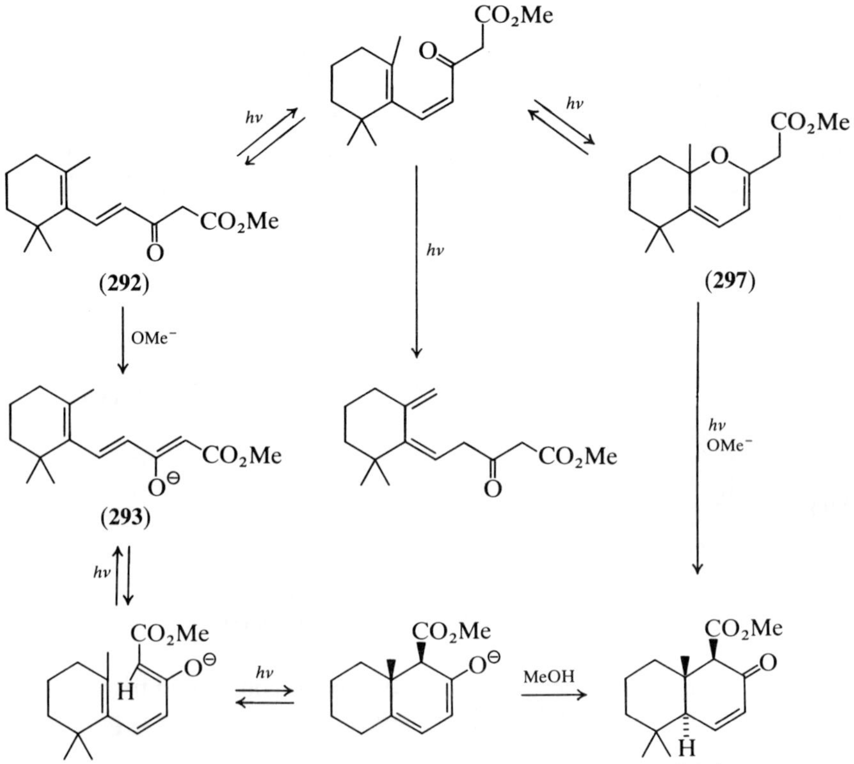

(292) (293) (294) (295) (296) (297)

(297) → hν OMe⁻ → (298)

(299) → hν OMe⁻ → (300)

THP =

involving the enolate anion during the course of studies on the photochemistry of β-ketoesters [245, 246].

In the presence of sodium ethoxide, the β-ketoester (**292**) was converted into the enolate (**293**), which includes a conjugated triene system and undergoes photocyclization to give the ring-closed β-ketoester (**296**) in 65% yield. The same product (**296**) was also synthesized in good yield by irradiation of the pyran (**297**), which was readily prepared by irradiation of the β-ketoester under neutral conditions [245].

This photochemical formation of the cyclohexane derivatives (**296**) from the enolated β-ketoester (**292**) can be explained as follows. The enolated β-ketoester (**293**) isomerizes photochemically to the *cis*-triene (**294**), which is subjected to photochemical electrocyclization in the conrotatory fashion to give the unstable dienolate (**295**). Finally, (**295**) irreversibly accepts a proton from the methanol solvent to give the product (**296**). Similarly, photocyclization of the β-ketoesters (**297**) and (**299**) proceeds smoothly to give the cyclohexenes (**298**) and (**300**) which are potential key synthetic intermediates for some sesquiterpenes.

Analogous photocyclizations giving naphthalene and phenanthrenes have also been reported [247].

EtONa; (1) hν (2) HCl [43%]

(1) hν OEt⁻ (2) H⁺ [48%]

Another potential application of photochemistry to the synthesis of six-membered rings is the preparation of anthracene derivatives in which regioselective photocyclization of enolated *o*-methoxytrienes is a key reaction [248].

ClOC, MeO, Bu$^t$ONa, OMe, $h\nu$, H, OMe, [51%]

### 6.4.2 Preparation of six-membered rings by the Norrish type II reaction followed by cycloaddition

Among many applications of the Norrish type II reaction of the excited carbonyl group to organic synthesis, two transformations involving photoenolization of an aromatic ketone followed by cycloaddition are very useful routes to naphthalene derivatives. Although on irradiation benzophenone was readily reduced in the presence of hydrogen donors, 2-alkylbenzophenones were not. For example, the ketones (**301**) and (**303**) in benzene undergo intramolecular photoenolization as a result of the Norrish type II reaction, and the photoenols (**302**) and (**304**) thus formed reverted back to (**301**) and (**303**) in the dark [249]. On the other hand, in the presence of dimethyl acetylenedicarboxylate, the ketone (**303**) readily underwent photocyclization followed by cycloaddition to give the adduct (**305**) in excellent yield [249].

$h\nu$ / dark; $CCO_2Me \equiv CCO_2Me$

(**301**) R = Ph
(**303**) R = H

(**302**) R = Ph
(**304**) R = H

(**305**) R = H

(303) $\xrightarrow{h\nu}$ $\underset{\longleftarrow}{\xrightarrow{h\nu}}$ $\xrightarrow{O_2}$ (306)

Henderson and Ullman [250] have further investigated the same type of reactions and succeeded in the preparation of 9-anthrone (**306**).

One of the most elegant applications of photochemistry to organic synthesis is the total synthesis of estrone by Quinkert's group [251]. Irradiation of the ketone (**307**) in pyridine and mesitol led to photoenolization as a result of photochemical hydrogen abstraction from the *o*-methyl group; this was followed by intramolecular Diels–Alder reaction of the resulting enol (**308**) to give the tetracyclic steroidal compound (**309**) together with a small amount of the isomer (**310**). Further three-step functionalization of (**309**) completed an elegant total synthesis of (±)-estrone (**311**).

Recently, Quinkert *et al.* [252] have also succeeded in the asymmetric synthesis of (+)-estrone via a route involving the same reaction.

(**307**) $\xrightarrow{h\nu}$ (**308**) [61%] (**310**) (**309**) (**311**) (±)-estrone

**Experimental 6.17** (±)-9$\beta$-Hydroxy-3-methoxy-1,3,5(10)-estratrien-17-one and its 9$\alpha$ isomer [251]

A solution of (**307**) (1.84 g, 6.12 mmol), sublimed mesitol (7.1 g, 52 mmol) and pyridine (dried over molecular sieves, 44.6 ml, 560 mmol) in methylcyclohexane

(1.3 l) was irradiated with a Rayonet reactor (3500A) through the filter solution (0.01% aqueous 2,3-dimethyl-3,6-diaza-1,6-cycloheptadiene for the light filter 340 nm) at 98 °C for 24 h. The reaction vessel was maintained at 98 °C by standing in a heating jacket. The reaction was monitored by GC (Hewlett-Packard 5730 with integrator 3370A) which showed a starting material to product ratio of 1 : 10 at the end of irradiation. The reaction mixture was condensed on a rotary evaporator at 30 °C and the residue obtained was chromatographed on alumina (activity III) by using petroleum ether–ethanol (999 : 1) as eluent. Cooling of the same fractions at −20 °C gave crystals which were recrystallized from ether–dichloromethane (2 : 1). The mother liquor was further purified by preparative thin layer chromatography on silica gel (silica gel-P, UV 254 +366, Merck) with cyclohexane–ethanol (95 : 5). Overall yields: (**309**) 0.83 g, 45%, MP 172–174 °C (ether : pentane 2 : 1); (**310**) 0.29 g, 16%, MP 124–129 °C (ether : pentane 2 : 1).

## 6.5 MEDIUM-RING AND MACROCYCLIC COMPOUNDS

There are only a few examples of the photochemical synthesis of medium-ring and macrocarbocyclic compounds (the corresponding heterocyclic compounds have been prepared as described in Section 7.5). Here we describe some syntheses of medium-ring and macrocarbocyclic compounds by the following photochemical reactions [253]:

(1) $\alpha$-cleavage reactions of carbonyl compounds;
(2) photoelectrocyclic reactions;
(3) photoextrusion of small molecules.

### 6.5.1 Preparation of larger-rings by $\alpha$-cleavage reactions of carbonyl compounds

$\alpha$-Substituted cycloalkanones are known to undergo photochemical $\alpha$-cleavage to give ring-enlarged products as shown by Carlson's group

(**312**) $\xrightarrow{h\nu}$ [(**313**)] ⟶ (**314**)

(**314**) $\xrightarrow{h\nu}$ [ ] ⟶ (**315**) ⟶ (60%)

(**314**) $\xrightarrow{H_2}$ (20%)

[254]. Irradiation of the cyclopropyl ketone (**312**) resulted in $\alpha$-cleavage by the Norrish type I reaction to form the biradical (**313**), isomerization and recombination of which gave the nine- and twelve-membered cyclic compounds (**314**) and (**315**).

Similarly, the cyclophanes (**317**) were prepared by irradiation of the macrocyclic 2-phenylcycloalkanones (**316**) via $\alpha$-cleavage of the carbonyl function followed by recombination of radicals [255].

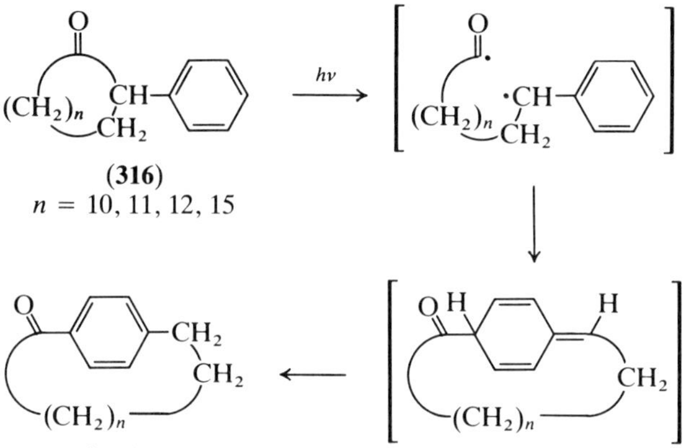

(**316**)
$n$ = 10, 11, 12, 15

(**317**)

### 6.5.2 Preparation of larger rings by photochemical electrocyclic reactions

Photochemical electrocyclic reactions involving C—C bond cleavage provide a useful route to medium-ring and macrocyclic compounds. They do not involve new C—C bond formation but only rearrangement of bond electrons. As an example of the synthesis of a natural product, dihydrocostunolide (**319**) was prepared by irradiation of santonin (**318**) [256].

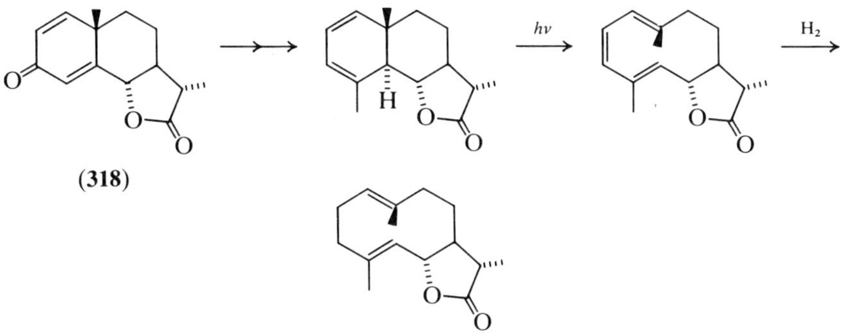

(**318**)

(**319**)

Similarly, methacyclophane (**321**) and [12]annulene (**323**) were prepared by photoisomerization of the 1,3-cyclohexadienes (**320**) and (**322**) respectively [257].

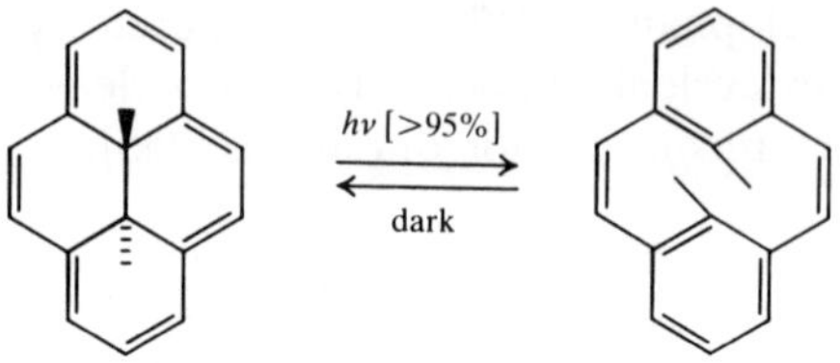

hν [>95%]
dark

(**320**) (**321**)

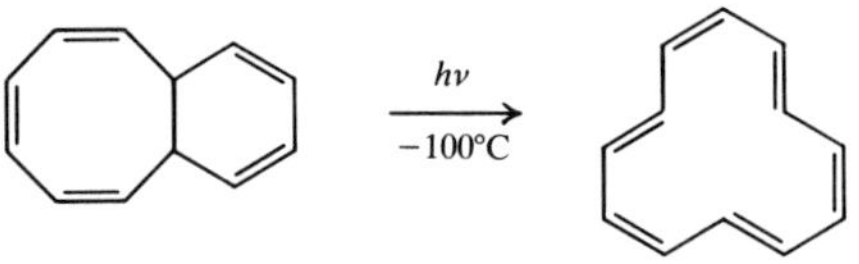

hν
−100°C

(**322**) (**323**)

Eight-, nine-, twelve- and sixteen-membered carbocyclic compounds have been prepared by similar photochemical disrotatory reactions of the respective cyclobutenes [259].

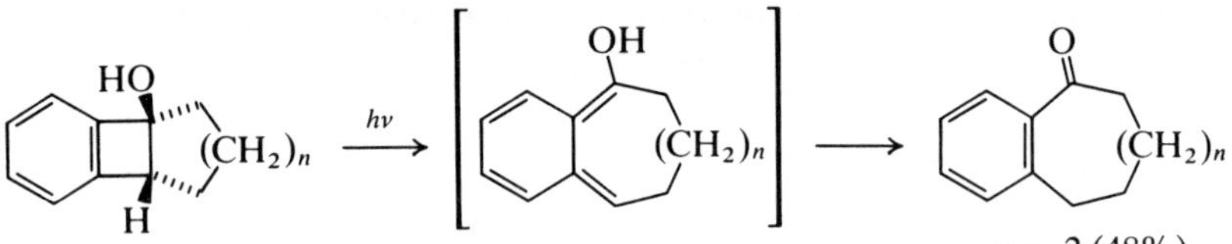

HO H $(CH_2)_n$ hν OH $(CH_2)_n$ O $(CH_2)_n$

$n = 2$ (48%)
$n = 3$ (23%)

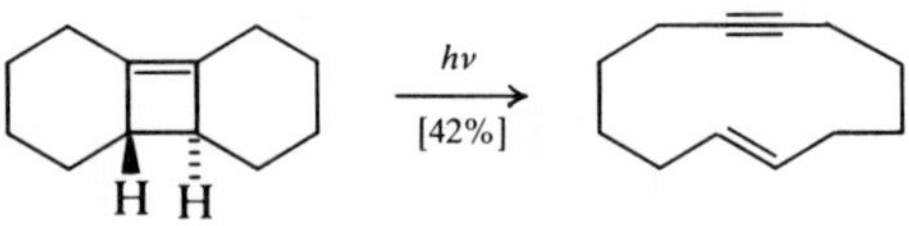

hν
[42%]
H H

### 6.5.3 Preparation of larger rings by photochemical extrusion of small molecules

Photoextrusion of $CO_2$ from cyclic lactones has been applied to the synthesis of certain cyclophanes [260].

Truesdale *et al.* [261,262] have shown that cyclophanes, an interesting class of compounds, can be synthesized by photodecarboxylation of cyclic esters. The cyclophanes, which are formed in very good yields (e.g. 70% for the paracyclophane (**325**) from (**324**)) most likely arise from a sequential photoextrusion of $CO_2$.

O O O O

$h\nu$ (254 nm) [79%]

(**324**) (**325**)

Photoextrusion of $SO_2$ from sulphones has found recent popularity in the synthesis of [2,2]cyclophanes by a number of groups [263] among which Rebafka and Staab [264] have reported the synthesis of a "stacked" donor–acceptor complex.

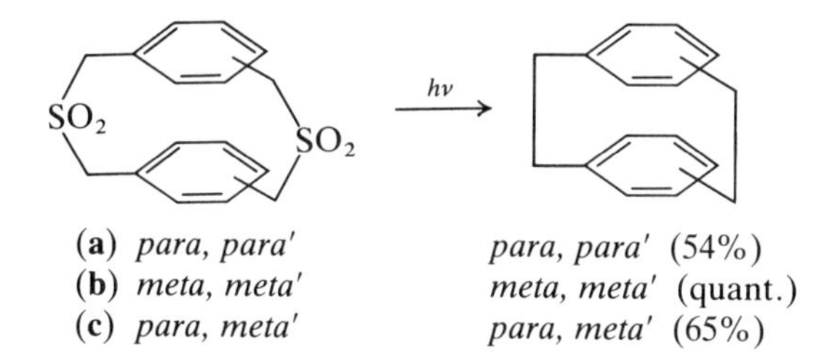


In addition to these examples of the photochemical synthesis of macrocyclic compounds, there are a number of syntheses of medium-ring and macrocyclic compounds by two-step reactions involving photochemical [2 + 2] cycloaddition of the appropriate enone to the appropriate olefin followed by ring opening of the resulting cyclobutanes under various conditions (Section 6.2.1.5).

## REFERENCES

1. H. Morrison, in *Organic Photochemistry* (ed. A. Padwa), Vol. 4, p. 143. Marcel Dekker, New York, 1979.
2. R. S. Givens, in *Organic Photochemistry* (ed. A. Padwa), Vol. 5, p. 227. Marcel Dekker, New York, 1981.
3. H. E. Zimmermann and G. L. Grunewald, *J. Am. Chem. Soc.* **88**, 183 (1979).

4. H. E. Zimmermann, B. W. Binkely, R. S. Givens and M. A. Sherwin, *J. Am. Chem. Soc.* **89**, 3932 (1967).
5. H. E. Zimmermann, R. W. Binkely, R. S. Givens, G. L. Grunewald and M. A. Sherwin, *J. Am. Chem. Soc.* **91**, 3316 (1969).
6. H. E. Zimmermann, *Acc. Chem. Res.* **4**, 272 (1971).
7. A. J. Barker, J. S. H. Kueh, M. Mellor, D.A. Otieno and G. Pattenden, *Tetrahedron Lett.* **1979**, 1881.
8. J. R. Edman, *J. Am. Chem. Soc.* **91**, 7103 (1969).
9. J. R. Edman, *J. Am. Chem. Soc.* **88**, 3454 (1968).
10. E. Ciganek, *J. Am. Chem. Soc.* **88**, 2882 (1966).
11. T. J. Katz and H. Acton, *J. Am. Chem. Soc.* **95**, 2738 (1973).
12. N. J. Turro, C. A. S. Renner, W. H. Waddell and T. J. Katz, *J. Am. Chem. Soc.* **98**, 4320 (1976).
13. M. Schneider, *Angew. Chem. Int. Ed. Engl.* **14**, 707 (1975).
14. P. G. Gassman and W. J. Greenlae, *J. Am. Chem. Soc.* **95**, 980 (1973).
15. J. J. Tufariello, A. C. Bayer and J. J. Spadaro, *J. Am. Chem. Soc.* **101**, 3309 (1979).
16. M. Frank-Newmann and C. D. Buchecker, *Tetrahedron Lett.* **1980**, 671.
17. W. J. Greenlee and R. B. Woodward, *J. Am. Chem. Soc.* **98**, 6075 (1976).
18. C. Berger, M. Franck-Newmann and G. Ourisson, *Tetrahedron Lett.* **1968**, 3451.
19. E. Piers, M. B. Gerghty, R. D. Smillie and M. Soucy, *Can. J. Chem.* **53**, 2849 (1975).
20. S. R. Wilson and R. B. Turner, *J. Org. Chem.* **38**, 2870 (1973).
21. M. Franck-Newmann, *Angew. Chem. Int. Ed. Engl.* **7**, 65 (1968).
22. H.-D. Scharf and W. Kusters, *Chem. Ber.* **104**, 3016 (1971).
23. H. Prinzbach, H. Hagemann, J. H. Hartenstein and R. Kitzing, *Chem. Ber.* **98**, 2201 (1965).
24. D. W. Boykin and R. E. Lutz, *J. Am. Chem. Soc.* **86**, 504 (1964).
25. D. E. McGreer, M. G. Vinje and R. S. McDaniel, *Can. J. Chem.* **43**, 1417 (1965).
26. J. Wiemann, N. Thoai and F. Weisbuch, *Tetrahedron Lett.* **1965**, 2983.
27. S. W. Baldwin, in *Organic Photochemistry* (ed. A. Padwa), Vol. 5, p. 123. Marcel Dekker, New York, 1981.
28. P. E. Eaton and T. W. Cole, *J. Am. Chem. Soc.* **86**, 3157 (1964).
29. E. J. Corey, R. B. Mitra and H. Uda, *J. Am. Chem. Soc.* **86**, 485 (1964).
30. J. D. White and D. N. Gupta, *J. Am. Chem. Soc.* **90**, 6171 (1968).
31. K. Wiesner, L. Poon, J. Jirkovsky and M. Fishman, *Can. J. Chem.* **47**, 433 (1969).
32. A. Kunai, T. Omori, K. Kimura and Y. Odaira, *Bull. Chem. Soc. Jpn.* **48**, 731 (1975).
33. P. E. Eaton and K. Lin, *J. Am. Chem. Soc.* **87**, 2052 (1965).
34. E. J. Corey, M. Tada, R. LaMathieu and L. Libit, *J. Am. Chem. Soc.* **87**, 2051 (1965).
35. E. J. Corey, J. D. Bass, R. LaMathieu and R. B. Mitra, *J. Am. Chem. Soc.* **86**, 5570 (1964).
36. P. E. Eaton and K. Lin, *J. Am. Chem. Soc.* **86**, 2087 (1964).
37. H. Shinozaki, S. Arai and M. Tada, *Bull. Chem. Soc. Jpn* **49**, 821 (1976).
38. R. Noyori, H. Inoue and M. Kato, *J. Chem. Soc. Chem. Commun.* **1970**, 1695.
39. H. Hart, B. Chen and M. Jeffares, *J. Org. Chem.* **44**, 2722 (1979).
40. G. Ciamician and P. Silber, *Chem. Ber.* **41**, 1928 (1908).
41. M. Zandomeneghi, M. Cavazza, L. Moi and F. Pietra, *Tetrahedron Lett.* **21**, 713 (1980).
42. O. L. Chapman and G. Lenz, in *Organic Photochemistry* (ed. O. L. Chapman), Vol. 1, p. 283. Marcel Dekker, New York, 1967.
43. O. L. Chapman and O. S. Weiss, in *Organic Photochemistry* (ed. O. L. Chapman), Vol. 3, p. 197. Marcel Dekker, New York, 1973.

44. N. J. Turro, in *Modern Molecular Photochemistry*, Chap. 11. Benjamin/Cummings, Menlo Park, California, 1974.
45. K. Nakanishi and H. Sato, in *Natural Products Chemistry* (ed. K. Nakanishi, T. Goto, S. Ito, S. Noyori and S. Nozoe), Vol. 2, p. 523. Academic Press, New York, 1975.
46. T. S. Cantrell, W. S. Haller and J. C. Williams, *J. Org. Chem.* **34**, 509 (1969).
47. T. S. Cantrell, *Tetrahedron* **27**, 1227 (1971).
48. J. C. Dalton, K. Dawes, N. J. Turro, D. S. Weiss, J. A. Bartlop and J. D. Coyle, *J. Am. Chem. Soc.* **93**, 7213 (1971).
49. P. J. Wagner and R. W. Spoerke, *J. Am. Chem. Soc.* **91**, 4437 (1969).
50. G. L. Lange, H. M. Campbell and E. Neidert, *J. Org. Chem.* **38**, 2117 (1973).
51. G. L. Lange, E. E. Neidert, W. J. Orrom and D. J. Wallace, *Can. J. Chem.* **56**, 1628 (1978).
52. P. G. Sammes, *Tetrahedron* **32**, 405 (1976).
53. M. Tada and K. Miura, *Bull. Chem. Soc. Jpn* **49**, 713 (1976).
54. Z. Valenta and H. J. Liu, *Org. Synth.* **57**, 113 (1977).
55. J. D. Shiloff and N. R. Hunter, *Can. J. Chem.* **57**, 3301 (1979).
56. G. Jones, in *Organic Photochemistry* (ed. A. Padwa), Vol. 5, p. 1. Marcel Dekker, New York, 1981.
57. O. L. Chapman, T. H. Koch, F. Klein, P. J. Nelson and E. L. Brown, *J. Am. Chem. Soc.* **90**, 1657 (1968).
58. P. J. Nelson, D. Osterm, J. D. Laddila and O. L. Chapman, *J. Org. Chem.* **34**, 811 (1969).
59. Z. Yoshida, M. Kimura and S. Yoneda, *Tetrahedron Lett.* **1975**, 1001.
60. V. Desobry and P. Margaretha, *Helv. Chim. Acta* **58**, 2161 (1975).
61. G. V. Thi and P. Margaretha, *Helv. Chim. Acta* **59**, 2236 (1976).
62. I. Altmayer and P. Margaretha, *Helv. Chim. Acta* **60**, 874 (1977).
63. P. E. Eaton, *J. Am. Chem. Soc.* **84**, 2454 (1962).
64. S. W. Baldwin and M. T. Crimmins, *J. Am. Chem. Soc.* **102**, 1198 (1980).
65. K. Tatsuta, K. Akimoto and M. Kinoshita, *J. Am. Chem. Soc.* **101**, 6116 (1979).
66. L. Duc, R. A. Mateer, L. Brassier and G. W. Griffin, *Tetrahedron Lett.* **1968**, 6173.
67. B. D. Challanel and P. de Mayo, *J. Chem. Soc. Chem. Commun.* **1966**, 982.
68. K. H. Lee and P. de Mayo, *J. Chem. Soc. Chem. Commun.* **1979**, 493.
69. M. Van Audenhove, D. Termont, D. de Keukeleire and M. Vandewalle, *Tetrahedron Lett.* **1978**, 2057.
70. D. Termont, D. de Keukeleire and M. Vandewalle, *J. Chem. Soc. Perkin Trans. 1* **1977**, 2349.
71. R. O. Loutfy and P. de Mayo, *Can. J. Chem.* **50**, 3465 (1972).
72. E. P. Serebryakos, S. D. Kulomzina-Pletneva and A. Kh. Margaryam, *Tetrahedron Lett.* **35**, 77 (1979).
73. P. de Mayo, J. Takeshita and A. B. M. A. Sattar, *Proc. Chem. Soc.* **1962**, 119.
74. P. de Mayo and H. Takeshita, *Can. J. Chem.* **41**, 440 (1963).
75. H. Hikino and P. de Mayo, *J. Am. Chem. Soc.* **86**, 3582 (1964).
76. P. de Mayo, *Pure Appl. Chem.* **9**, 597 (1964).
77. B. D. Challand, H. Hikino, G. Kornis, G. Lange and P. de Mayo, *J. Org. Chem.* **34**, 794 (1969).
78. R. G. Hunt, C. J. Potter, S. T. Reid and M. L. Roantree, *Tetrahedron Lett.* **1975**, 2327.
79. M. Umehara, T. Oda, Y. Ikebe and S. Hishide, *Bull. Chem. Soc. Jpn* **49**, 1075 (1976).
80. M. Gorodetsky, Z. Luz and Y. Mazur, *J. Am. Chem. Soc.* **89**, 1183 (1967).
81. H. Nozaki, M. M. Kurita, T. Mori and R. Noyori, *Tetrahedron* **24**, 1821 (1968).
82. E. W. Garbisch, *J. Am. Chem. Soc.* **85**, 1696 (1963).

83. A. A. Bothner-By and R. K. Harris, *J. Org. Chem.* **30**, 254 (1965).
84. W. O. George and V. G. Mansell, *J. Chem. Soc.* (B) **1968**, 132.
85. S. W. Baldwin, R. E. Gawley, R. J. Doll and K. H. Leung, *J. Org. Chem.* **40**, 1865 (1975).
86. S. W. Baldwin and R. E. Gawley, *Tetrahedron Lett.* **1975**, 3669.
87. See Ref. [27], p. 173.
88. M. Tada, T. Kokubo and T. Sato, *Bull. Chem. Soc. Jpn* **43**, 2162 (1970).
89. D. Verierov, T. Bercovici, E. Fischer, Y. Mazur and A. Yogev, *J. Am. Chem. Soc.* **95**, 8173 (1973).
90. R. C. Gueldner, A. C. Thompson and P. A. Hedin, *J. Org. Chem.* **37**, 1854 (1972).
91. H. Kosugi, S. Sekiguchi, R. Sekita and H. Uda, *Bull. Chem. Soc. Jpn* **49**, 520 (1976).
92. S. W. Baldwin and J. M. Wilkinson, *J. Am. Chem. Soc.* **102**, 3634 (1980).
93. See Ref. [27], p. 175.
94. R. E. Gawley, *Synthesis* **1976**, 777.
95. M. E. Jung, *Tetrahedron* **32**, 3 (1976).
96. S. Danishefsky and T. Kitahara, *J. Am. Chem. Soc.* **96**, 7807 (1974).
97. H. Takeshita and S. Tanno, *Bull. Chem. Soc. Jpn* **46**, 880 (1973).
98. H. Takeshita, H. Iwabuchi, I. Kouno, M. Iino and D. Monura, *Chem. Lett.* **1979**, 647.
99. H. Takeshita, R. Kikuchi and Y. Shoji, *Bull. Chem. Soc. Jpn* **46**, 690 (1973).
100. J. J. Partridge, N. K. Chandha and M. R. Uskokovic, *J. Am. Chem. Soc.* **95**, 532 (1973).
101. G. Buchi, J. A. Carlson, J. E. Powell and L.-F. Tietze, *J. Am. Chem. Soc.* **95**, 540 (1973).
102. L.-F. Tietze, *Chem. Ber.* **107**, 2499 (1974).
103. C. R. Hutchinson, K. C. Mattes, M. Nakane, J. J. Partridge and M. R. Uskokovic, *Helv. Chim. Acta* **61**, 1221 (1978).
104. M. Nakane, H. Gollman, L. R. Hutchinson and P. L. Knutson, *J. Org. Chem.* **45**, 2536 (1980).
105. S. W. Baldwin and M. T. Crimmins, *J. Am. Chem. Soc.* **104**, 1132 (1982).
106. S. W. Baldwin, M. T. Crimmins and V. I. Cheek, *Synthesis* **1978**, 210.
107. A. Kh. Margaryan, E. P. Sevebryyakov and V. E. Kucherov, *Izv. Akad. Nauk SSSR Ser. Khim.* **1976**, 840.
108. D. C. Owsley and J. J. Bloomfield, *J. Chem. Soc.* (C) **1971**, 3445.
109. Y. Sugihara, N. Norokoshi and I. Murata, *Tetrahedron Lett.* **1977**, 3887.
110. P. E. Eaton, *Tetrahedron Lett.* **1964**, 3695.
111. G. O. Schenck and R. Steinmetz, *Bull Soc. Chim. Belg.* **71**, 781 (1962).
112. R. Robson, P. W. Grubb and J. A. Bartlop, *J. Chem. Soc.* **1964**, 2153.
113. G. O. Schenck, W. Hartmann, S.-P. Mannsfeld, W. Metzner and H. Krauchi, *Chem. Ber.* **95**, 1642 (1962).
114. G. O. Schenck, W. Hartmann and R. Steinmetz, *Chem. Ber.* **96**, 498 (1963).
115. C. Kaneko and T. Naito, *Chem. Pharm. Bull.* **27**, 2254 (1979).
116. C. Kaneko, T. Naito and M. Somei, *J. Chem. Soc. Chem. Commun.* **1979**, 804.
117. T. Naito, N. Nakayama and C. Kaneko, *Chem Lett.* **1981**, 423.
118. C. Kaneko, T. Naito and T. Ohashi, *Heterocycles* **20**, 1275 (1983).
119. T. Naito and C. Kaneko, *Tetrahedron Lett.* **22**, 2691 (1981).
120. W. L. Dilling, *Photochem. Photobiol.* **25**, 605 (1977).
121. M. Brown, *J. Org. Chem.* **33**, 162 (1968).
122. K. Yoshihara, Y. Ohta, T. Sakai and Y. Hirose, *Tetrahedron Lett.* **1969**, 2263.
123. J. H. Tumlinson, R. C. Guedner, D. D. Hardee, A. C. Thompson, P. A. Hedin and J. P. Miryard, *J. Org. Chem.* **36**, 2616 (1971).
124. R. Zurfluh, L. L. Dunham, V. L. Spain and J. B. Siddall, *J. Am. Chem. Soc.* **92**, 425 (1979).
125. W. A. Ayer and L. M. Browne, *Can. J. Chem.* **52**, 1352 (1974).

126. R. L. Cargill and B. W. Wright, *J. Org. Chem.* **40**, 120 (1975).
127. W. E. Billups, J. H. Cross and C. V. Smith, *J. Am. Chem. Soc.* **95**, 3438 (1973).
128. K. Mori, *Tetrahedron* **34**, 915 (1978).
129. P. Ho, S. F. Lee, D. Chang and K. Wiesner, *Experientia* **27**, 1377 (1971).
130. D. E. Bergstrom and W. C. Agosta, *Tetrahedron Lett.* **1974**, 1087.
131. B. D. Challand, G. Kornis, G. L. Lange and P. de Mayo, *J. Chem. Soc. Chem. Commun.* **1967**, 704.
132. M. Mellor, D. A. Otieno and G. Pattenden, *J. Chem. Soc. Chem. Commun.* **1978**, 138.
133. S. W. Baldwin and J. M. Wilkinson, *Tetrahedron Lett.* **1979**, 2657.
134. See Ref. [27], p. 198.
135. P. A. Wender and J. C. Lechleiter, *J. Am. Chem. Soc.* **100**, 4321 (1978).
136. P. A. Wender and J. C. Lechleiten, *J. Am. Chem. Soc.* **99**, 267 (1977).
137. S. R. Wilson, L. R. Phillips, Y. Pelisten and J. C. Huffman, *J. Am. Chem. Soc.* **101**, 7373 (1979).
138. J. R. Williams and J. F. Callahan, *J. Chem. Soc. Chem. Commun.* **1979**, 404.
139. J. R. Williams and J. F. Callahan, *J. Chem. Soc. Chem. Commun.* **1979**, 405.
140. G. L. Lange and F. C. McCarthy, *Tetrahedron Lett.* **1978**, 4749.
141. E. J. Corey and S. Nozoe, *J. Am. Chem. Soc.* **86**, 1652 (1964).
142. K. Hayano, Y. Ohfune, H. Shirahama and T. Matsumoto, *Chem. Lett.* **1978**, 1301.
143. D. K. M. Duc, M. Fetizon and M. Kone, *Tetrahedron* **34**, 3513 (1978).
144. M. Yanagiya, K. Kaneko, T. Kaji and T. Matsumoto, *Tetrahedron Lett.* **1979**, 1761.
145. H. Suginome, K. Kobayashi, M. Itoh and A. Furuhashi, *Chem. Lett.* **1985**, 727.
146. H. Suginome, C. F. Liu and A. Furusaki, *Chem. Lett.* **1985**, 27.
147. H. Suginome, C. F. Liu and M. Tokuda, *J. Chem. Soc. Chem. Commun.* **1984**, 334.
148. H. Suginome, C. F. Liu, M. Tokuda and A. Furusaki, *J. Chem. Soc. Perkin Trans. 1* **1985**, 327.
149. A. C. Weedon, in *Synthetic Organic Photochemistry* (ed. W. M. Horspool), p. 61. Plenum Press, New York, 1984.
150. H. Suginome, *J. Synth. Org. Chem. Jpn* **44**, 1029 (1986).
151. H. J. Liu and Z. Valenta, *J. Chem. Soc. Perkin Trans. 1* **1970**, 1116.
152. H. J. Liu, Z. Valenta, J. S. Wilson and T. T. J. Yu, *Can. J. Chem.* **47**, 509 (1969).
153. K. Wiesner, P. T. Ho, D. Cheng, Y. K. Lam, C. S. J. Pan and W. Y. Ren, *Can. J. Chem.* **51**, 3978 (1973).
154. K. Mori and M. Sasaki, *Tetrahedron Lett.* **1979**, 1329.
155. P. A. Wender and J. L. Hubba, *J. Org. Chem.* **45**, 365 (1980).
156. M. Yanagida, K. Kaneko, T. Kaji and T. Matsumoto, *Tetrahedron Lett.* **1979**, 1761.
157. P. A. Wender and L. J. Letendre, *J. Org. Chem.* **45**, 367 (1980).
158. H. Minato and I. Horibe, *J. Chem. Soc.* (C) **1967**, 1576.
159. P. A. Wender and J. C. Lechleiter, *J. Am. Chem. Soc.* **102**, 6340 (1980).
160. A. I. Meyers and S. A. Fleming, *J. Am. Chem. Soc.* **108**, 306 (1986).
161. M. Demuth, *Pure Appl. Chem.* **58**, 1233 (1986).
162. K. Wiesner, V. Musil and K. J. Wiesner, *Tetrahedron Lett.* **1968**, 5643.
163. R. Srinivasan and K. H. Carlough, *J. Am. Chem. Soc.* **89**, 4396 (1967).
164. R. S. Liu and G. S. Hammond, *J. Am. Chem. Soc.* **89**, 4936 (1967).
165. W. C. Agosta and S. Wolff, *J. Org. Chem.* **45**, 3139 (1980) and references therein.
166. M. C. Pirrung, *J. Am. Chem. Soc.* **103**, 82 (1981).
167. R. L. Cargill, J. R. Dalton, S. O'Connor and P. G. Michels, *Tetrahedron Lett.* **1978**, 4465.
168. C. R. Johnson and N. A. Meanwell, *J. Am. Chem. Soc.* **103**, 7667 (1981).
169. G. Mehta, A. Srikrihna, A. V. Reddy and M. S. Nair, *Tetrahedron* **37**, 4543 (1981).
170. G. Mehta and A. V. Reddy, *J. Chem. Soc. Chem. Commun.* **1981**, 756.

171. W. Oppolzer, F. Zutterman and K. Battig, *Helv. Chim. Acta* **66**, 522 (1983).
172. W. Oppolzer, *Acc. Chem. Res.* **15**, 135 (1982).
173. K. Tatsuta, K. Akimoto and M. Kinoshita, *J. Am. Chem. Soc.* **101**, 6116 (1979).
174. B. W. Disanayaka and A. C. Weedon, *J. Chem. Soc. Chem. Commun.* **1985**, 1282.
175. W. Oppolzer and T. Godel, *J. Am. Chem. Soc.* **100**, 2583 (1978).
176. W. Oppolzer and R. D. Wylie, *Helv. Chim. Acta* **62**, 1198 (1980).
177. J. K. Whitesell, R. S. Matthews and A. M. Helbling, *J. Org. Chem.* **43**, 784 (1978).
178. H. Takeshita, I. Kunno, M. Iino, H. Iwabuchi and D. Nomura, *Bull. Chem. Soc. Jpn* **53**, 3641 (1980).
179. H. Takeshita, S. Hatta and T. Hatsui, *Bull. Chem. Soc. Jpn* **57**, 619 (1984).
180. C. Kaneko and T. Naito, *Heterocycles* **19**, 2183 (1982).
181. C. Kaneko, Y. Momose, T. Maeda, T. Naito and M. Somei, *Heterocycles* **20**, 2169 (1983).
182. T. Naito and C. Kaneko, *J. Synth. Org. Chem. Jpn* **42**, 51 (1984).
183. T. Naito, Y. Momose and C. Kaneko, *Chem. Pharm. Bull.* **30**, 1531 (1982).
184. W. Oppolzer, *Synthesis* **1978**, 793.
185. T. Kametani and K. Fukumoto, *Heterocycles* **3**, 29 (1975).
186. T. Naito and C. Kaneko, *Chem. Pharm. Bull.* **28**, 3150 (1980).
187. N. V. Kirby and S. T. Reid, *J. Chem. Soc. Chem. Commun.* **1980**, 150.
188. J. H. M. Hill and S. T. Reid, *J. Chem. Soc. Chem. Commun.* **1983**, 501.
189. M. Ikeda, K. Ohno, T. Uno and Y. Tamura, *Tetrahedron Lett.* **1980**, 3403.
190. P. D. Davis, J. R. Blount and D. C. Neckers, *J. Org. Chem.* **45**, 462 (1980).
191. S. T. Reid and D. De Silva, *Tetrahedron Lett.* **24**, 1949 (1983).
192. T. Sano, Y. Horiguchi and Y. Tsuda, *Heterocycles* **16**, 355 (1981).
193. T. Sano, Y. Horiguchi, S. Kanbe and Y. Tsuda, *Heterocycles* **16**, 363 (1981).
194. T. Sano, Y. Horiguchi and Y. Tsuda, *Heterocycles* **12**, 1427 (1979).
195. C. Kaneko, N. Shimomura, Y. Momose and T. Naito, *Chem. Lett.* **1983**, 1239.
196. F. I. Sonntag and R. Srinivasan, *Org. Photo. Synth.* **1**, 39 (1971).
197. R. Srinivasan *J. Am. Chem. Soc.* **84**, 4141 (1962).
198. W. G. Dauben and C. D. Poulthr, *Tetrahedron Lett.* **1967**, 3021.
199. W. G. Dauben and G. J. Fonken, *J. Am. Chem. Soc.* **81**, 4060 (1959).
200. K. Dimroth, *Chem. Ber.* **70**, 1631 (1937).
201. W. G. Dauben and H. J. Cargill, *Tetrahedron* **12**, 186 (1961).
202. O. L. Chapman, D. J. Paston, G. W. Borden and A. A. Griswold, *J. Am. Chem. Soc.* **84**, 1220 (1962).
203. G. J. Fonken, *Chem. Ind.* **1961**, 1575.
204. E. J. Corey and J. Streith, *J. Am. Chem. Soc.* **86**, 950 (1964).
205. F. W. Fowler, *J. Org. Chem.* **37**, 1321 (1972).
206. L. A. Paquett and G. Slomp, *J. Am. Chem. Soc.* **85**, 765 (1963).
207. E. C. Taylor and R. O. Kan, *J. Am. Chem. Soc.* **85**, 776 (1963).
208. M. Pomeranz and G. W. Gruber, *J. Am. Chem. Soc.* **93**, 6615 (1971).
209. L. A. Paquette and D. E. Kuhla, *J. Org. Chem.* **34**, 2885 (1969).
210. O. L. Chapman and E. D. Hoganson, *J. Am. Chem. Soc.* **86**, 498 (1964).
211. L. A. Paquette, *J. Am. Chem. Soc.* **86**, 500 (1964).
212. J. Brennan, *J. Chem. Soc. Chem. Commun.* **1981**, 880.
213. T. Kametani, T. Mochizuki and T. Honda, *Heterocycles* **19**, 89 (1982).
214. C. Kaneko, K. Shiba, H. Fujii and Y. Momose, *J. Chem. Soc. Chem. Commun.* **1980**, 1177.
215. C. Kaneko, T. Naito and A. Saito, *Tetrahedron Lett.* **25**, 1591 (1984).
216. C. Kaneko, N. Katagiri, M. Sato, M. Muto, T. Sakamoto, S. Saikawa, T. Naito and A. Saito, *J. Chem. Soc. Perkin Trans. 1* **1986**, 1283.

217. C. Kaneko, M. Sato and N. Katagiri, *J. Synth. Org. Chem. Jpn* **44**, 1058 (1986).
218. R. B. Gagosian, J. C. Dalton and N. J. Turro, *J. Am. Chem. Soc.* **92**, 4752 (1970).
219. R. B. Gagosian, J. C. Dalton and N. J. Turro, *J. Am. Chem. Soc.* **97**, 5189 (1975).
220. R. R. Saners, M. Gordetsky, J. A. Whittle and C. K. Hu, *J. Am. Chem. Soc.* **93**, 5520 (1971).
221. W. H. Urry and D. J. Trecken, *J. Am. Chem. Soc.* **84**, 118 (1962).
222. P. J. Wagner, R. G. Zepp, K. C. Liu, M. Thomas, T. J. Lee, N. J. Lee and N. J. Turro, *J. Am. Chem. Soc.* **98**, 8125 (1976).
223. H. Aoyama, S. Suzuki, T. Hasegawa and Y. Omote, *J. Chem. Soc. Chem. Commun.* **1979**, 899.
224. H. Aoyama, T. Hasegawa and Y. Omote, *J. Am. Chem. Soc.* **101**, 5343 (1979).
225. A. Hassner, A. W. Coulter and W. S. Seese, *Tetrahedron Lett.* **1962**, 759.
226. M. P. Cava, R. L. Litle and D. R. Napier, *J. Am. Chem. Soc.* **80**, 2257 (1958).
227. M. P. Cava and E. Moroz, *J. Am. Chem. Soc.* **84**, 115 (1962).
228. O. L. Chapman and D. J. Pasto, *J. Am. Chem. Soc.* **82**, 3642 (1960).
229. W. G. Dauben, K. Koch and W. E. Thiessen, *J. Am. Chem. Soc.* **81**, 6087 (1959).
230. T. Miyoshi, M. Nitta and T. Mukai, *J. Am. Chem. Soc.* **93**, 3441 (1971).
231. W. G. Dauben, K. Koch, O. L. Chapman and S. L. Smith, *J. Am. Chem. Soc.* **85**, 2616 (1963).
232. K. B. Wiberg, B. R. Lowry and T. H. Coby, *J. Am. Chem. Soc.* **83**, 3998 (1961).
233. L. Horner and E. Spietschka, *Chem. Ber.* **88**, 934 (1955).
234. P. J. Wagner and C. Chiu, *J. Am. Chem. Soc.* **101**, 7134 (1979).
235. E. V. Blackburn and C. J. Timmons, *Q. Rev. Chem. Soc.* **23**, 482 (1969).
236. See Ref. [42], p. 247.
237. F. B. Mallory and C. W. Mallory, in *Organic Reactions* (ed. W. G. Dauben), Vol. 30, p. 1. Wiley, New York, 1984.
238. C. S. Wood and F. B. Mallory, *J. Org. Chem.* **29**, 3373 (1974).
239. M. V. Sargent and C. J. Timmons, *J. Am. Chem. Soc.* **85**, 2186 (1964).
240. A. Buquet, A. Couture and A. L. Combier, *J. Org. Chem.* **44**, 2300 (1979).
241. A. H. A. Tinnemans and W. H. Laarhoven, *J. Am. Chem. Soc.* **96**, 4611 (1974).
242. H. Havinga and J. L. M. A. Schlatmann, *Tetrahedron* **16**, 146 (1961).
243. G. J. Fonken, *Tetrahedron Lett.* **1962**, 549.
244. R. Ramage and A. Sattar, *Chem. Commun.* **1970**, 173.
245. T. H. Kim, Y. Hayase and S. Isoe, *Chem. Lett.* **1983**, 651.
246. J. D. White and R. W. Skeean, *J. Am. Chem. Soc.* **100**, 6296 (1978).
247. N. C. Yang, L. C. Lin, A. Shai and S. S. Yang, *J. Org. Chem.* **34**, 1845 (1969).
248. Y. Tamura, S. Fukumori, S. Kato and Y. Kita, *J. Chem. Soc. Chem. Commun.* **1974**, 285.
249. N. C. Yang and C. Rivas, *J. Am. Chem. Soc.* **83**, 2213 (1961).
250. W. A. Henderson and E. F. Ullman, *J. Am. Chem. Soc.* **87**, 5424 (1965).
251. G. Quinkert, W.-D. Weber, U. Schwartz, H. Stark, H. Baier and G. Durner, *Liebigs Ann. Chem.* **1981**, 2335.
252. G. Quinkert, U. Schwartz, H. Stark, W.-D. Weber, F. Adam, H. Baier, G. Frank and G. Durner, *Liebigs Ann. Chem.* **1982**, 1999.
253. K. Maruyama and Y. Kubo, *J. Synth. Org. Chem. Jpn* **38**, 351 (1980).
254. R. G. Carlson and W. S. Mardis, *J. Org. Chem.* **40**, 817 (1975).
255. X.-G. Lei, C. E. Doubleday, M. B. Zimmt and N. J. Turro, *J. Am. Chem. Soc.* **108**, 2444 (1986).
256. E. J. Corey and A. G. Hortmann, *J. Am. Chem. Soc.* **85**, 4033 (1963).
257. H.-R. Blattmann, D. Meuche, E. Heilbronner, R. J. Molyneux and V. Boekelheide, *J. Am. Chem. Soc.* **87**, 130 (1965).

258. J. F. M. Oth, H. Rottele and G. Schroder, *Tetrahedron Lett.* **1970**, 61.
259. J. Saltiel and L.-S. N. Lim, *J. Am. Chem. Soc.* **91**, 5404 (1964).
260. See Ref. [2], p. 309.
261. M. L. Kaplan and E. A. Truesdale, *Tetrahedron Lett.* **1976**, 3665.
262. E. A. Truesdale, *Tetrahedron Lett.* **1978**, 3777.
263. See Ref. [2], p. 322.
264. W. Rebafka and H. A. Staab, *Angew. Chem. Int. Ed. Engl.* **12**, 776 (1973).

# – 7 –

# Preparation of Heterocyclic Compounds

## 7.1 THREE-MEMBERED HETEROCYCLES

The aziridine ring has been prepared by the photochemical addition of an azidoformate to a C=C double bond. Irradiation of a dilute cyclohexene solution of ethyl azidoformate (**1**) gave the aziridine (**2**) in 56% yield in addition to small amounts of olefinic carbamates [1]. Lowering the temperature to −75 °C improved the yield of (**2**) to 75%.

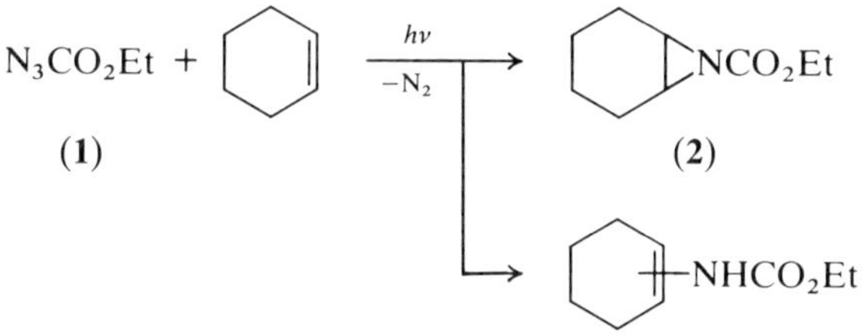

$N_3CO_2Et$ + cyclohexene $\xrightarrow[-N_2]{h\nu}$ (**2**) $NCO_2Et$ + $NHCO_2Et$ (olefinic carbamates)

(**1**) (**2**)

In some cases an aziridine derivative is a key intermediate for the synthesis of azepine derivatives, as shown below. When ethyl azidoformate (**1**) was dissolved in benzene and irradiated, a 70% yield of ethyl 1*H*-azepine-1-carboxylate (**4**) was obtained [2]. The ring-enlargement reaction is formulated as proceeding via the aziridine (**3**) in close analogy to the reaction of benzene with diazomethane to yield cycloheptatriene.

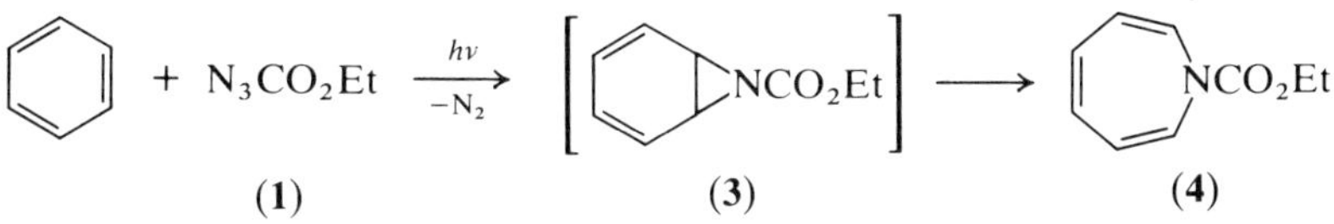

benzene + $N_3CO_2Et$ $\xrightarrow[-N_2]{h\nu}$ [ $NCO_2Et$ ] ⟶ $NCO_2Et$

(**1**) (**3**) (**4**)

The photochemically and thermally induced extrusion of stable neutral molecules from five-membered heterocycles has been used as a route to the aziridine, 1-azirine and 2-azirine ring systems. $\Delta^2$-1,2,3-Triazolines (**5**), prepared by the 1,3-dipolar cycloaddition of organic azides to allenes or less commonly by the addition of diazoalkanes to imines, serve as precursors to aziridines [3]. The photochemical route is generally preferred since by-product formation (e.g. isomeric imines) is minimized. The $\Delta^2$-triazoline photodecomposition process is consistent with an initial homolysis of the $N^1$—$N^2$ bond, leading to a singlet diradical (**6**), which then loses nitrogen to give a diradical (**7**). The photochemical extrusion of nitrogen from monocyclic $\Delta^2$-triazolines generally displays a high degree of stereoselectivity in which the triazoline stereochemistry is preserved in the triazoline ring. Monocyclic, fused and spiroaziridines (**9**) have been prepared by this route [4].

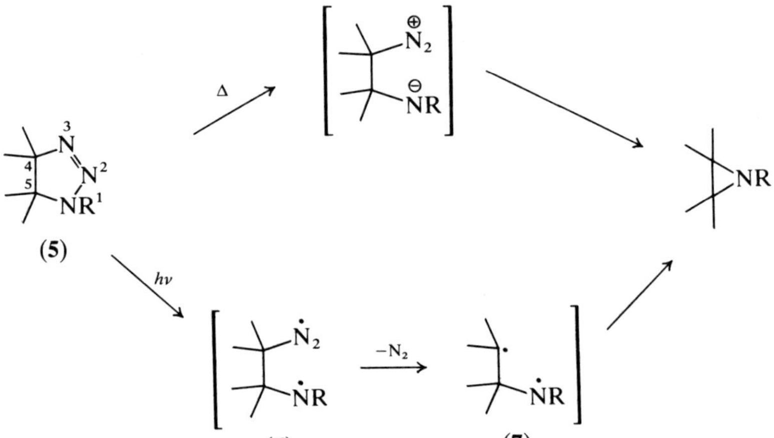

Δ

(**5**)

$h\nu$

$-N_2$

(**6**) (**7**)

$PhN_3$

$h\nu$ [90%]

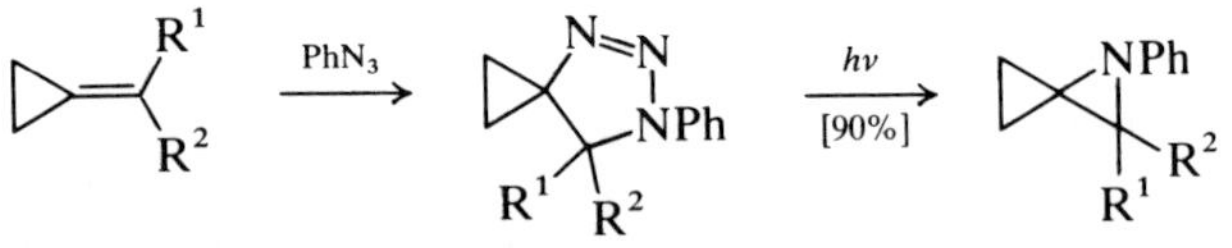

(**8**) (**9**)

**Experimental 7.1** 1-Phenyl-1-azaspiro[2,2]pentane [4]

A solution of the triazoline (**8**) (100 mg) in dichloromethane (10 ml) was irradiated at 0 °C through a Pyrex test tube in a Rayonet reactor using 3100-A bulbs until gas evolution ceased (ca. 2 h). Removal of the solvent under vacuum gave 75 mg (90%) of the spiroaziridine (**9**) as a yellow oil which was purified by column chromatography on basic alumina for analysis.

Aziridinimines (**11**) have been prepared quantitatively by photolysis of tetrazolines (**10**) [5].

Photochemical extrusion reactions of nitrogen and carbon dioxide are useful synthetic routes to oxiranes (**12**) and (**13**) and diaziridines (**14**) [6, 7].

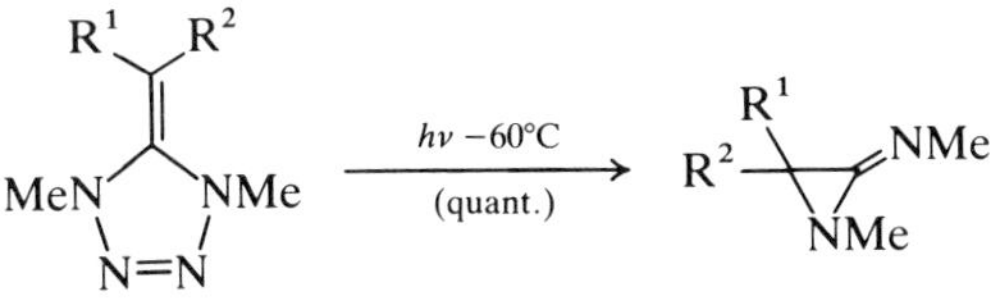

$h\nu$ −60°C (quant.)

**(10ab)** **(11ab)**

(**a**) $R^1 = R^2 = Me$

(**b**) $R^1 = H, R^2 = Me$

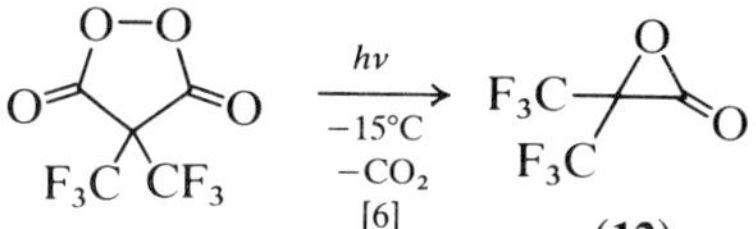

$h\nu$, −15°C, $-CO_2$ [6]

**(12)**

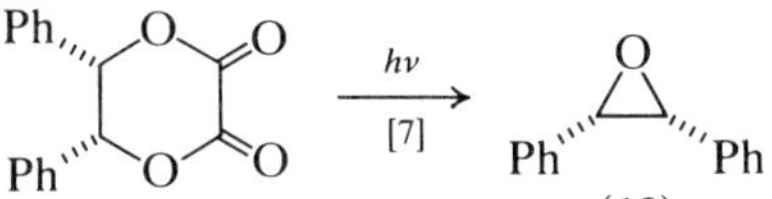

$h\nu$ [7]

**(13)**

$h\nu$, $-N_2$

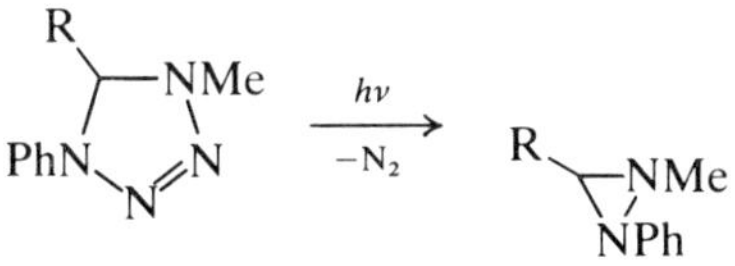

**(14)**

Thiirenes (**17**) have been isolated in an argon matrix at 8 K by photolysis of 1,2,3-thiadiazoles (**15**) or vinylene trithiocarbonates (**16**) [8].

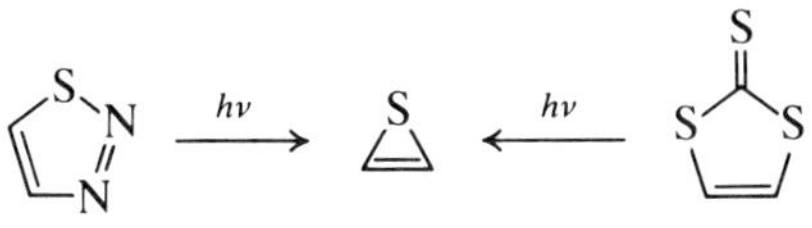

**(15)** $\xrightarrow{h\nu}$ **(17)** $\xleftarrow{h\nu}$ **(16)**

Oxaziridines can be prepared by photoisomerization of nitrones [9]. Irradiation of the t-butyl nitrones of benzaldehyde and *p*-nitrobenzaldehyde yielded about 90% of the oxaziridines (**18**) and (**19**).

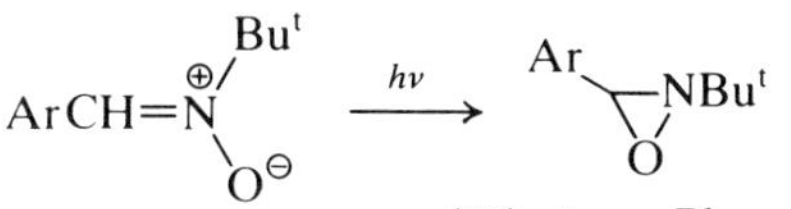

$ArCH{=}N^{\oplus}(Bu^t)O^{\ominus} \xrightarrow{h\nu}$ oxaziridine

(**18**) Ar = Ph

(**19**) Ar = $C_6H_4NO_2$-p

**Experimental 7.2** Preparation of oxazirane by irradiation of nitrone [9]

A solution of $\alpha$-(*p*-nitrophenyl)-*N*-t-butylnitrone (10 mg) in ethanol (70 ml) was irradiated with a mercury arc (Hanovia, type 16200) for 25 min. Evaporation of the solvent at room temperature gave a residue which was dissolved in methylene chloride and then washed with water, cold 15% aqueous ammonia, and finally with 10% sulphuric acid. The organic extract was dried over magnesium sulphate, and most of the solvent was distilled off at atmospheric pressure through a short column packed with glass helices. During this distillation the pot temperature was not allowed to exceed 60 °C. It was necessary to carry out this distillation fairly rapidly since some thermal isomerization of the oxazirane to nitrone took place. The residue was recrystallized from petroleum ether to give the oxazirane (**19**), MP 58–60 °C in 40% isolated yield.

## 7.2 FOUR-MEMBERED HETEROCYCLES

### 7.2.1 Preparation of oxetanes

Photochemical preparation of oxetanes has potential uses, as suggested by a number of patents [10]. These are for a pyridylcarboxaldehyde–furan adduct as a fungicide, the 2 : 1 benzophenone–furan adduct (**20**) as an anti-bacterial agent, and the photoadduct (**21**) as an insecticide [11].

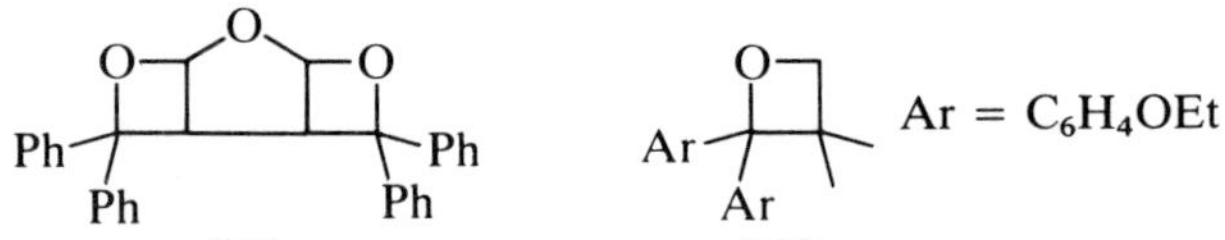

Ph Ph Ph Ph

(**20**)

Ar Ar $Ar = C_6H_4OEt$

(**21**)

Perhaps more interesting to synthetic chemists is the role of oxetanes as intermediates. Electrophiles and nucleophiles may bring about ring opening, regioselective cracking can take place, and heterolytic cleavage to provide an intermediate for reaction with a dipolarophile is also possible.

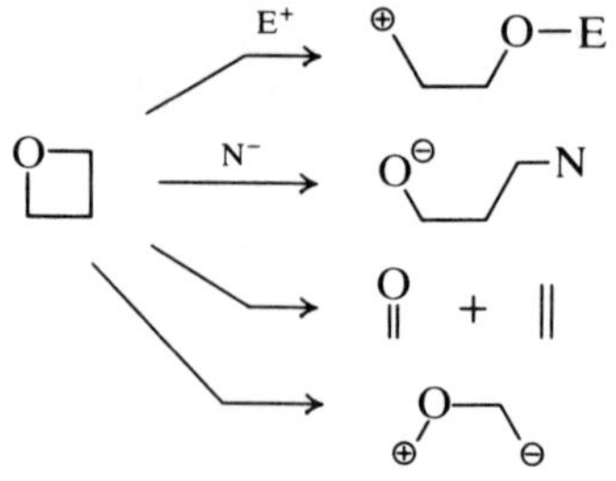

$E^+$ O–E

$N^-$ $O^\ominus$ N

#### 7.2.1.1 Preparation of oxetanes by the Paterno–Büchi reaction

Photo[2 + 2]cycloaddition of either aldehydes or ketones to C=C double bonds is known as the Paterno–Büchi reaction [12]. There are two types of this reaction: (i) the reaction between carbonyl compounds and electron-rich alkenes, and (ii) the reaction between carbonyl compounds and electron-deficient alkenes.

##### *(i) The Paterno–Büchi reaction with electron-rich alkenes*

*(a) Stereochemistry* Irradiation of several alkanals in the presence of high concentrations of alkenes (>5 M) produced photoadducts in good yield. Irradiation of butanal and 2-methyl-2-butene gave the regioisomers (**22**) and (**23**) in a 4 : 1 ratio. The stereochemistry of addition was revealed in the adducts of acetaldehyde and 2-butenes. Thus, on irradiation with *trans*-2-butene, the oxetane isomers (**24**) and (**25**) predominated over the all-*cis* oxetane (**26**). High stereoselectivity was maintained for the addition to *cis*-2-butene, with a preference (8 : 1) for the adducts with apparent retention of stereochemistry (**25**) and (**26**) [13].

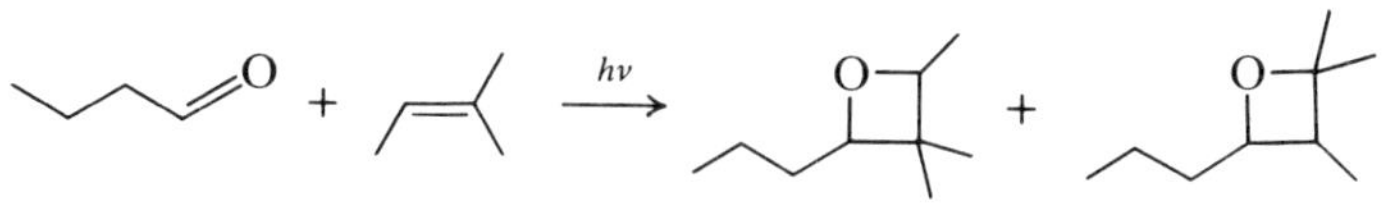

O + hν ⟶ O + O

(**22**) (**23**)

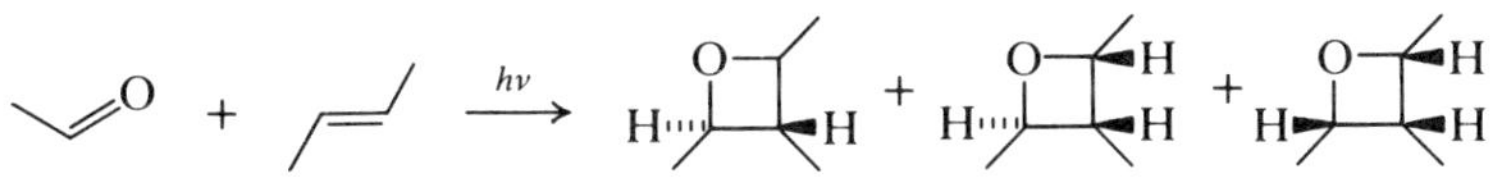

O + hν ⟶ O H H'''' H + O H H'''' H + O H H H

(**24**) (**25**) (**26**)

The addition of aldehydes to conjugated alkenes has been studied with consistent results [14]. With *trans*-piperylene, four oxetanes (**27**)–(**30**) were obtained in the ratio 51 : 16 : 25 : 8. There was high stereoselectivity in addition (97% retention of alkene configuration), a 65% regioselectivity in attack at the more substituted double bond, virtually complete regio-selectivity in the formation of 3- versus 2-enyl oxetanes, and 77% stereo-selectivity in favour of the *trans* adduct for the pair (**29**) and (**30**). Similar results were obtained for the addition of acetaldehyde to *cis*-piperylene. Under the condition of these irradiations (a 1.0 M benzene solution at 300 nm) the formation of dimers and diene isomerization were not important.

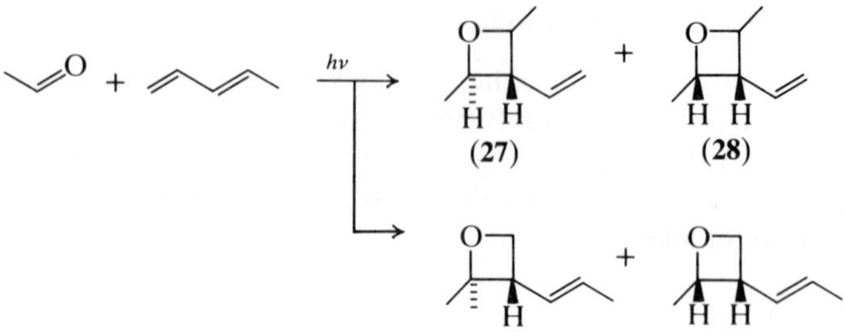

(27) (28) (29) (30)

*(b) Concentration dependence* The Paterno–Büchi reaction between carbonyl compounds and cyclic alkenes results in the formation of bicyclic oxetanes, which are useful intermediates in a general synthetic strategy, although several restrictions must be taken into account [12].

The salient features that determine the distribution of oxetanes and other products and the stereochemistry of photoaddition to simple alkenes and dienes involve the nature of the addends and the concentration of the alkene component in the following ways [12]:

(1) alkenals show a higher tendency for stereoselective addition and in many cases yield oxetanes more readily than alkanones;
(2) oxetane formation becomes predominant when the alkene concentration is very high (>2 M);
(3) at lower concentrations of alkene a number of reactions compete with photocycloaddition such as alkene dimerization, hydrogen abstraction (carbonyl photoreduction) and alkene isomerization;
(4) conjugated dienes are better addends (at moderate concentrations) than simple alkenes.

From the mechanistic point of view, with regard to the electron donor–acceptor interaction of two components in the Paterno–Büchi reaction, it is not surprising that the combination of electrophilic carbonyl excited states with unsaturated systems such as enol ethers and electron-rich heterocycles is useful in producing photoadducts. Schroeter and Orlando [15] found that photolysis of the aliphatic carbonyl compounds acetone, cyclohexanone and propanal with vinyl ethers (with the latter as solvent or at very high concentration) gave oxetanes in isolated yields of 50–70%. Irradiation of acetone and ethyl vinyl ether led to the adducts (**31**) and (**32**) in a 70 : 30 ratio.

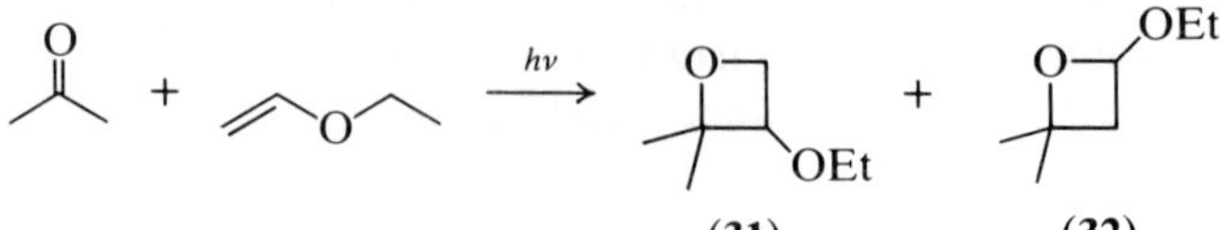

(31) (32)

**Experimental 7.3** The Paterno–Büchi reaction between acetone and ethyl vinyl ether [15]

A solution of 43.5 g (0.75 mol) of acetone in 300 ml of ethyl vinyl ether was irradiated at room temperature for 50 h. Irradiation was carried out under nitrogen in an internally water-cooled reactor at 15–25 °C with a 450-W medium-pressure mercury lamp in a quartz reactor with a Vycor 7910 glass filter to eliminate wavelengths below 250 nm. Excess ethyl vinyl ether was carefully removed at 50 mmHg and the residue was distilled to afford 59 g (60.5%) of a mixture of ethoxyoxetanes, BP 60–80 °C (70 mmHg). Gas chromatographic analysis on a 2-foot Apiezon column (temperature-programmed 5750 F & M research chromatograph) and the NMR spectrum showed this to be a 25 : 75 mixture of the isomers of 4,4-dimethyl-3-ethoxyoxetanes. Distillation through an 80-cm spinning band column afforded 35 g of 2,2-dimethyl-3-ethoxyoxetane (**31**), BP 67–68 °C (65 mmHg) and 4,4-dimethyl-2-ethoxyoxetane, BP 54 °C (62 mmHg).

Photocycloaddition of aliphatic carbonyl compounds and heterocycles has been used for the preparation of oxetanes. A wide variety of aldehydes add to furan with yields that are relatively high in some cases (70–80%) when recovered starting materials are accounted for [16, 17]. Propanal and furan, for example, gave the adduct (**33**) in 80% adjusted yield. Ketones are reluctant photoaddends—acetone and furan gave an oxetane in only 2% yield.

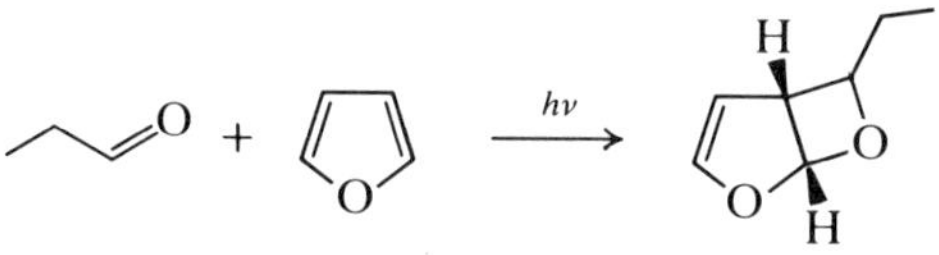

(**33**)

The combination of alkanals or alkanones with pyrrole has been reported [18]. The products actually isolated were not oxetanes, but their role as primary photoproducts can be inferred. Thus photolysis of butanal and *N*-methylpyrrole gave the alcohol adduct (**34a**) in 60% yield. Evidence for the intermediacy of a labile oxetane (**35**), which rearranged thermally to the heterocyclic alcohol (**34b**), was obtained by NMR analysis of the photolysis mixture from irradiation of acetone and *N*-methylpyrrole.

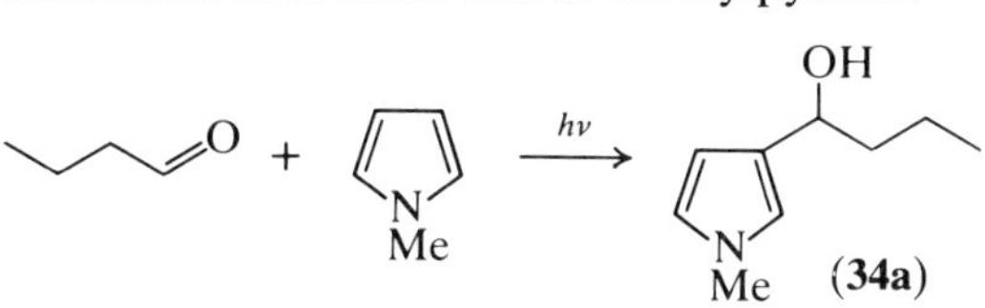

(**34a**)

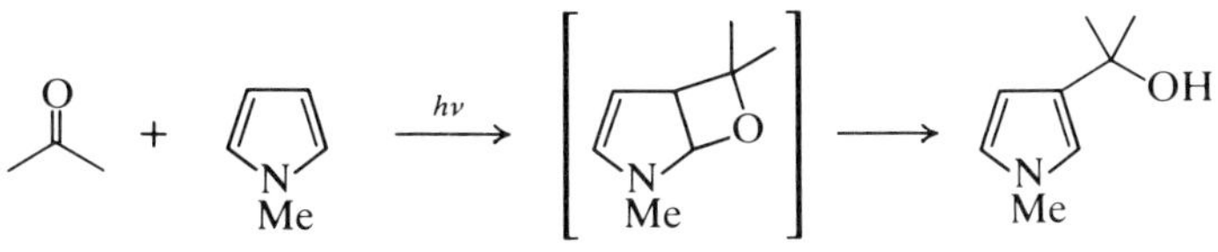

(**35**) (**34b**)

The regiochemistry of carbonyl addition to heterocycles is uniform and consistent with attack of the electron-deficient carbonyl oxygen at a nucleophilic site (normally the 2- or 5-position) on the heterocyclic ring. This pattern of electrophilic attack is analogous to that observed generally in ground-state reactions and appears to be predictable on the basis of the electron-density distribution in frontier molecular orbitals [19].

### *(ii) Paterno–Büchi reaction with electron-deficient alkenes*

Under photochemical conditions, aliphatic ketones add to electron-deficient alkenes to give oxetanes in good yields. Irradiation of an acetone solution of fumaronitrile gave the 1 : 1 adducts **(36)** and **(37)** in addition to maleonitrile [20].

NC–CH=CH–CN (trans) + acetone —$h\nu$→ **(36)** (64%) + **(37)** (23%)

$h\nu$ ⇅ $h\nu$

NC–CH=CH–CN (cis) + acetone —$h\nu$→ **(37)** + **(36)**

**Experimental 7.4** Irradiation of fumaronitrile and acetone [20]

A solution of 90 g of fumaronitrile in 1350 ml of acetone was irradiated with a Hanovia mercury arc lamp (450 W) with water cooling and nitrogen-gas stream agitation. The course of the reaction was followed by GLC assay at regular intervals. After 56 h the disappearance of fumaronitrile had slowed sufficiently the irradiation was stopped. Removal of the acetone gave an oil that assayed by GLC as follows: fumaronitrile, 14%; maleonitrile, 12%; *trans*-oxetane **(36)**, 64%; and *cis*-oxetane **(37)**, 22.5%. The oil was distilled through a 36-inch spinning band column at a reflux ratio of 5 : 1.

Fraction 1: BP 83–86 °C (4 mmHg), 26.7 g, consisted largely of fumaronitrile and maleonitrile by GLC assay. Fraction 2: BP 88–91 °C (1 mmHg), was crystallized from an ether–hexane mixture to give 45 g of *trans*-oxetane **(36)**, MP 41–42.3 °C. Fraction 3: BP 117–118 °C (1 mmHg), was crystallized from ether to give 19.7 g of *cis*-oxetane **(37)**, MP 59.5–60.2 °C.

Reactivities of electron-deficient alkenes substituted with various groups increase in proportion to their electron-withdrawing character as follows: MeCOO–<MeOOC–<NC–.

### *(iii) Intramolecular Paterno–Büchi reaction*

Carbonyl compounds with intramolecular C=C double bonds undergo intramolecular Paterno–Büchi reactions to give bicyclic oxetanes. This reaction has provided a useful synthetic method for polycyclic compounds that are not easily prepared by other methods. One of the earliest examples of this type of Paterno–Büchi reaction is the report by Srinivasan of the isomerization of 5-hexen-2-one to (**38**) [21].

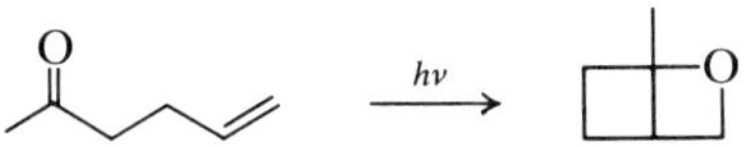

$h\nu$

(**38**)

The salient features of intramolecular addition are (1) the success of this reaction of the carbonyl and alkene moieties compared with other intramolecular reactions under a variety of conditions; and (2) the enhanced reactivity of linked addends compared with their intermolecular equivalents. For example, the naphthyl ketone (**39**) and the enedione (**40**) both cyclize with quantitative yields [22], while acetonaphthone and biacetyl are normally reluctant addends.

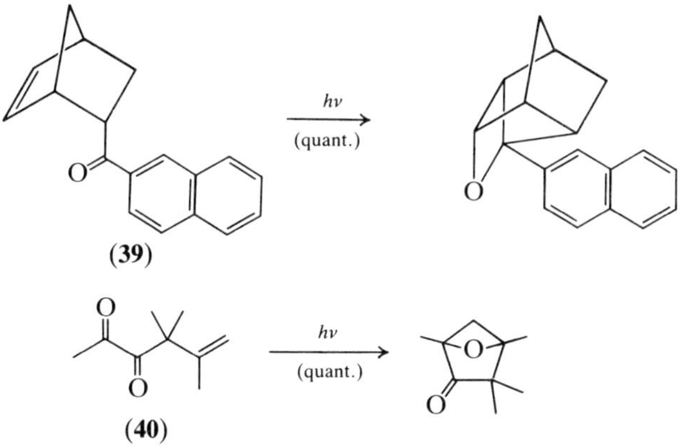

$h\nu$ (quant.)

(**39**)

$h\nu$ (quant.)

(**40**)

### *(iv) Synthetic applications*

Although the Paterno–Büchi reaction is potentially synthetically useful it has not been so widely exploited as enone photoaddition. The transformation of photogenerated oxetanes is described below, with reactions organized in terms of mode of decomposition and the type of photoadduct. The well-known pathways involve oxetane cracking, ring-chain isomerization, and ring enlargement. Each of these paths depends for its driving force on the controlled release of oxetane ring strain, and each is successful in generating useful new functionalities for further modification.

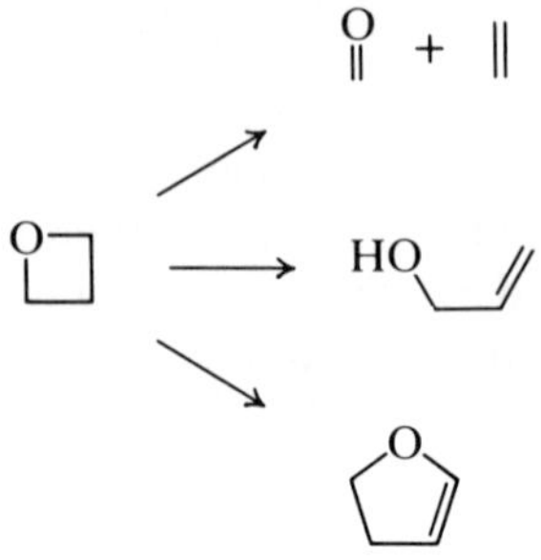

*(a) Carbonyl–alkene metathesis* Büchi *et al.* [23] have reported the fragmentation of oxetane photoadducts into carbonyl–alkene pairs. Of special interest was the regioselectivity in fragmentation of (**41**) which apparently produced a new carbonyl–alkene pair (acetaldehyde was identified).

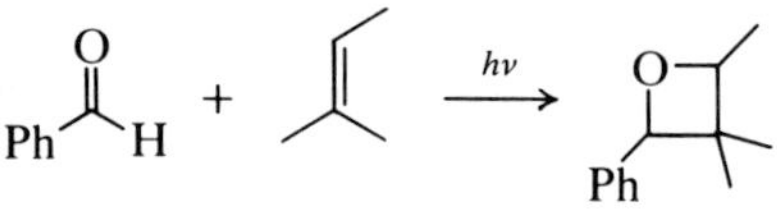

(**41**)

Other examples of selective cleavage of photoadducts can be found in Arnold's review [24]. The potential synthetic utility of coupled photochemical and thermal cracking reactions involving oxetane intermediates was proposed in 1973 [25], as shown below.

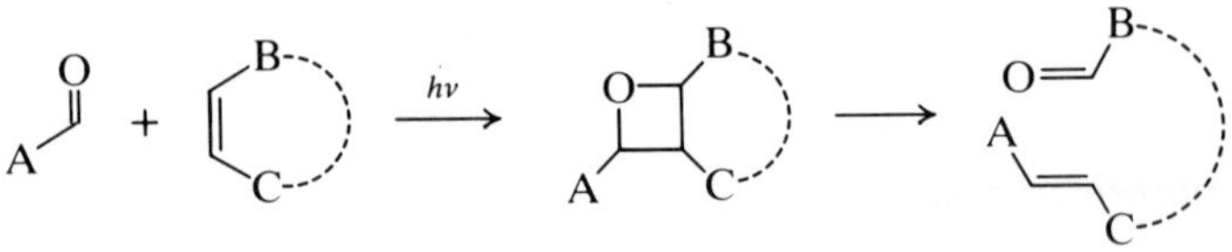

A number of studies concerning both catalysed and uncatalysed fragmentation of oxetanes offer some direction in evaluating the pyrolysis step, with attention also given to the stereochemistry of decomposition. Fragmentation reactions involving transition-metal catalysts have also been reported [26, 27]. Thermal decomposition of oxetanes proceeds along the following homogeneous paths:

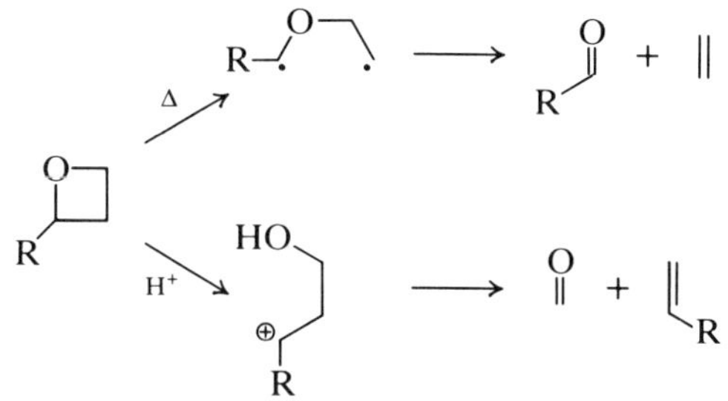

Homogeneous pyrolysis gives biradicals with selectivity suggesting C—C bond cleavage. Biradicals either as intermediates or as transition states are stereochemically randomized, which is consistent with experiment. On the other hand, catalysed cleavage leads to a more stable carbonium ion, which is then fragmented on trapping. The course of metal-catalysed ring opening is less certain.

The structural features of carbonyl–alkene metathesis products are ideally suited for conversion to a large class of natural products, including pheromones and insect sex attractants (for reviews see [28, 29]). Several examples of pheromone synthesis using carbonyl–alkene metathesis are described here. Lithium aluminum hydride reduction of (**43**), which was prepared from the oxetane (**42**), gave *trans*-6-nonen-1-ol (**44**), an attractant of the Mediterranean fruit fly (*Ceratitus capitata*) [30].

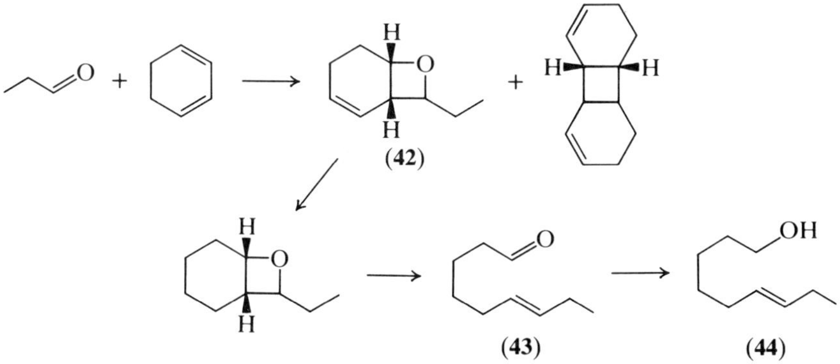

A "7 + 5" adduct from cycloheptene and valeraldehyde was used for the synthesis of a pheromone [31]. Photolysis in acetonitrile gave the adducts (**45**) (30%) as a mixture of three stereoisomers. An enal (**46**) was obtained (30%) by pyrolysis at 550 °C, and conventional steps gave 7-dodecen-1-ol acetate (**47**). Comparison of the acetate with authentic samples showed that it consisted of approximately 1 : 1 stereoisomers. The *E* isomer corresponds to the pheromone of the false colling moth (*Cryptophlebia leucotreta*), whereas the *Z* isomer is the attractive substance for the cabbage looper moth (*Trichoplusia ni*).

hν

(**45**)

O

OAc

(**46**)

(**47**)

*(b) Synthesis from enol ether adducts* Among Paterno–Büchi reactions, the addition of enol ethers to a wide variety of carbonyl compounds proceeds in highest yield. Ring-opening reactions for these oxetanes are selective, favouring decomposition of 2-alkoxyoxetane isomers under conditions in which the 3-alkoxy derivatives are left unchanged [32]. For example, treatment of the acetone adduct (**48**) with ethanol under reflux gave an acetal (**49**) (96%). Reaction with water at room temperature brought about selective ring opening to give aldol (**50**). Nucleophilic addition also leads to the selective decomposition of oxetanes, giving 1,3-diol derivatives. Treatment of (**48**) with alkyl magnesium bromide gave (**51**) (84%) and unreacted (**48**). Lithium aluminium hydride reduction of (**48**) gave the 1,3-diol (**52**).

OH $OR^1$ $OR^2$

$R^2OH$

(**49**)

OH CHO

$OR^1$ $H_2O$

(**50**)

$R^2MgX$

OH $OR^1$

(**48**)

$LiAlH_4$

$R^2$

(**51**)

OH $OR^1$

(**52**)

The synthetically useful glycerate (**54**) was obtained from the 3-alkoxyoxetane (**53**) on vigorous hydrolysis with $H_2O$–HCl at 25 °C.

$H_3O^+$

(**53**) (**54**)

*(c) Synthesis from heterocyclic adducts* The decomposition of heterocyclic adducts displays large directing effects on catalytic cleavage of the oxetane ring. This is best explained by the stability of the ring-opened carbonium-ion species.

3(4)-Substituted heterocyclic systems can be prepared from heterocyclic photoadducts, particularly with aliphatic carbonyl compounds. These systems are usually difficult to prepare by non-photochemical methods. The first synthetic application employing a furan adduct was the preparation of 3-furfuryl alcohol (**56**) [33]. Treatment of the photoadducts with *p*-TsOH in ethanol at room temperature gave (**56**) (58–73%) presumably via the protonated oxetane (**55**). Perillaketone (**58**) was synthesized by photochemical oxetane formation followed by acidic rearrangement and Jones oxidation of the resulting alcohol (**57**).

$H^+$

(**55**) (**56**)

R = Me, $COOC_4H_9$, $CH_2OCOMe$, $CH_2CH_2CHMe_2$, Ph

(1) *hν* (2) $H^+$

(**57**)

(**58**)

Nitrogen heterocyclic adducts are more labile than their furan counterparts. Thus irradiation gives 3-substituted pyrroloalcohols directly. Photoadducts of *N*-methylpyrrole with aliphatic carbonyl components were isolated as their 3-substituted pyrroles as follows [18, 19]. The photoadduct (**59**) was prepared in 38% isolated yield and used for the preparation of the 3-substituted *N*-methylpyrroles (**60**)–(**62**).

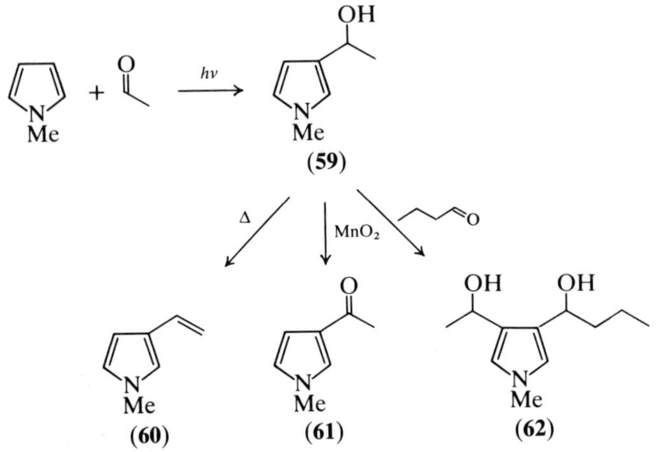

The *threo*-aldols (**65**) and (**67**) were stereoselectively prepared from the oxetane (**64**), which was readily prepared by a Paterno–Büchi reaction of the furan (**63**) and an aldehyde. Acid hydrolysis of the adduct (**64**) gave the 1,4-carbonyl *threo*-aldol (**65**), while introduction of several electrophiles to the adduct (**64**) and ring-opening reaction gave the substituted aldols (**67**).

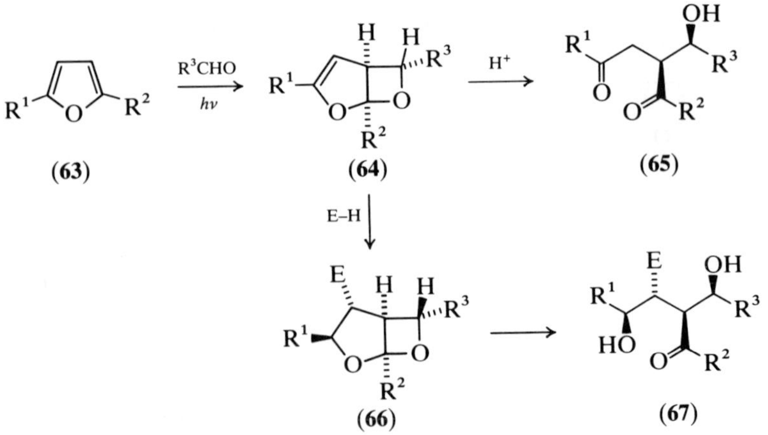

The natural $\alpha$-pyrone compound ($\pm$)-asteltoxin (**68**) was also synthesized in total 3.0% yield by applying the Paterno–Büchi reaction between 3,4-dimethylfuran and $\beta$-benzyloxypropanal and a subsequent 15-step reaction sequence [34].

Me Me + H OBn O → hν, Vycor filter → Me Me OBn O O H (63%)

Bn = $CH_2Ph$

OH OH O H O O O OMe

(**68**) asteltoxin

*(d) Synthesis by catalytic ring-opening reactions* Biologically active aromatic prostaglandin analogues have been prepared by the catalytic ring opening of aryloxetanes. On the basis of a model study in which the adduct (**70**) (prepared from benzaldehyde and (**69**) in 40% yield) was subjected to reductive cleavage with Na–BuOH to give (**71**) in 95% yield. A similar adduct (**72**) (30%) was transformed into the prostanoid (**73**) via aldehyde liberation, the Wittig reaction, and vinyl cyclopropane ring-opening reactions [35].

*(e) Synthesis by intramolecular photoaddition* Preparation of oxetanes by intramolecular photoaddition is a potentially important synthetic tool because highly strained and unusual polycyclic ring systems can be readily prepared and used as synthetic intermediates [36].

This procedure is exemplified by the synthesis of azulenes via oxetane formation from $\gamma,\delta$-unsaturated ketones. Irradiation of (**74**) gave the oxetane (**75**) as a major adduct, which was subjected to thermal decomposition at 140–250 °C to give a mixture of the alcohol (**76**) and the diene product (**77**) [37]. This mixture was converted into azulene by dehydration–dehydrogenation over palladium on alumina in 25% yield.

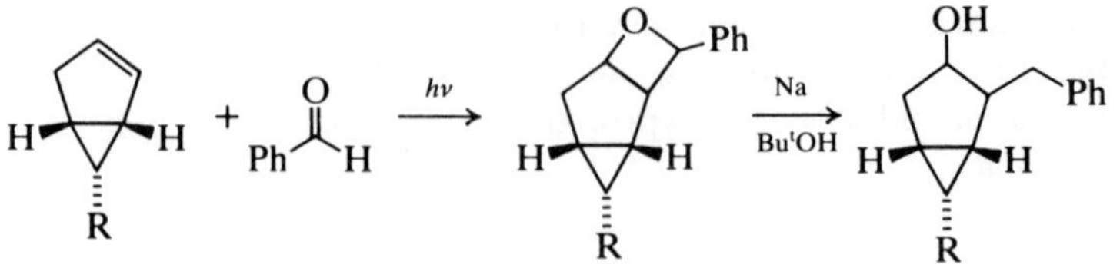

H H R + Ph O H $\xrightarrow{h\nu}$ O Ph H H R $\xrightarrow[Bu^tOH]{Na}$ OH Ph H H R

(**69**) (**70**) (**71**)

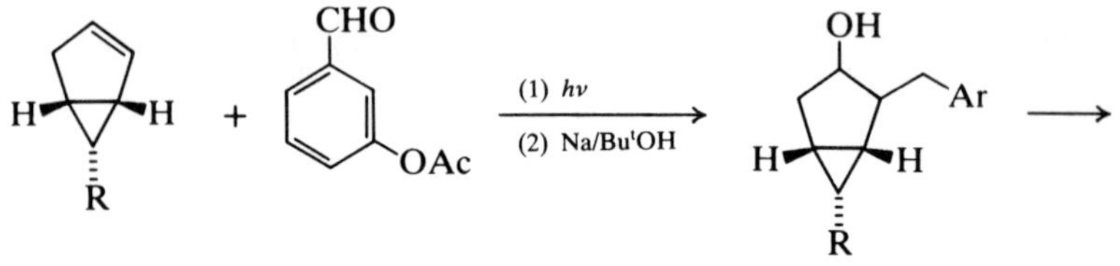

H H R + CHO OAc $\xrightarrow[(2)\ Na/Bu^tOH]{(1)\ h\nu}$ OH Ar H H R $\longrightarrow$

(**72**)

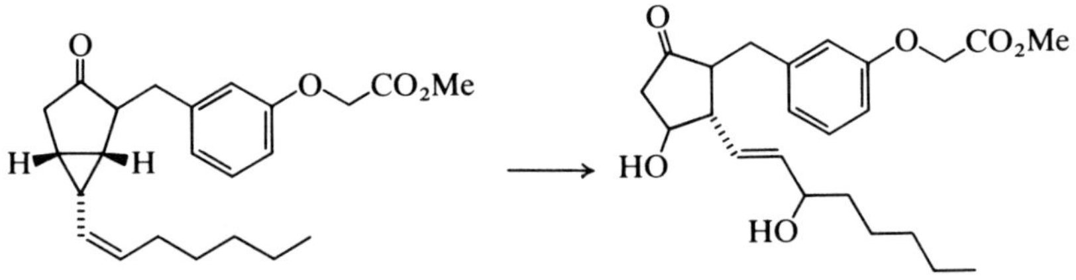

O H H O $CO_2Me$ $\longrightarrow$ O HO O $CO_2Me$ HO

(**73**)

R = O O

O $\xrightarrow{h\nu}$ O $\xrightarrow{\Delta}$

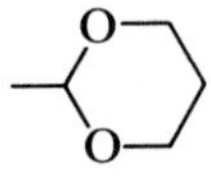

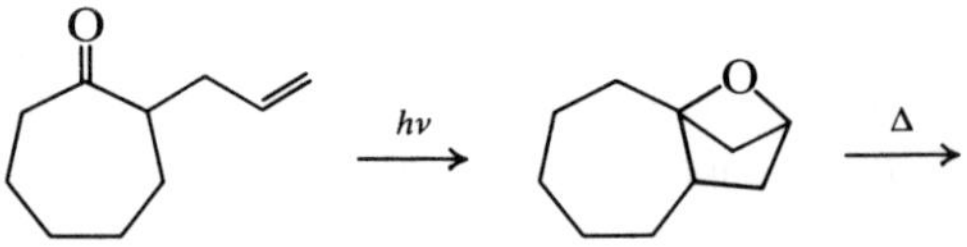

(**74**) (**75**)

H OH + $\longrightarrow$

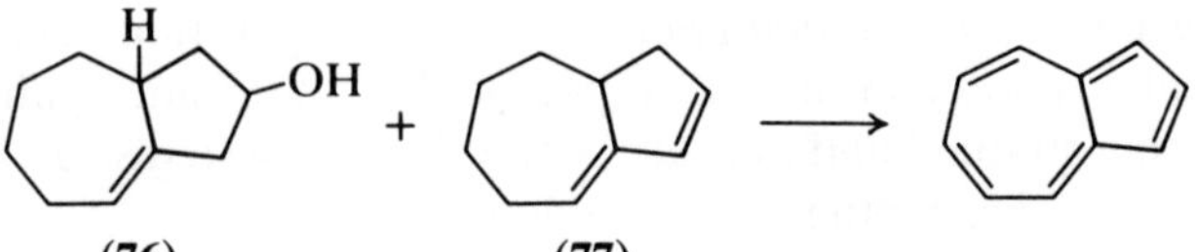

(**76**) (**77**)

Another example is the synthesis of the large-ring lactone (**80**) [38]. Photolysis of the benzophenone analogue (**78**) in $CCl_4$ gave the bicyclic oxetane (**79**) (83%), and subsequent treatment on silica gel resulted in ring cleavage to give (**80**) (90%).

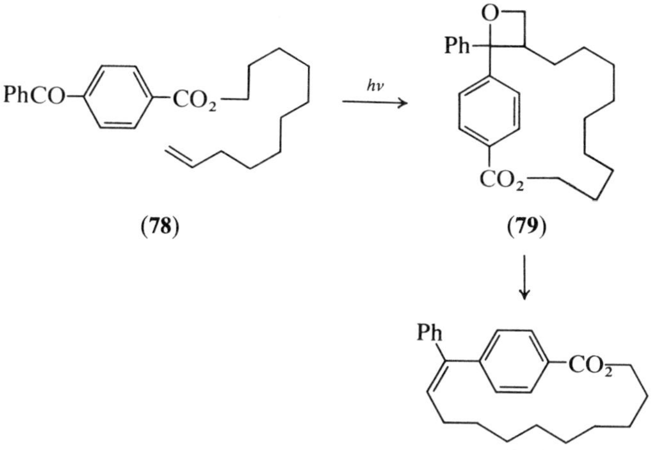

PhCO — — $CO_2$ — hν → Ph — $CO_2$

(78) (79)

↓

Ph — $CO_2$

(80)

An elegant synthesis of 1-hydroxycholesterol was completed by the application of intermolecular oxetane formation [39]. The steroidal cyclodecenone (**81**) was first prepared from cholesterol in four steps (56%) and then irradiated in acetone to give the oxetane (**82**) in 40% yield. Acid treatment of (**82**) with HI in acetic acid resulted in selective ring opening of the oxetane to give the glycol (**83**).

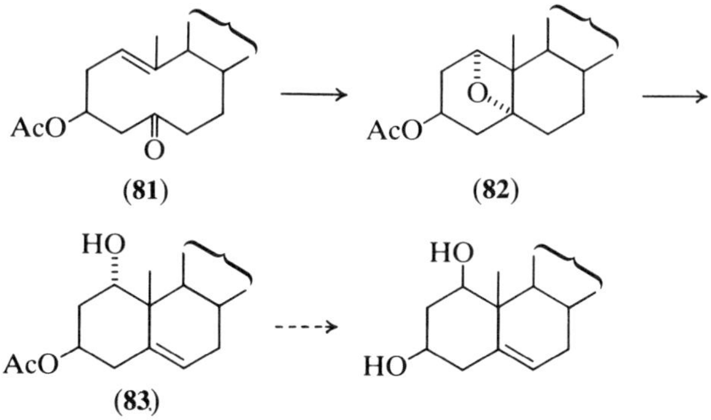

AcO — O → AcO — O →

(81) (82)

HO — AcO ----→ HO — HO

(83)

**Experimental 7.5** 6,6-Diphenyl-2,7-dioxabicyclo[3.2.0]hept-3-ene*

A solution of benzophenone (4 g) in furan (180 ml) freshly distilled over triphenylphosphine was irradiated with a Philips HPK 125-W high-pressure mercury lamp at 12 °C. During irradiation the solution was stirred magnetically under nitrogen. After 44 h illumination, the furan was evaporated, whereupon the product (5.2 g, 94% remained, MP 105–106 °C.

* G. O. Schenk, W. Hartmann and R. Steinmetz, *Chem. Ber.*, **96**, 498 (1963).

#### 7.2.1.2 Preparation of oxetanes by photochemical extrusion reactions

There are a few such reactions, which have been found to yield oxetanes but have not been developed as synthetic methods. The thermolysis and photolysis of $\gamma$-methyl-$\alpha$-peroxyvalerolactone gave 2,2-dimethyloxetane, presumably by way of a diradical intermediate [40].

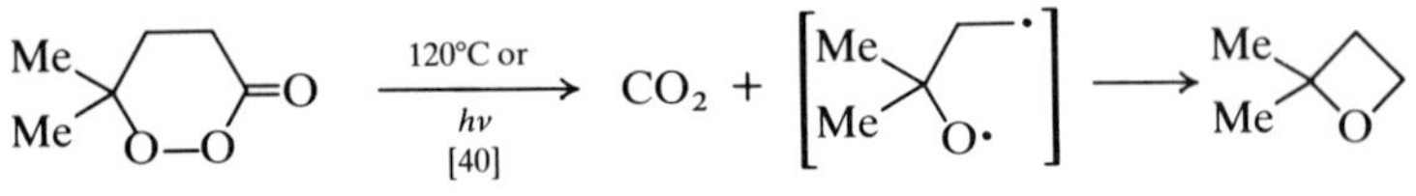

120°C or $h\nu$ [40] → $CO_2$ +

#### 7.2.1.3 Preparation of oxetanes by the Norrish type II reaction

The Norrish type II photocycloaddition of $\alpha$-alkoxyketones provides a photochemical synthesis of oxetanes [41].

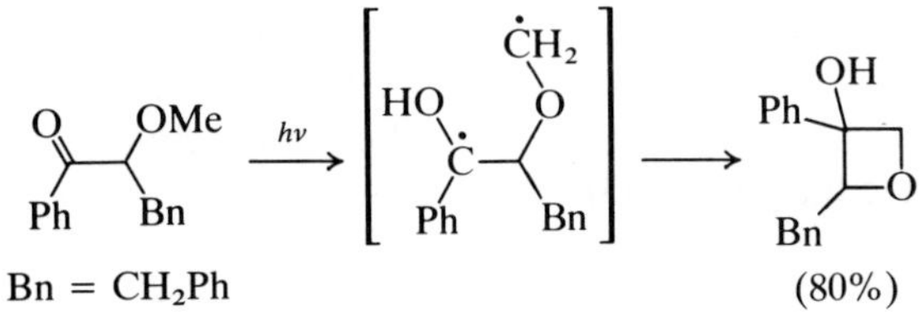

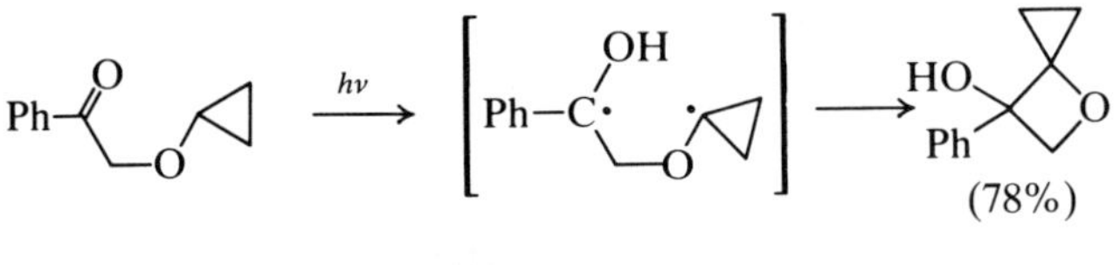

Bn = $CH_2Ph$ (80%)

(78%)

### 7.2.2 Preparation of azetidines

#### 7.2.2.1 Preparation of azetidines by the Norrish type II reaction

Intramolecular hydrogen abstraction in aromatic (*N*-acylamino)methyl ketones is an effective synthetic route to azetidine derivatives [42, 43]. This behaviour differs considerably from that of aromatic $\alpha$-aminoketones, which split to give two components, an aromatic ketone and an amine, on irradiation.

As in the preparation of cyclobutanes by the Norrish type II reaction of $\alpha$-diketones (section 6.2.3), $\gamma$-hydrogen abstraction from a $\alpha$-diketone system can be extended to $\alpha$-ketoamide systems. The products obtained

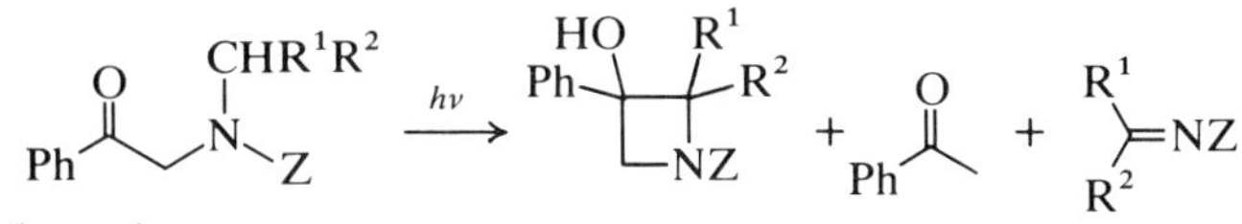

$R^1 = R^2 = H$, Z = *p*-Ts
$R^1 = Me$, $R^2 = H$, Z = COPh
$R^1 = R^2 = Me$

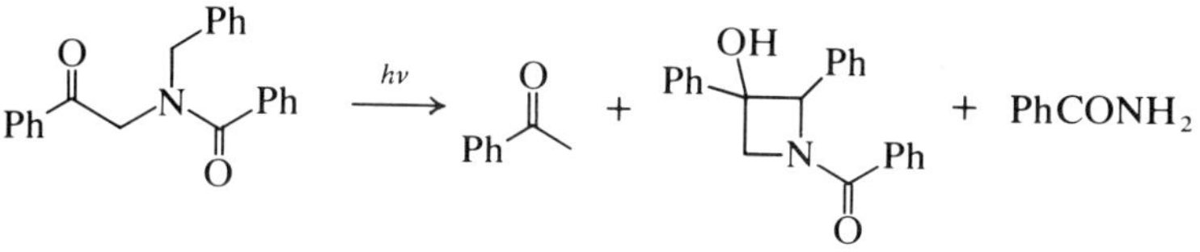

+ $PhCONH_2$

are 2-azetidinones, which are related to penicillins and other β-lactam antibiotics [44, 45].

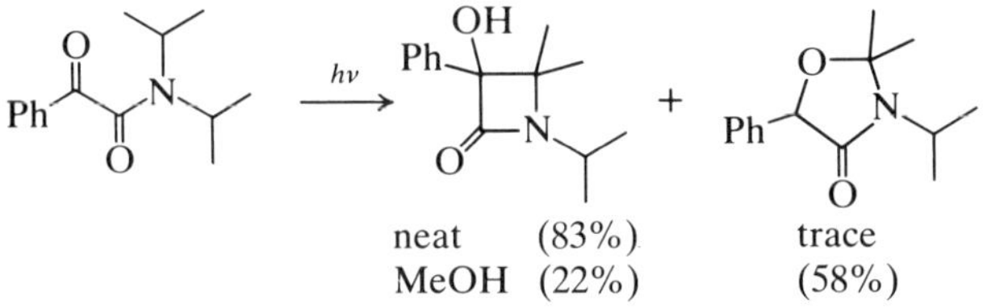

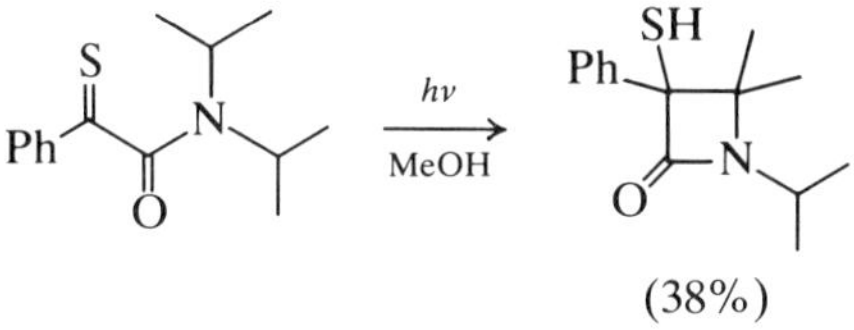

neat (83%) trace
MeOH (22%) (58%)

(38%)

**Experimental 7.6** General procedure for solid-state photoreaction of α-ketoamides [45]

The recrystallized α-ketoamide (100–300 mg) was sandwiched between a pair of Pyrex plates (thickness 2 mm) and put into a polyethylene envelope. The envelope was sealed and placed in a cold medium and irradiated with a high-pressure mercury lamp (100 W) for 1–10 h. The lamp was used with a vessel for immersion-type irradiation that contained dry air. The products were isolated by column chromatography on silica gel.

### 7.2.2.2 Preparation of azetidines by cycloaddition reactions

Ketenes formed as intermediate products in the photolysis of diazoketones can add to imines to give 2-azetidinones [46].

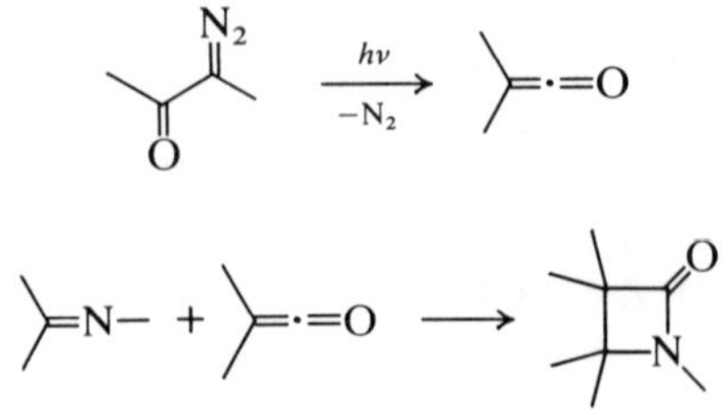

$h\nu$, $-N_2$

Ketoketenes are known to add to imines in the dark, while the formation of $\beta$-lactams from aldoketenes and imines can only take place photochemically. In general, partially aliphatic-substituted imines show less reactivity than their completely aromatic-substituted counterparts. Experiments with diazopyruvate esters ($N_2$—CO—COOR) did not result in addition to imines.

Similarly, the ketone formed as an intermediate in the photolysis of phenylpyruvic acid esters is known to give the 2-azetidinone (**84**) in good yield [47].

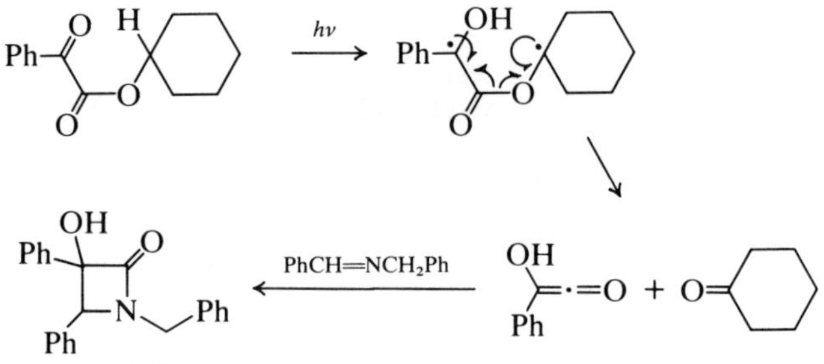

$h\nu$

PhCH=NCH$_2$Ph

(**84**) (76%)

#### 7.2.2.3 Preparation of azetidines by photochemical pericyclic reactions

For a time the only known photochemical reaction of 2-pyridones was the formation of the corresponding photodimer [48, 49]. However, Corey and Streith [50] were able to prepare photopyridone (2-azabicyclo[2.2.0]hex-5-en-3-one) by irradiation of *N*-methyl-2-pyridone. Recently, the application of this photochemical pericyclic reaction of 2-pyridones to the synthesis of 2-azetidinones related to the biologically active $\beta$-lactam antibiotics has been investigated by three groups, Brennan [51], Kametani *et al.* [52] and Kaneko *et al.* [53–55] according to the following strategy. Brennan [51] and Kametani *et al.* [52] have investigated the synthetic potentiality of photopyridones from (**85**) and (**87**) by preparing the *cis*- and *trans*-3,4-disubstituted azetizin-2-ones (**86**) and (**88**) via a route involving ring opening of the cyclobutenes.

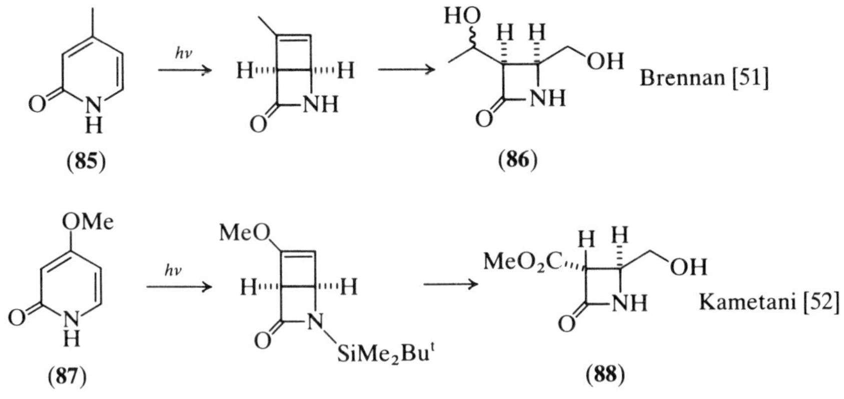

Brennan [51]

(**85**) (**86**)

Kametani [52]

(**87**) (**88**)

On the basis of these results, Kaneko *et al.* found an efficient photochemical preparation of photopyridones from 4-oxy-2-pyridones, which undergo unexceptional pericyclic reactions with no dimerization, followed by an effective ring-opening reaction of the resulting cyclobutanol by a retro-aldol reaction [54, 55].

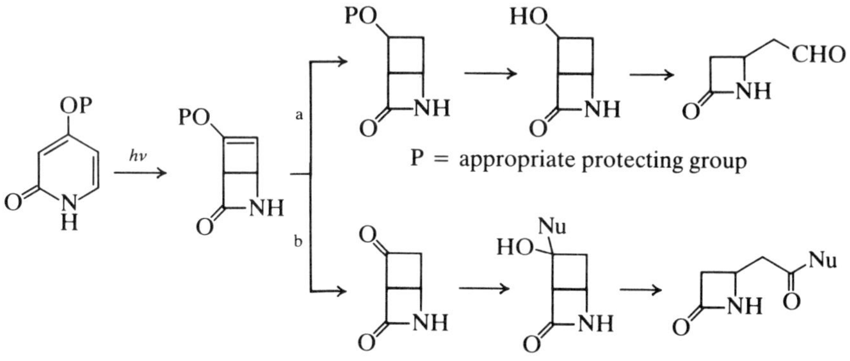

P = appropriate protecting group

Irradiation of the 4-acetoxy-2-pyridone (**89**) gave the photopyridone (**90**), which was catalytically reduced to give exclusively the dihydro derivative (**91**) as a result of hydrogenation from the *exo* side. Deacetylation of (**91**) by $K_2CO_3$–MeOH in the presence of $NaBH_4$ led to the 2-azetidinone (**93**) in good yield, while in the absence of the borohydride the 2-pyridone (**94c**) was obtained, possibly via the intermediate shown in brackets [56].

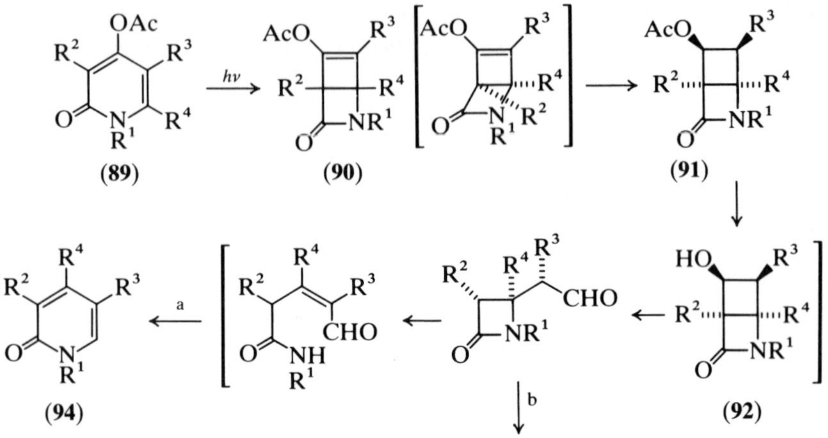

(89) (90) (91) (92) (93) (94)

| | $R^1$ | $R^2$ | $R^3$ | $R^4$ |
|---|---|---|---|---|
| **a** | H | H | H | H |
| **b** | H | H | H | Me |
| **c** | Me | H | H | Me |
| **d** | H | Pr | H | H |
| **e** | H | H | Me | H |

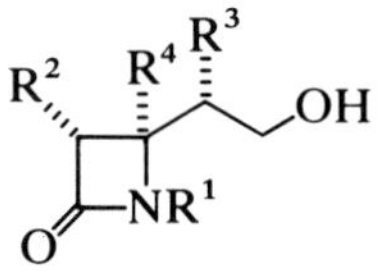

(a) $K_2CO_3$/MeOH
(b) $NaBH_4$–$K_2CO_3$/MeOH

An intermediate (**98**) was isolated by using the 4-benzyloxypyridone (**95**) as the starting pyridone of the photolysis. Catalytic reduction of the photopyridone (**96**) resulted in hydrogenation of the double bond, while leaving intact the benzyl group, which was removed by hydrogenolysis in the presence of 0.3% hydrochloric acid. The azetidinone (**98**) is equivalent to

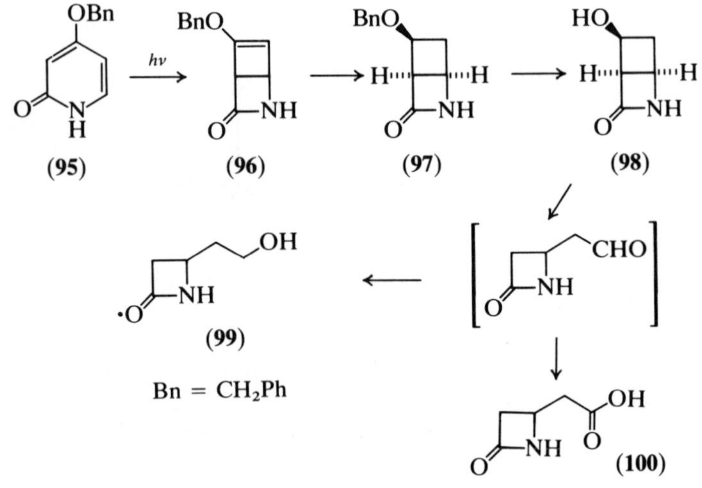

(95) (96) (97) (98) (99) (100)

Bn = $CH_2Ph$

4-(2-hydroxyethyl)azetidin-2-one (**99**) under basic conditions and was converted to the alcohol (**99**) by treatment with $K_2CO_3$–MeOH in the presence of $NaBH_4$ and to the carboxylic acid (**100**) in the presence of $KMnO_4$.

The more stable crystalline 2-azetidinone (**102**) was prepared from the photopyridone (**101**) by hydrolysis of the enol ether moiety. Thus two versatile synthetic intermediates (**92**) and (**102**) for the biologically important carbapenem were prepared via a route involving photochemical pericyclic reactions [57, 58].

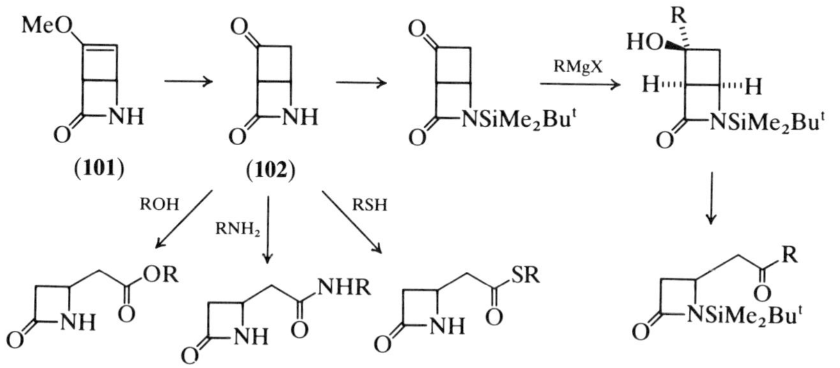

(**101**) (**102**) RMgX ROH $RNH_2$ RSH

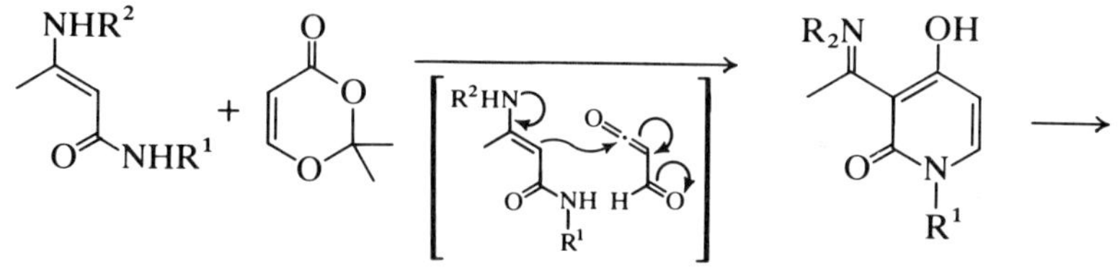

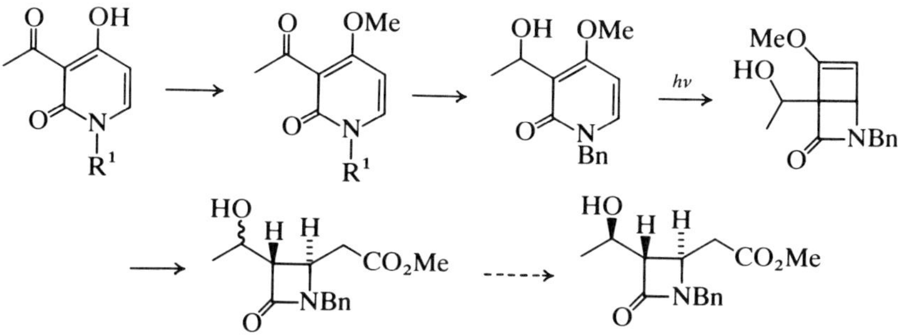

(**103**)

*hν*

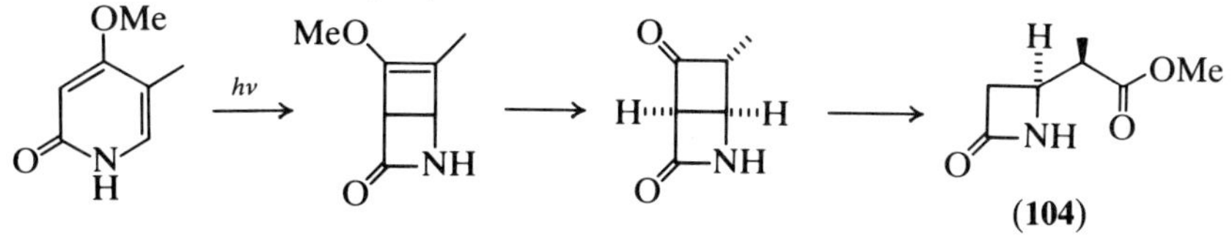

(**104**)

Applications of this route to the synthesis of the key intermediates (**103**) [59] and (**104**) [53] to thienamycin and 1$\beta$-methylcarbapenem were also carried out and asymmetric syntheses of optically active congeners have been developed [53, 60].

**Experimental 7.7** 5-Acetoxy-2-azabicyclo[2.2.0]hex-5-en-3-one

A solution of 4-acetoxy-2-pyridone (158 mg, 1.03 mmol) in a mixture of acetonitrile (37.5 ml) and ether (132.5 ml) under argon was irradiated by a high-pressure mercury lamp (Ushio 450 W, Pyrex filter) for 1 h. The solvent was evaporated off under reduced pressure to give the product (**90a**) (148 mg, 93%) as prisms from dichloromethane–ether, MP 103–105 °C.

### 7.2.3 Preparation of thietanes

As in the case of benzophenone, thiobenzophenone can add photochemically to C═C double bonds to form thietanes. Several additions to alkenes are of preparative value, but require more than usual care in selection of reagents and conditions and are further limited by the high thermal and photochemical reactivity of ground-state thione molecules.

Irradiation of thiobenzophenone and acrylonitrile with a high-pressure mercury lamp gave regioselectively an adduct (**105**) in 90% yield [61].

Ph Ph C=S —hv, >300 nm→ Ph Ph C=S* —NC-CH=CH₂→ [Ph Ph C=S … CH₂=CH-CN] → (**105**) (NC, S, Ph, Ph thietane) ↛ (NC, S, Ph, Ph thietane)

Irradiation of thiobenzophenone and *cis*-1,2-dichloroethylene with a mercury lamp resulted in the formation of *cis*-2,2-diphenyl-3,4-dichlorothietane (**106**) in 83% yield. With *trans*-1,2-dichloroethylene, the corresponding *trans* isomer (**107**) was formed in 90% yield [61]. For electron-deficient alkenes substituted with electron-withdrawing groups, the reactivity as acceptors increases in proportion to the electron-withdrawing character in the order MeCOO–<MeOOC–<NC–.

Irradiation of thiobenzophenone at long wavelengths (>400 nm) in the presence of electron-deficient alkenes and enol ethers led to adducts regioselectively but non-stereoselectively [62]. Irradiation at about 300 nm,

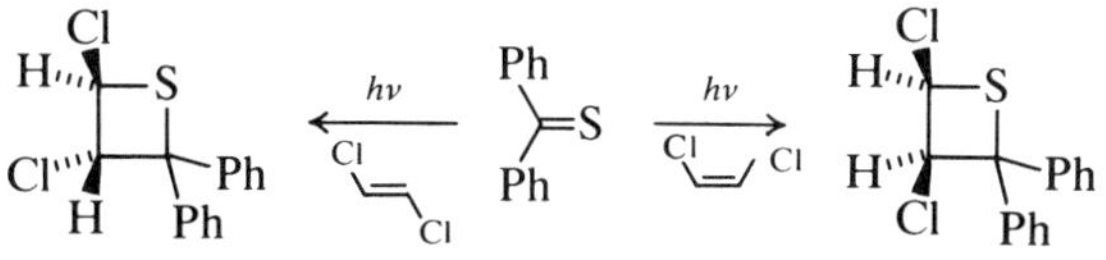

(107) (106)

especially with electron-deficient alkenes (acrylate, acrylonitrile and vinyl chloride derivatives) gave the adducts (**108**) (85%) and (**109**) (93%) with high stereoselectivity.

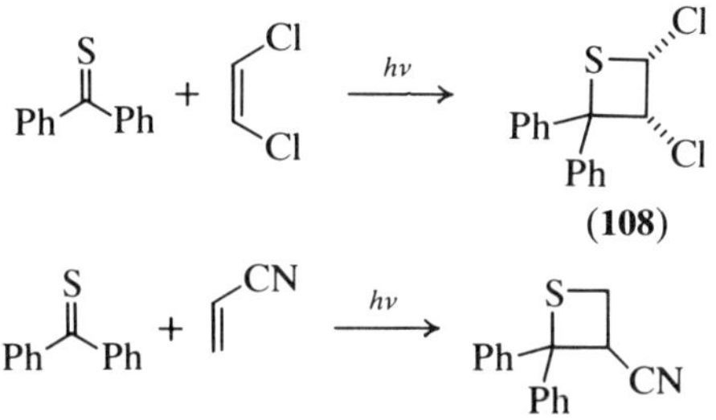

(**108**)

(**109**)

Intermolecular photochemical addition of thioimides to alkenes proceeds smoothly to give spirothietanes, which are key intermediates for various types of compounds [63]. Irradiation of the *N*-methylthioimides (**110**) and the 2,3-dimethyl-2-butenes (**111**) gave the spirothietanes (**113**) in good yield irrespective of the structure of the thioimide and the alkene. This reaction proceeds via the biradical intermediate (**112**) [15].

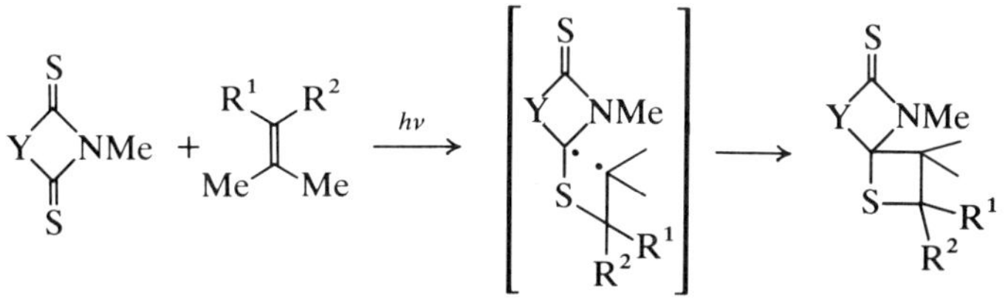

(**110**) (**111**) (**112**) (**113**)

| | Y | $R^1$ | $R^2$ | Yield (%) |
|---|---|---|---|---|
| **a** | $(CH_2)_2$– | Me | Me | 83 |
| **b** | $(CH_2)_2$– | H | H | 58 |
| **c** | $(CH_2)_3$– | Me | Me | 86 |
| **d** | $CH_2OCH_2$– | Me | Me | 76 |
| **e** | $CH_2SCH_2$– | Me | Me | 42 |

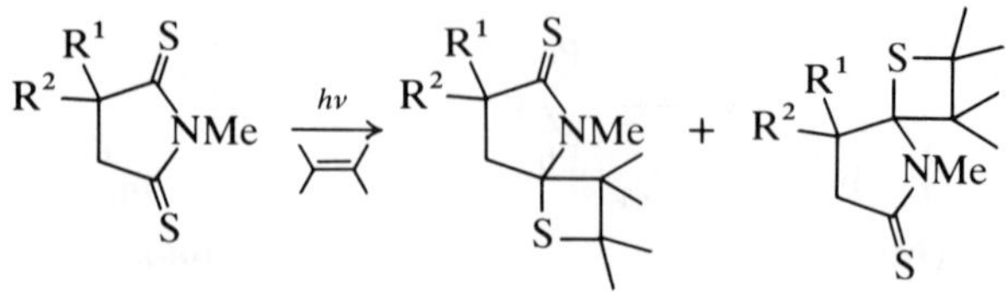

(114) (115) (116)

| | $R^1$ | $R^2$ | **115** (%) | **116** (%) |
|---|---|---|---|---|
| **a** | Me | H | 22 | 5 |
| **b** | Me | Me | 54 | — |
| **c** | $-(CH_2)_5-$ | $-(CH_2)_5-$ | 50 | — |

In the reactions of the unsymmetric thioimides (**114**) the major products were (**115**) rather than (**116**) as a result of steric hindrance by the substituents [63].

Photolysis of the aromatic thioamide (**117**) and the alkenes (**118**) conjugated with an aromatic ring resulted in the formation of the 1,2-dithianes (**120**) in addition to the thietanes (**119**) via the proposed reaction pathways shown below [64, 65].

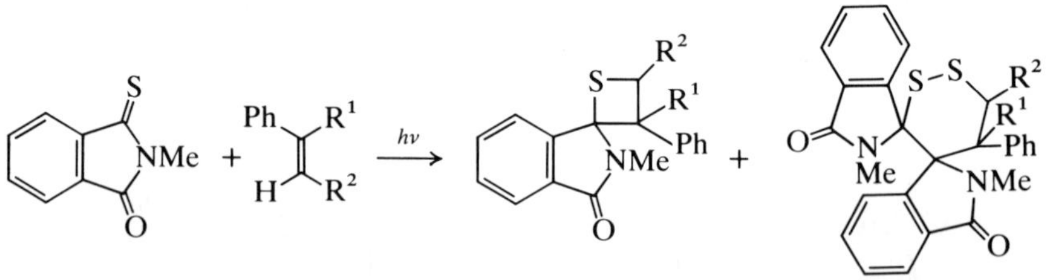

(117) (118) (119) (120)

| $R^1$ | $R^2$ | **119** (%) | **120** (%) |
|---|---|---|---|
| H | Me | 53 | 37 |
| Ph | H | 34 | — |
| H | Ph | 80 | 15 |
| H | CN | 46 | 43 |
| H | COOMe | 58 | 35 |

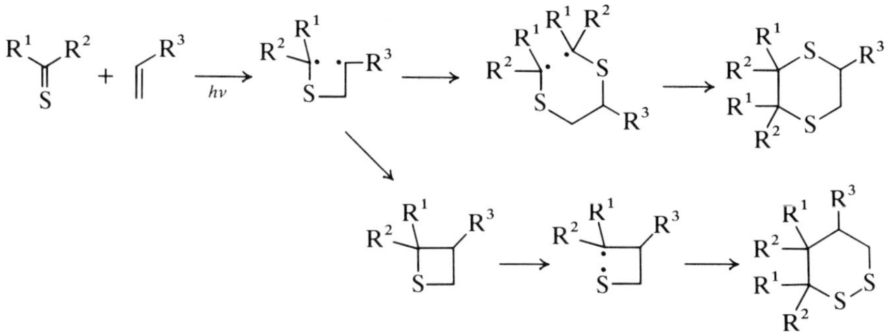

Photoaddition of the thioimides (**121**) to alkenes proceeded intramolecularly to give the tricyclic thietanes (**122**) [66].

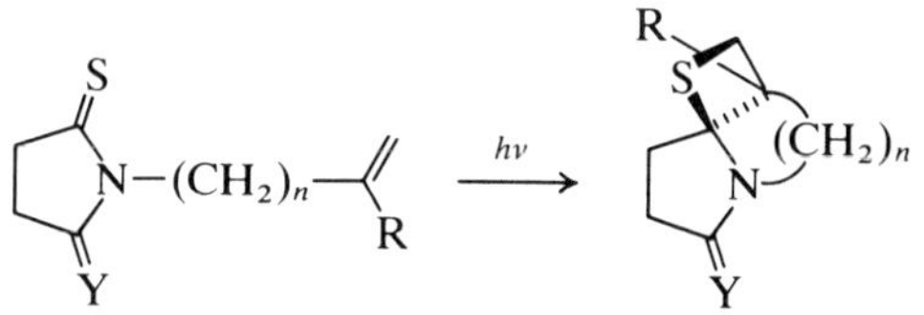

(**121**) (**122**)

| | Y | $n$ | R | **122** (%) |
|---|---|---|---|---|
| **a** | S | 1 | H | 31 |
| **b** | S | 1 | Me | 82 |
| **c** | S | 2 | Me | 65 |
| **d** | O | 1 | Me | 42 |

Similarly, the highly strained polycyclic thietanes (**123**) and (**124**) were prepared by the same intramolecular reaction [67].

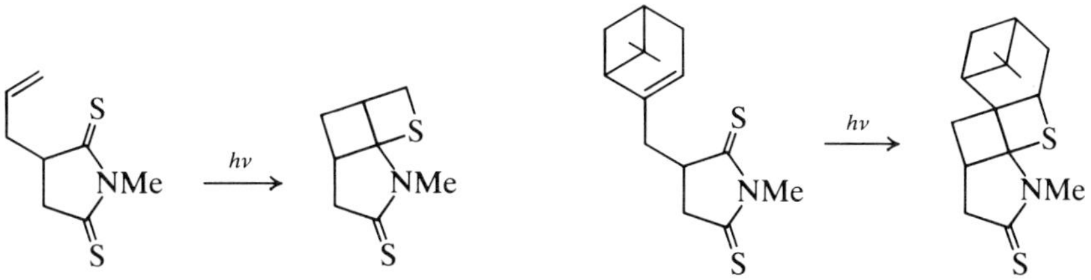

(**123**) (42%)

(**124**) (45%)

## 7.3 FIVE-MEMBERED HETEROCYCLES

### 7.3.1 Pyrroles and their benzo derivatives

#### 7.3.1.1 Preparation of pyrroles by photocyclization

*N*-Substituted diphenylamines (**125**) can be regarded as analogues of stilbene with the C=C bond replaced by the –NR group, and undergo a stilbene-type photocyclization involving six $\pi$-electrons to give the carbazole derivatives (**126**); the partially hydrogenated systems (**127**) similarly give (**128**) [68–71].

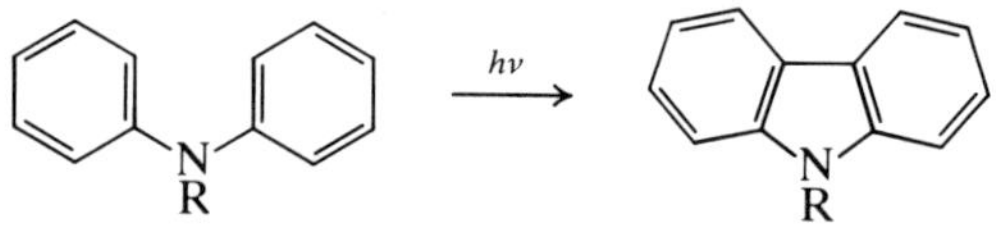

hν

N R (**125**) → N R (**126**)

hν

N Me (**127**) → H, N Me H (**128**)

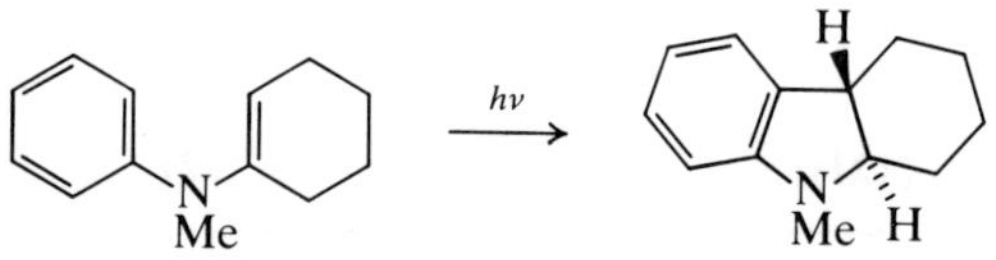

The synthesis of carbazoles by this photocyclization is best carried out with the use of oxygen as an oxidant. The three isomeric *N*-pyridylanilides behave analogously, giving the expected azacarbazoles in 70–81% yield. Non-oxidative photocyclization of the enamine (**129**) gave the *trans*-indoline (**130**) [72, 73]. Asymmetric photocyclization of (**129**) with circularly polarized light gave (**130**) with slight enantiomeric excess [74].

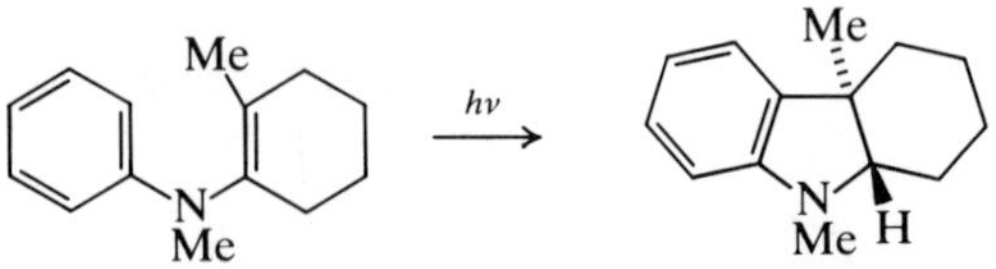

hν

Me, N Me (**129**) → Me, N Me H (**130**) (70%)

A mixture of two indolines (**132**) and (**133**) was obtained by irradiation of the ketone (**131**) via two possible pathways [75].

The *o*-halogeno- or *o*-cyano-aniline derivatives (**134**) and (**136**) are also

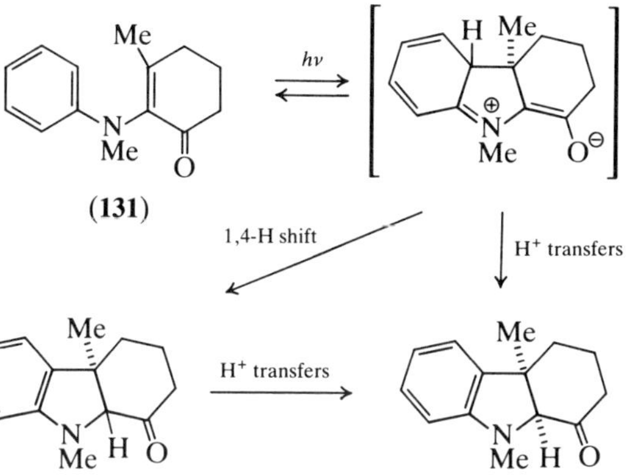

(131)

1,4-H shift

$H^+$ transfers

$H^+$ transfers

(132) (7%)

(133) (90%)

good starting compounds for the photochemical synthesis of the carbazoles (**135**) and (**137**).

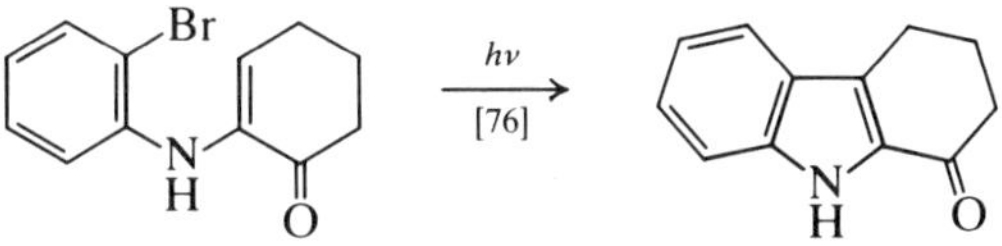

(134)

(135) (80%)

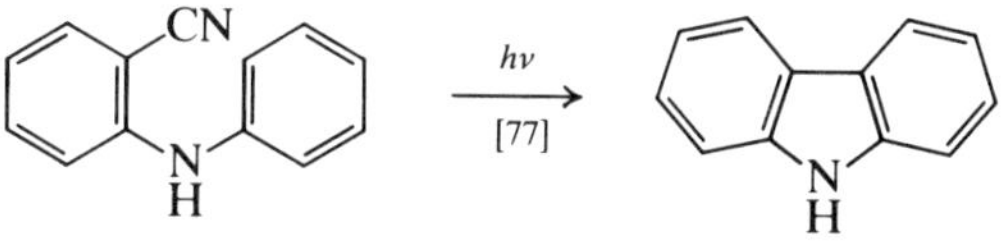

(136)

(137) (87%)

Interestingly, the enolated aniline derivative (**138**) underwent smooth photocyclizations to give (**139**) under non-acidic conditions and the dehydration product (**140**) under acidic conditions.

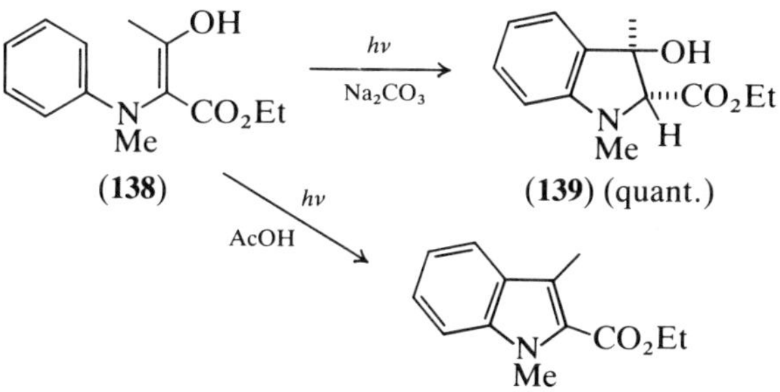

(138)

(139) (quant.)

(140) (92%)

**Experimental 7.8** *trans*-9-Methyl-1,2,3,4,4*a*,9*a*-hexahydrocarbazole [72]

A solution of 1-(*N*-methylanilino)cyclohexene (4.18 g) in diethyl ether (300 ml) was purged with argon for 30 min and then irradiated with a Hanovia 550 W mercury lamp in a Pyrex immersion well. The progress of the reaction was monitored by TLC on silica gel with ethyl acetate in Skelly B petroleum ether. After 6 h irradiation, the solvent was evaporated from the reaction mixture. The resulting oil was crystallized from 95% ethanol to give *trans*-9-methyl-1,2,3,4,4*a*,9*a*-hexahydrocarbazole (**128**) (2.3 g, 55%), MP 58–60 °C (picrate: MP 125–126 °C).

### 7.3.1.2 Preparation of pyrroles by other photochemical reactions

Reactions of *o*-haloanilines with enolate anions in liquid ammonia to form indoles may be initiated photochemically and proceed through radical nucleophilic substitution ($S_{RN}1$) of halogen [78,79].

Br NH$_2$ + $^{\ominus}$O R —hν, liq. NH$_3$→ NH R

Photo-stimulated intramolecular substitution by the anions of *o*-halo-anilides is valuable for oxindole synthesis, and the mechanism is proposed to be of the electron-transfer type $S_{RN}1$ rather than a classical addition–elimination mechanism. The reaction is effective when R = H if two equivalents of the base are used to generate the dianion [80].

Cl R$^2$ N R$^1$ O —hν, −78°C, LiNR$_2$→ R$^2$ =O NR$^1$

**Experimental 7.9** Irradiation of the acetamide [80]

To a solution of lithium diisopropylamide (10 mmol) in 30 ml of THF, maintained at −78 °C under an argon atmosphere, was added a solution of *N*-propionyl-*o*-chloro-aniline (0.46 g, 2.5 mmol) in 20 ml of THF. After the addition was complete, the solution of the dianion was allowed to rise to 25 °C and was then irradiated for 3 h in a Rayonet Model RPR-240 photoreactor equipped with four 12.5 W lamps emitting at 350 nm. The reaction mixture was quenched with water, acidified to pH 1 with 6 M HCl, and extracted with ether. The extracts were dried over magnesium sulfate, filtered, and concentrated. Purification of the crude product by medium-pressure chromatography (diisopropylamine–hexane 1 : 3) followed by recrystallization from ether–hexane afforded 0.27 g (73%) of the oxindole, MP 120–121 °C.

The ring-opening reaction of 2-*H*-azirines to yield vinyl nitrenes on photolysis also leads to pyrrole formation. Some examples proceeding via

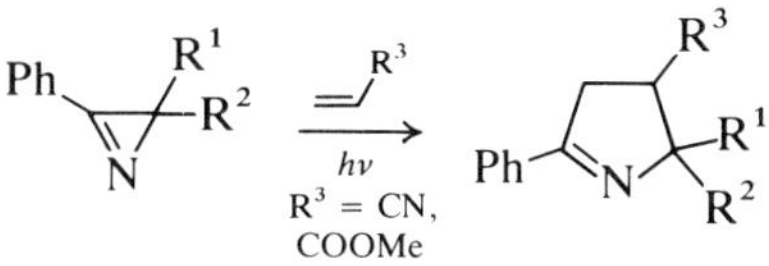

[81]

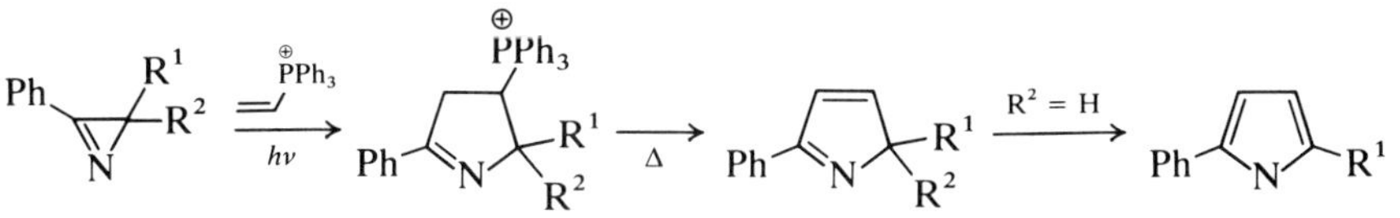

[82]

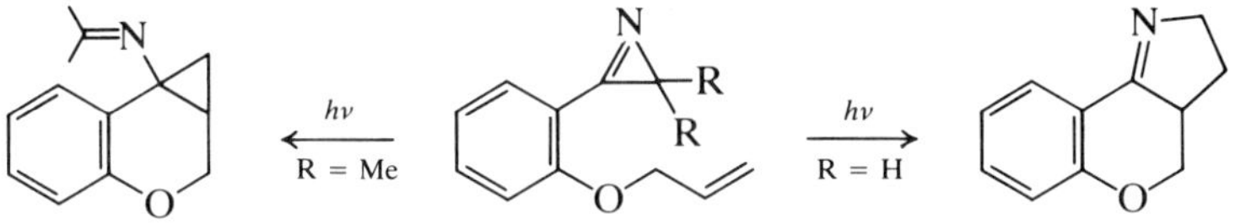

[83]

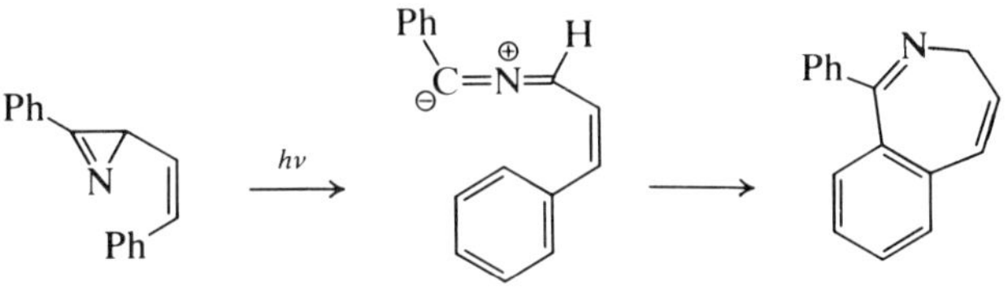

[84]

nitrile ylides are shown above. The consequences of attempts to carry out such reactions intramolecularly depend not only upon the spatial relationship of the double bond and the nitrile ylides but also upon the substituents on the azirine moiety, since these can determine whether the resulting ylide is linear or bent [81–84].

Photochemical rearrangement of pyridine *N*-oxide and its analogues is a synthetic route to pyrrolecarboxaldehydes; it is promoted by copper sulphate, giving yields of 30–40% [85].

Photochemical ring expansion of quinoline *N*-oxides (**141**) to benzoxazepines (**142**), followed by hydrolytic or thermal ring contraction, is another route to indoles and can be useful for certain specific syntheses [86, 87]. When the quinoline *N*-oxide bears an aryl or cyano substituent the benzoxazepine can usually be isolated in good yield. Two separate pathways for the cyclization have been observed, one leading to 2-substituted indole-3-carboxaldehydes and the other to 2,3-unsubstituted indoles. The precise course of these reactions depends upon the substitution pattern. To put the effectiveness of the method as a route to indoles in perspective, several examples are given here [88–90].

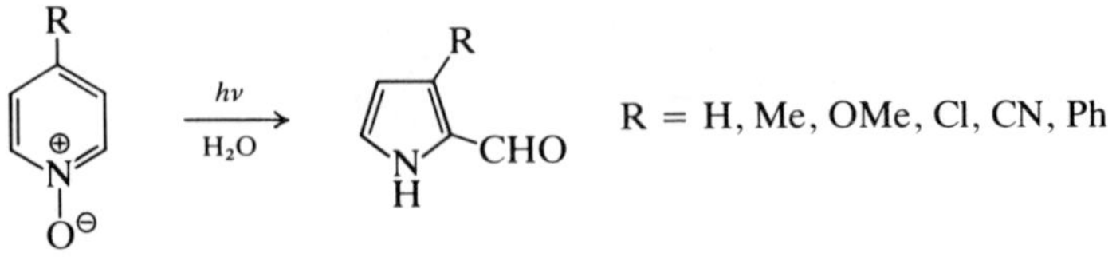

R = H, Me, OMe, Cl, CN, Ph

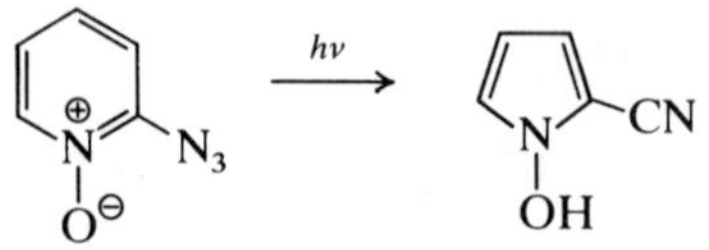

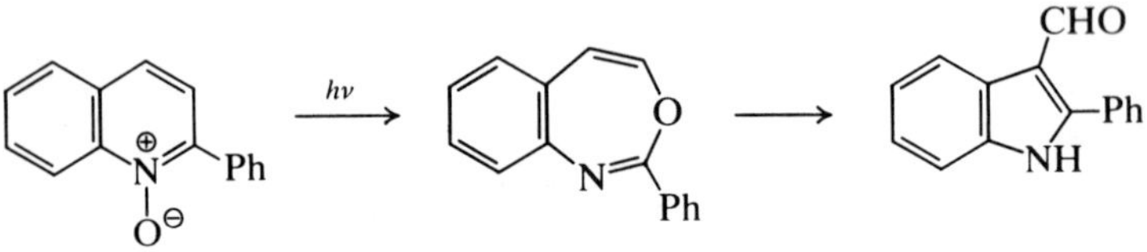

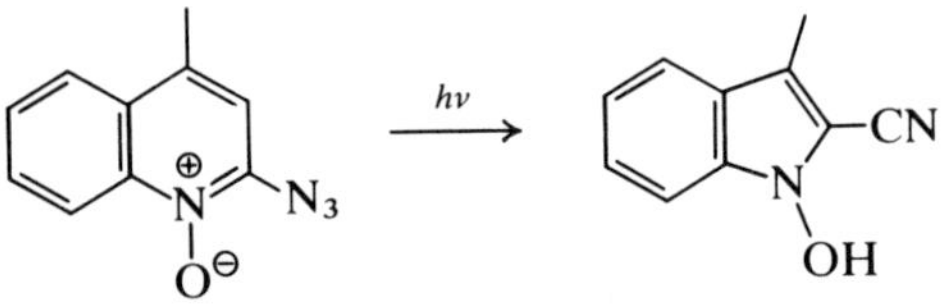

Photochemical routes for the synthesis of carbazoles via nitrenes are generally high-yield processes. Either nitro compounds or azides can serve as the intermediate carbazole precursors. *o*-Biphenylazide (**143**) and ring-

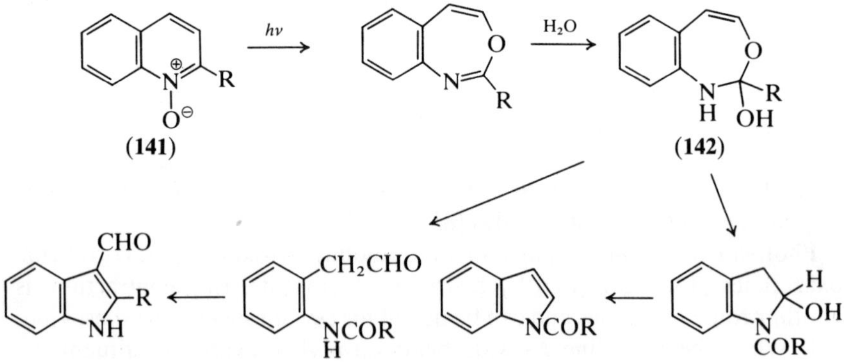

(**141**) (**142**)

| Quinoline substituent | Indole substituent | Yield (%) | Ref. |
|---|---|---|---|
| 2-Ph | 2-Ph, 3-CHO | 40 | [88] |
| 2-Me, 4-COOH | 2-Me, 3-CHO | 40 | [89] |
| 2-Ph, 4-COOMe | 2-Ph, 3-COOMe | 96 | [90] |

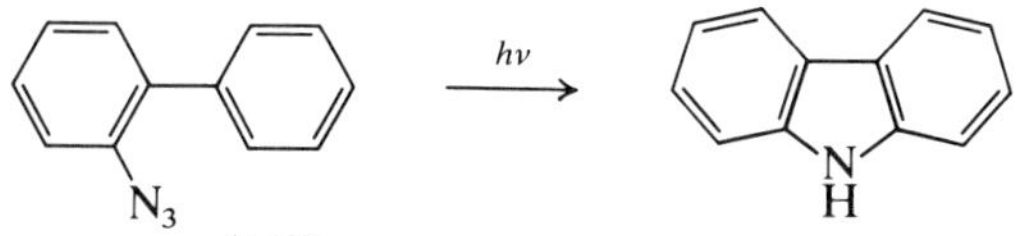

$N_3$ $h\nu$ N H

(**143**)

substituted analogues gave carbazoles in good yield under photolytic conditions [91].

### 7.3.2 Furans, thiophenes and their benzo derivatives

When either oxygen or sulphur replaces the central double bond of the stilbene system, analogous photocyclizations involving a six $\pi$-electron system proceed to provide a new synthetic route to either furans or thiophenes. Although 1-phenoxynaphthalene fails to undergo oxidative photocyclization, the *o*-chloro and *o*-methoxy derivatives undergo eliminative photocyclization. 4-Phenoxy-2,3,5,6-tetrachloropyridine (**144**) [92] and 2-(*o*-chlorophenoxy)naphthalene (**145**) (93) photocyclized with loss of hydrogen chloride.

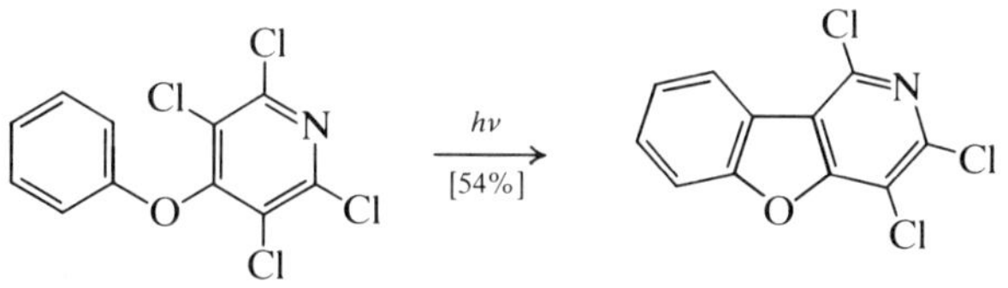

Cl Cl N Cl O Cl $h\nu$ [54%] Cl N Cl O Cl

(**144**)

Cl O $h\nu$ [45%] O

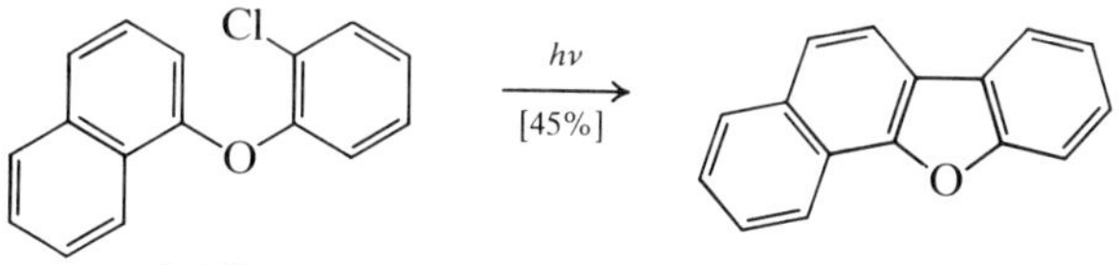

(**145**)

Photocyclization of diarylethers with elimination of an *o*-methoxyl group proceeds in very low yield except in certain multisubstituted systems such as (**146**) [94].

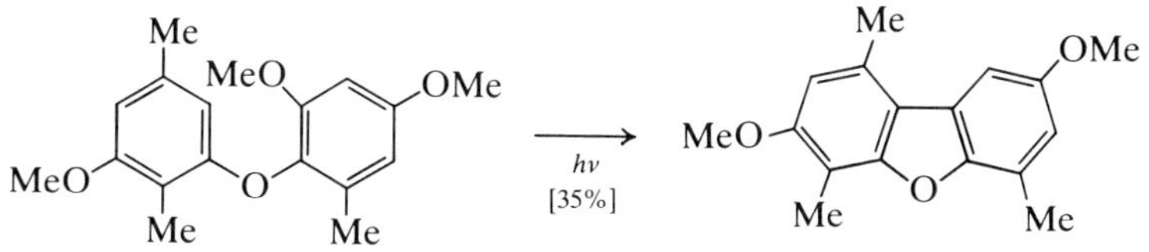

Me MeO OMe MeO O Me Me $h\nu$ [35%] Me OMe MeO O Me Me

(**146**)

The ether (**147**) and sulphide (**149**) behave analogously, giving the *cis* photocyclization products (**148**) and (**150**) respectively on irradiation in a methanol–benzene mixture, but giving mostly the less stable *trans* photocyclization products in 27–30% yields (along with polymeric material) on irradiation in benzene [95].

(**147**) (X = O)
(**149**) (X = S)

(**148**) (X = O) (88%)
(**150**) (X = S) (91%)

**Experimental 7.10** Photocyclization of 2-phenoxy-3,5,5-trimethyl-2-cyclohexen-1-one [95]

A solution of (**147**) (20 g, 0.087 mol, 0.043 M) in benzene–methanol–acetic acid (1 : 1 : 1, 2000 ml) was purged with argon for 30 min prior to and during irradiation with Pyrex-filtered light. After 22 h the reaction was complete (vapour-phase chromatographic analysis, 5% SE-30 on Chromosorb W, 80–100 mesh). The solvent was evaporated and ether (200 ml) was added to the residue. The ether solution was washed with 1 N sodium hydroxide ($3 \times 50$ ml) and saturated sodium chloride ($2 \times 50$ ml), and dried over magnesium sulphate. Evaporation of solvent gave pure (**148**) which crystallized on standing (17.6 g, 88%). Recrystallization from ether–petroleum ether gave the analytically pure (**148**) (15.9 g, 80%), MP 85–87 °C.

Interestingly, the *m*-methyl- or methoxy-substituted ethers (**151ab**) underwent regioselective photocyclization at the *ortho* carbon adjacent to the substituents [95]. Thus the methyl-substituted ether (**151a**) gave a 3 : 1 mixture of products (**152a**) and (**153**), while the methoxy analogue (**151b**) gave (**152b**) as the only detected product.

(**151ab**)
(**a**) R = Me
(**b**) R = OMe

(**152a**) (75%)
(**152b**) (90%)

(**153**) (25%)

When the sulphide (**154**) was irradiated in the presence of *N*-phenylmaleimide, the photochemically generated dihydro intermediate (**155**) was trapped, leading to the adduct (**156**) in 81% yield. In the absence of

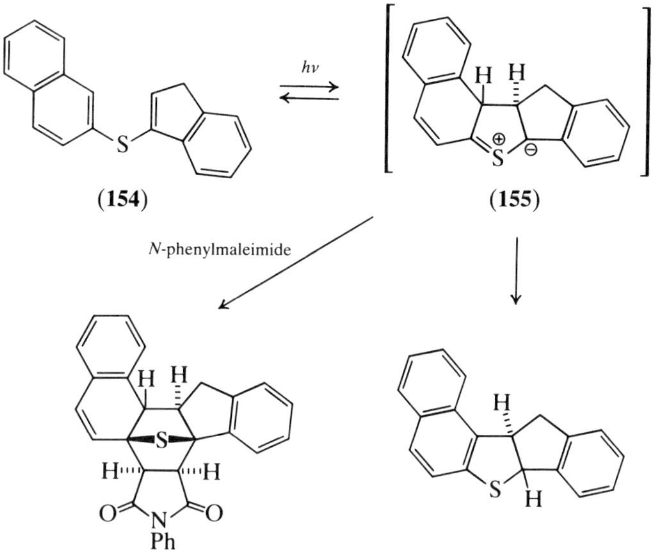

(156) (157)

the maleimide the intermediate (**155**) rearranged by a 1,4-hydrogen shift to give the *trans* product (**157**) in 78% yield [96].

Upon irradiation, the thioketone (**158**) gave the thiophene derivative (**159**) as a photocyclized product [97].

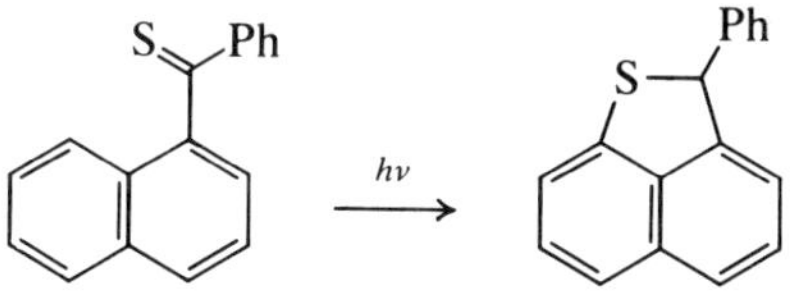

(158) (159) (51%)

Other photochemical syntheses of furan, thiophene, selenophene and thiazole and their benzo derivatives are described in reference [98].

Irradiation of allyloxy-ketones and aldehydes has led, by $\delta$-hydrogen

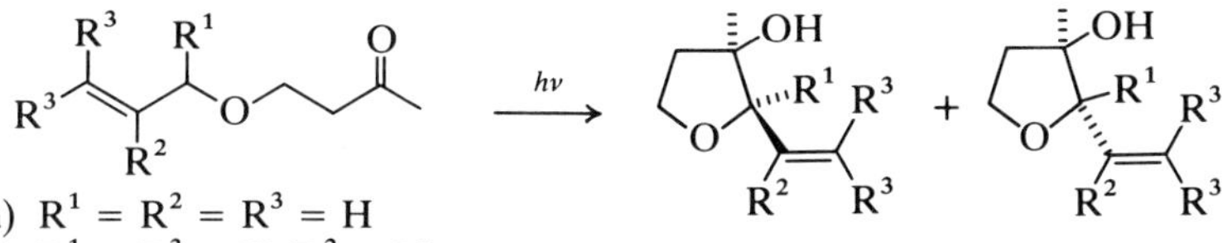

(**a**) $R^1 = R^2 = R^3 = H$
(**b**) $R^1 = R^3 = H, R^2 = Me$
(**c**) $R^1 = Me, R^2 = R^3 = H$
(**d**) $R^1 = R^2 = H, R^3 = Me$

abstraction, to a useful high-yield synthesis of 2-alkenyl-3-hydroxytetrahydrofurans. This substitution pattern occurs in the natural mycotoxins citreoviridin and asteltoxin [99].

Similarly, photochemical abstraction of $\delta$-hydrogen was employed in the preparation of 2,3-diphenyl-6-methoxybenzo[*b*]furan (**160**) [100].

Bn = $CH_2Ph$

(**160**)

### 7.3.3 Other five-membered heterocycles with two or more heteroatoms

Although five-membered heterocycles with two or more heteroatoms can be prepared by photochemical methods, their photochemical processes are not simple, and those interested in synthetic methods for these heterocycles should refer to the comprehensive review [98].

## 7.4 SIX-MEMBERED HETEROCYCLES

### 7.4.1 Pyridines and their benzo derivatives

#### 7.4.1.1 Preparation by photocyclization

As an extension of the photocyclization of stilbene and related compounds that provides a useful synthetic method for polycyclic compounds, Schiff's bases were studied and also found to be good precursors for the photochemical synthesis of phenanthridines and related heterocyclic compounds.

Irradiation of the imine (**161**) under normal oxidative conditions in organic solvents fails to produce 9-azaphenanthrene (**162**) [101, 102].

$h\nu$ $O_2$, organic solvent

(**161**) (**162**)

(163) ⇌ (161) [$h\nu$; $h\nu$ or dark]

This failure of (**161**) to undergo photocyclization in organic solvents can be attributed (at least in part) to its thermal conversion to the related *trans* isomer (**163**) [101].

However, the photocyclization of (**161**) to (**162**) did proceed in 98% sulphuric acid [103, 104]. Photocyclization involves the formation of the protonated imine (**164**), and the dihydro intermediate (**165**) is trapped directly as (**164**) or by sulphuric acid or oxygen.

(**164**) ⇌ [(**165**)] → (38%) [$h\nu$; oxidant]

In sulphuric acid the thermal *cis–trans* isomerization of the protonated imine is very slow, as exemplified by the oxidative photocyclization of (**166**), with two phenyl groups *cis* to one another, in cyclohexane [101].

(**166**) → (46%) [$h\nu$, $I_2$]

As an exception to the general rule that acidic conditions are required, the imine (**167**) is reported to undergo oxidative photocyclization in ether [105].

(**167**) → (56%) [$h\nu$, $O_2$]

An alternative synthesis of phenanthridines that avoids the problem of imine photocyclization is illustrated by the three-step conversion of the

(**168**) $\xrightarrow{BF_3}$ (**169**) $\xrightarrow[I_2]{h\nu}$ (95%) $\xrightarrow{LiAlH_4}$

hydroxamic acid (**168**) via the cyclic borate (**169**) to a phenanthridine, in which the last step is reduction of the photocyclized product with lithium aluminium hydride [106].

A number of photochemical syntheses of substituted phenanthridines and related polycyclic heterocycles are described in the excellent review [98].

As azastilbene analogues, styrylpyridinium salts (**170**) have been found to be precursors for the photochemical synthesis of the benzo[*a*]quinolizinium

(**170**) $\xrightarrow{h\nu}$ (**171**) $\xrightarrow[I_2]{h\nu}$ (**172**)

TABLE 7.1

Irradiation of styrylpyridinium salts (**170**)

| $R^1$ | $R^2$ | $R^3$ | $R^4$ | $R^5$ | (**172**) Yield (%) |
|---|---|---|---|---|---|
| — | — | — | — | — | 60 |
| Me | Me | — | — | — | 47 |
| Ph | Ph | — | — | — | 50 |
| — | — | Me | — | — | 56 |
| — | — | — | — | Me | 66 |
| — | — | OCOPh | — | — | 43 |
| — | — | — | — | Cl | 60 |
| — | — | OCOPh | OCOPh | — | 25 |
| — | — | OCOPh | — | OCOPh | 50 |

ion and some of its derivatives (**172**) [107, 108]. Irradiation of (**170**) in the presence of iodine as an oxidant gave the photocyclized tricyclic compounds (**172**) via the *cis* intermediates (**171**). In view of this, if an improved method for making the styrylpyridinium salts (**170**) can be found, this photochemical method appears to offer an advantageous route to the benzo[*a*]-quinolizinium compounds (**172**).

Another related photochemical synthesis of 4*H*-benzo[*a*]quinolizin-4-one (**175a**) and its 7-phenyl derivative (**175b**) is from an *N*-styrylpyridinone (**174ab**), which is prepared by *N*-alkylation of pyridone with a styryl bromide (**173ab**) [109].

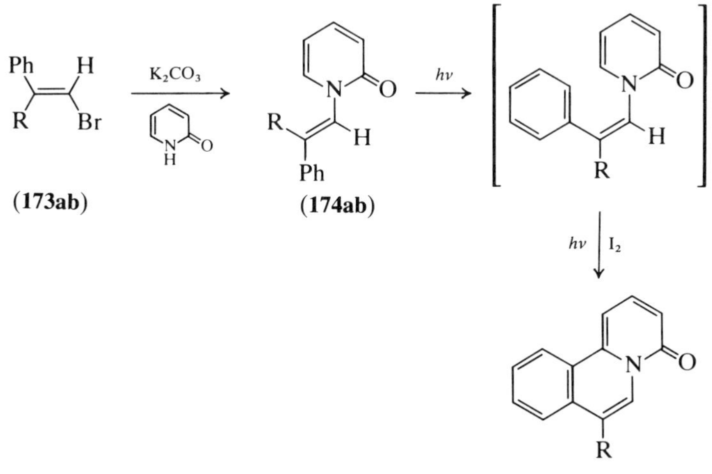

Ph H
R Br
$K_2CO_3$
N H O
R H Ph
N O
$h\nu$
N O H R
$h\nu$ $I_2$
N O R

(**173ab**)

(**174ab**)

(**175a**) R = H
(**175b**) R = Ph

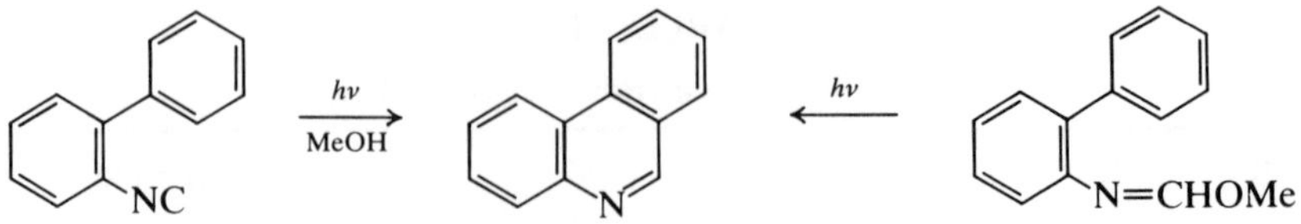

hν MeOH → NC (162) hν ← N=CHOMe (176)

Aryl isocyanides can also be photochemically ring closed to a suitably placed aromatic ring, as in the synthesis of phenanthridine (**162**), where it was established that solvent interaction was required; the methoxy imine (**176**) also gave phenanthridine on irradiation [110].

Much more work has been done on the photochemical cyclization of amides, mainly enamides [111–116]. Irradiation of the enamides (**177**) can give either tetrahydrophenanthridones or hexahydrophenanthridones, depending upon the reaction condition employed. Without added iodine, a mixture of the *cis*- and *trans*-lactams (**178**) and (**179**) was obtained, the ratio varying with the solvent used: in non-polar solvents the *trans* : *cis* ratio can be 15.6 : 1. If iodine was added to the solution to be irradiated, the quinol-2-ones (**180**) were obtained [117].

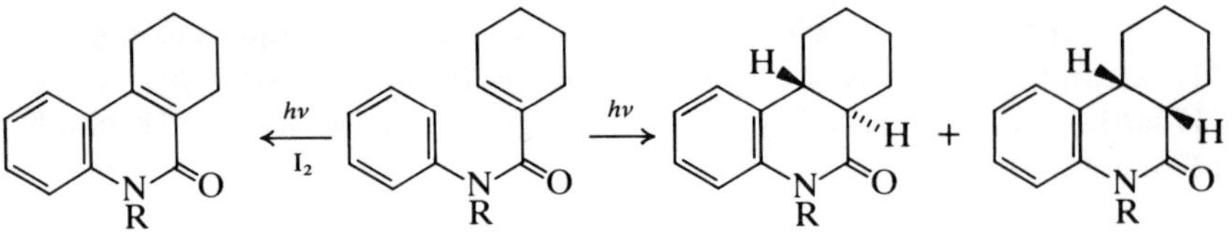

(180) ← hν, $I_2$ — (177) — hν → (178) + (179)

R = Me, $CH_2Ph$

The mechanism has been discussed and is shown below [113]: enolization is followed by conrotatory photochemical electrocyclization and then a 1,5-sigmatropic hydrogen shift. There was evidence of some incorporation of $^2H$ into the 3-position of the dihydroquinol-2-one if a [2,4,6-$^2H_3$]anilide was used [118].

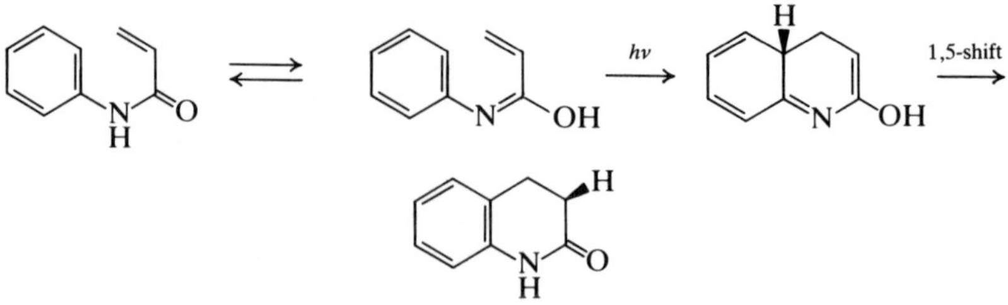

⇌ hν → 1,5-shift →

Much information is available on the cyclization of the *N*-cyclohexenylbenzamides (**181**), which give either hexahydro- or tetrahydrophenanthridones, depending upon the reaction conditions [119]. From the enamides (**181**) only one isomer (**182**), with *trans* ring fusion, was obtained; this was also the case in the synthesis of the polycycles (**183**) [120]. The reaction is not regioselective unless an *ortho* substituent is present. Such an *ortho* substituent, if it is a good leaving group (halogen or methoxy), can permit the production of the tetrahydrophenanthridones without added iodine, as shown here for the synthesis of the benzo[*c*]phenanthridone (**185**) from the enamide (**184**). A better yield was obtained in the cyclization of the

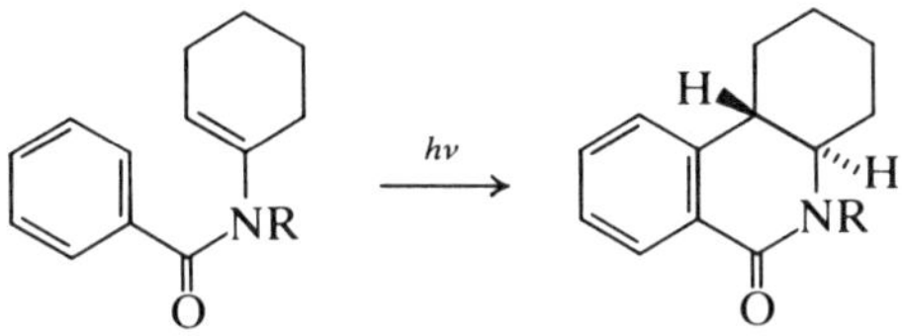

(**181**) $\xrightarrow{h\nu}$ (**182**)

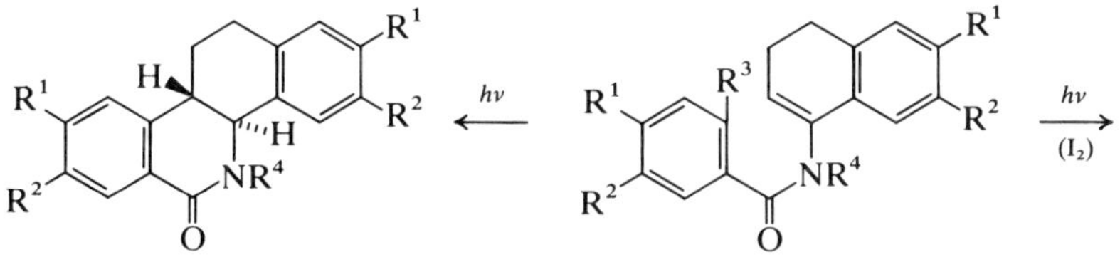

(**183**) (52%)
$R^1 = R^2 = H, R^4 = Me$

(**184**) $R^3 = Br$
(**186**) $R^3 = OMe$

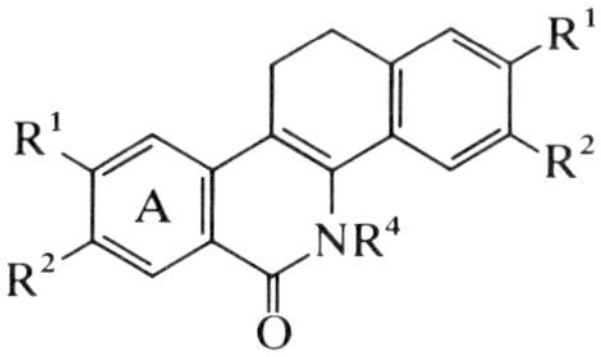

$h\nu$ ($I_2$)

(**185**) $R^1 = R^2 = H$
(**187**) $R^1, R^2 = OCH_2O$

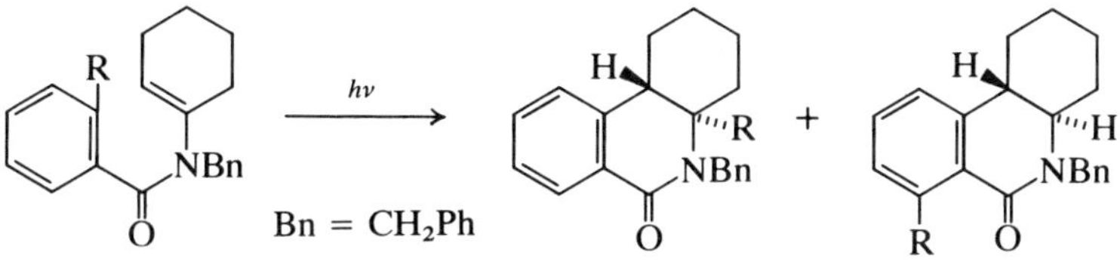

(**188**) $\xrightarrow{h\nu}$

Bn = $CH_2Ph$

R = OMe, COOMe, $CONH_2$

R = $NH_2$

*o*-methoxy substituted enamides (**186**), with control of the regiochemistry of the product (**187**); without the methoxy group, but using iodine, both regioisomers in ring A were obtained. An *ortho* substituent can isomerize among the *ortho* substituents; in the enamide (**188**) the methoxy group and the methoxycarbonyl group migrated, while the amino group directed cyclization to the *ortho* position [121, 122].

Benzanilides can be cyclized to phenanthridones when irradiated in the presence of iodine [123]. In the examples given, phenanthridone itself (**193**, $R^3 = H$) has been prepared from benzanilide (**189**), from *o*-iodobenzoylaniline (**190**) and also from benzoyl-*o*-iodoaniline (**191**); in the last two cases no iodine was necessary. A variety of *o*-methoxybenzanilides (**192**) have been cyclized in moderate to excellent yields, if the benzene ring A contains at least one extra methoxy group [124]. As expected, benzo[*c*]-phenanthridones such as (**194**) can be similarly prepared [125]. The phenanthridones can be converted into phenanthridines by standard procedures.

(**189**) $R^1 = R^2 = R^3 = H$ (**193**) (9–48%)
(**190**) $R^1 = I, R^2 = R^3 = H$
(**191**) $R^2 = I, R^1 = R^3 = H$
(**192**) $R^1 = OMe, R^2 = H, R^3 =$ various groups

(**194**) $R^1 = R^2 = OCH_2O$ (70%)

In the course of investigations into the reactivity of enamides as unique synthons and also in the evaluation of enamide photocyclization as a useful synthetic tool, Ninomiya and coworkers [126] discovered the reductive photocyclization of enamides in the presence of a hydride reagent. In addition to oxidative and non-oxidative photocyclizations, this reductive photocyclization disclosed a broad aspect of the cyclizability of enamides.

Irradiation of *N*-cyclohexenylbenzamides in the presence of sodium borohydride or borodeuteride gives a mixture of the dienes (**196**) and (**197**). The labelling produced by the use of borodeuteride has provided evidence for the intermediacy of the zwitterionic structures (**195**).

Thus the reductive photocyclization of enamides has established photocyclization as a simple and useful method for the construction of various six-membered lactams with various oxidation levels [126]. Reductive photocyclization of enamides has recently been used in the total synthesis of a number of alkaloids, including an asymmetric synthesis

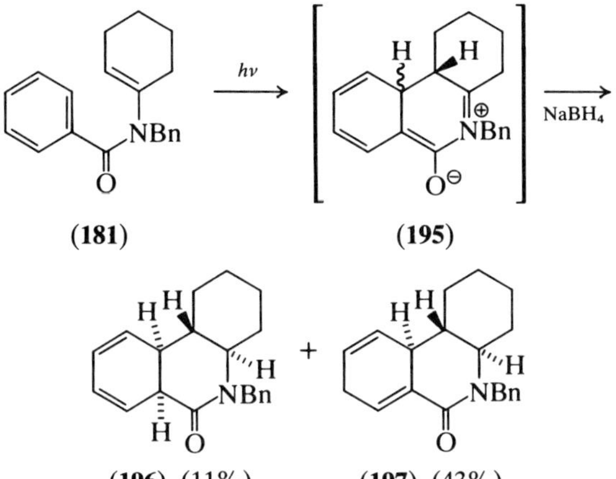

hν

NaBH$_4$

(181) (195)

(196) (11%) (197) (43%)

[127–141]. The common strategy in these syntheses consists in the construction of the basic structure of the target alkaloids and subsequent functionalization to the final alkaloids. Among a number of indole alkaloids, ergot and yohimbine alkaloids were smoothly synthesized as follows.

*Ergot alkaloids*. Reductive photocyclization of the enamide (**199**), prepared from the tricyclic ketone (**198**), gave the lactam (**200**) in good

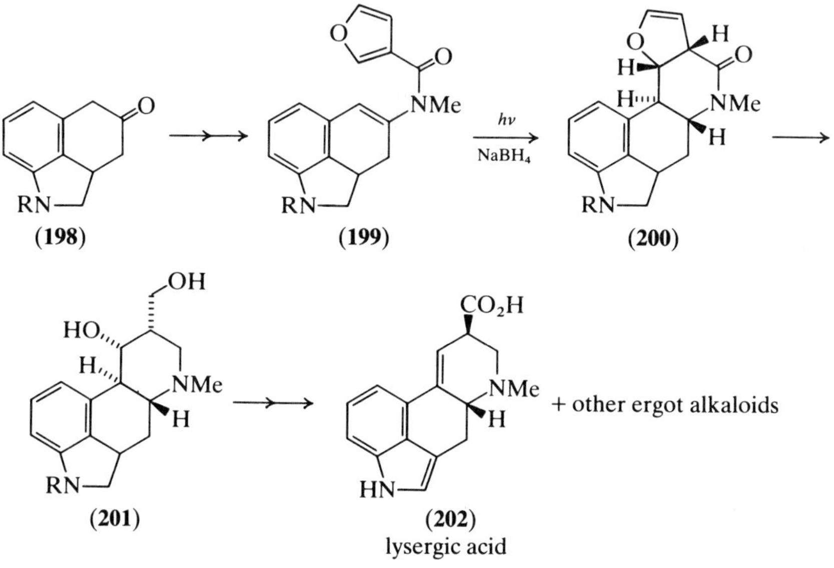

hν
NaBH$_4$

(198) (199) (200)

(201) (202)
lysergic acid

+ other ergot alkaloids

yield. The dihydrofuran ring in the lactam (**200**) was oxidatively cleaved to give an aminoglycol (**201**), which was used as the common key intermediate for the total synthesis of ergot alkaloids including (±)-lysergic acid (**202**) [127–131].

*Yohimbine alkaloids*. Reductive photocyclization of the *p*-methoxy substituted enamide (**203**), prepared from harmalane, proceeded smoothly in acetonitrile–methanol (9 : 1) to give the hydrogenated lactam (**204**) in quantitative yield; this is a basic structure of the yohimbine alkaloids and was successively converted into (±)-yohimbine (**206**), (±)-alloyohimbine (**207**) [132], (±)-19,20-dehydroyohimbines (**208**) [133] and (±)-deserpidine (**209**) [134] via a common key intermediate (**205**).

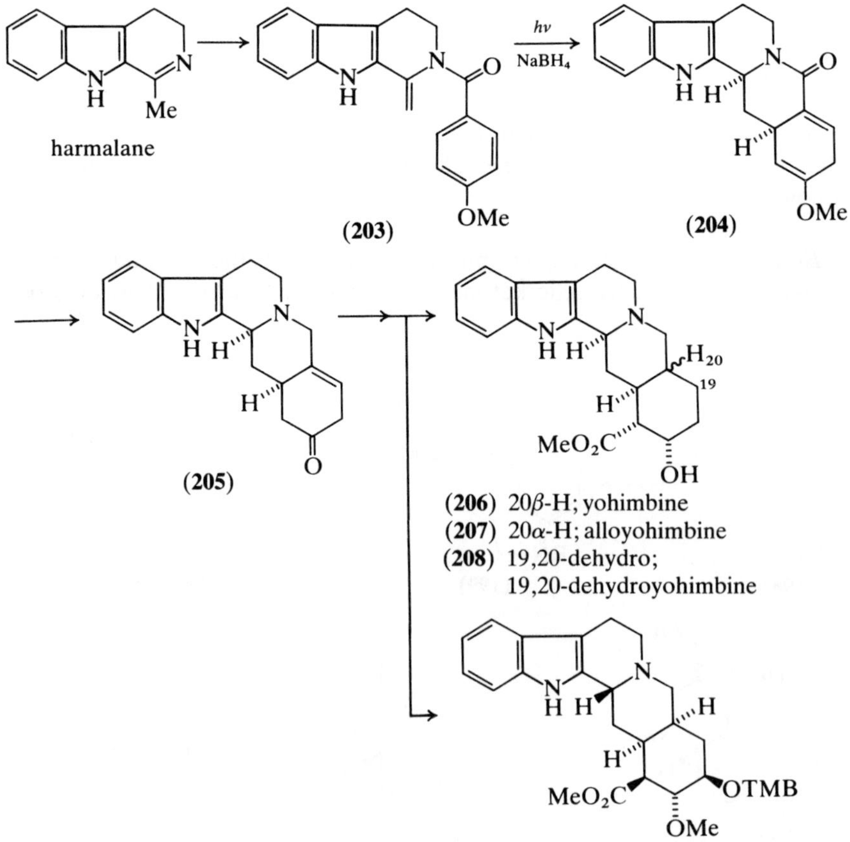

harmalane

(**203**)

$h\nu$, $NaBH_4$

(**204**)

(**205**)

(**206**) 20β-H; yohimbine
(**207**) 20α-H; alloyohimbine
(**208**) 19,20-dehydro; 19,20-dehydroyohimbine

(**209**) deserpidine
(TMB = 3,4,5-trimethoxybenzoyl)

*Corynantheine–heteroyohimbine alkaloids*. Using reductive photocyclization of the enamides (**210ab**) with a furan ring, Ninomiya and coworkers have succeeded in an efficient total synthesis of the corynantheine–heteroyohimbine alkaloids (±)-ajmalicine (**213**) and (±)-corynantheine (**214**) via the hydrogenated lactam (**211**) [135, 136], and (±)-hirsuteine (**215**) and (±)-geissoschizine (**216**) [137] via the lactam (**212**).

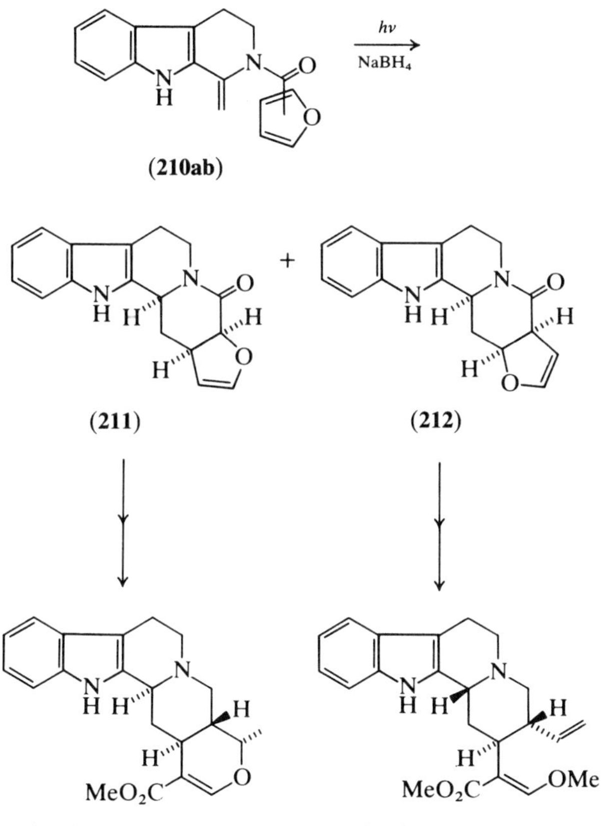

hν
NaBH$_4$

(**210ab**)

(**211**) + (**212**)

(**213**) ajmalicine and
(**214**) corynantheine

(**215**) hirsuteine and
(**216**) geissoschizine

*Ipecac alkaloids*. In a procedure similar to the total synthesis of corynantheine–heteroyohimbine alkaloids using furoylenamine type enamides, (±)-emetine (**220**) was synthesized via the two hydrogenated lactams (**218**) and (**219**), which were prepared by the reductive photocyclization of the enamides (**217a**) [135] and (**217b**) [138].

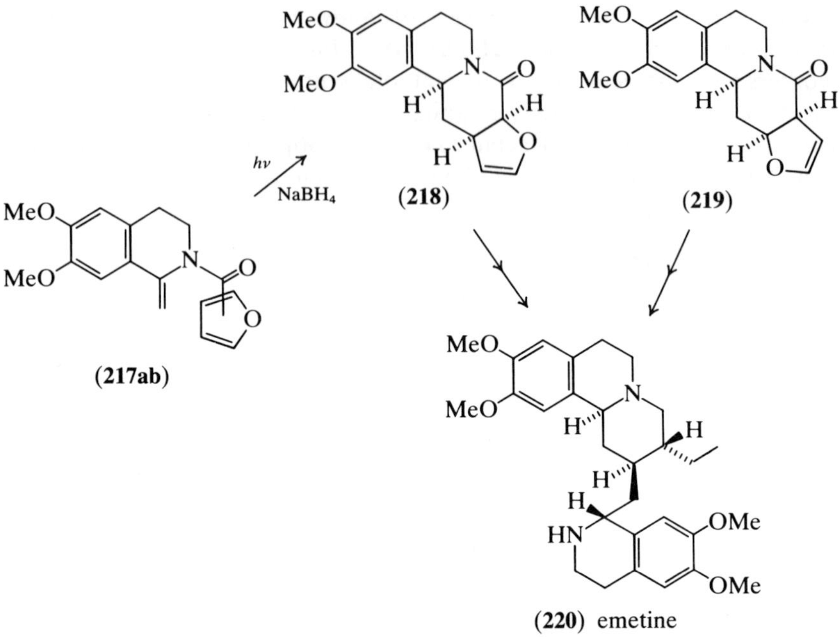

MeO MeO N O H H H O (218)

MeO MeO N O H H H O (219)

$h\nu$ NaBH$_4$

MeO MeO N O O (217ab)

MeO MeO N H H H H HN OMe OMe

(220) emetine

*Asymmetric synthesis of (−)-xylopinine*. Reductive photocyclization of the enamide (**221**) in the presence of a chiral metal-hydride complex, prepared from lithium aluminium hydride and quinine, gave the optically active lactam (**222**), which was used for the asymmetric synthesis of (−)-xylopinine (**223**) with 37% e.e. [139, 140]. Kametani *et al.* [141] have also succeeded in the asymmetric synthesis of the same alkaloid by diastereoselective photocyclization of the chiral enamide (**224**).

Photocyclization of the unsaturated amide (**226**), which contains a six $\pi$-electron system involving a double bond of the aromatic ring and the unsaturated amide group, has provided another synthetic route for the quinolones (**227**) [142].

Photocyclization of the *o*-thiocarboxamidostyrenes (**228**) [112] gave the quinolines (**229**) as a result of elimination of the thio group [143].

**Experimental 7.11** *cis*- and *trans*-6*a*,7,8,9,10,10*a*-Hexahydro-5-methylphenanthridin-6(5*H*)-one [117]

A solution of the anilide (**177**) (1.6 g) in ether (400 ml) was irradiated externally with a low-pressure mercury lamp (120 W, Eikosha PIL-120) at room temperature under nitrogen atmosphere for 40 h. The solvent was removed *in vacuo* at room temperature to yield a residue which was chromatographed on silica gel (silica gel 60,

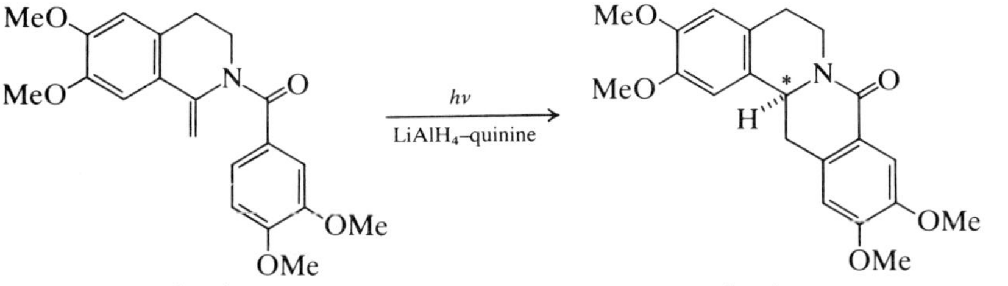

MeO MeO N O OMe OMe (221) $h\nu$ $LiAlH_4$–quinine MeO MeO H * N O OMe OMe (222)

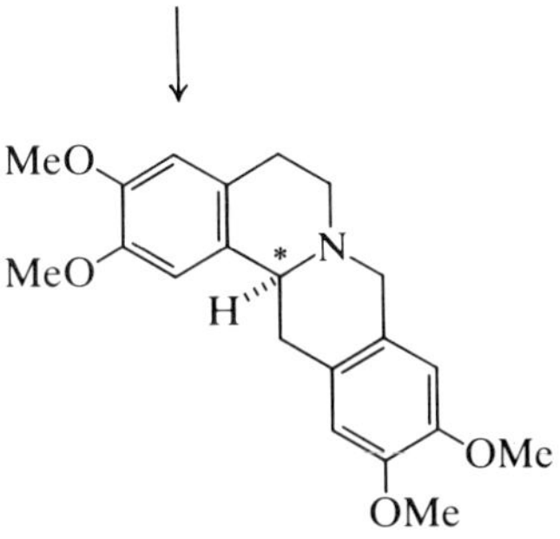

MeO MeO H * N OMe OMe (223) (−)-xylopinine

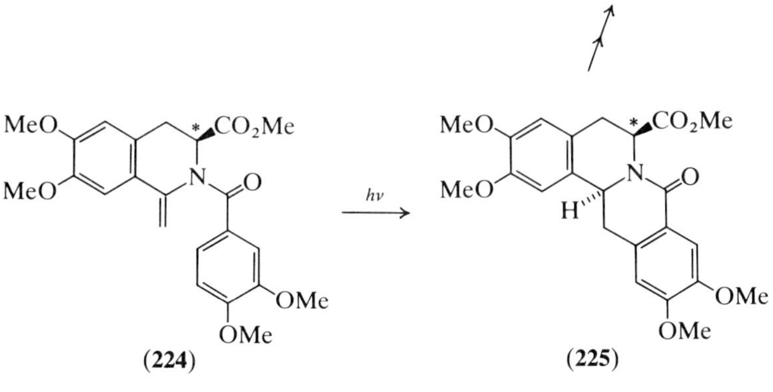

MeO MeO * $CO_2Me$ N O OMe OMe (224) $h\nu$ MeO MeO * $CO_2Me$ H N O OMe OMe (225)

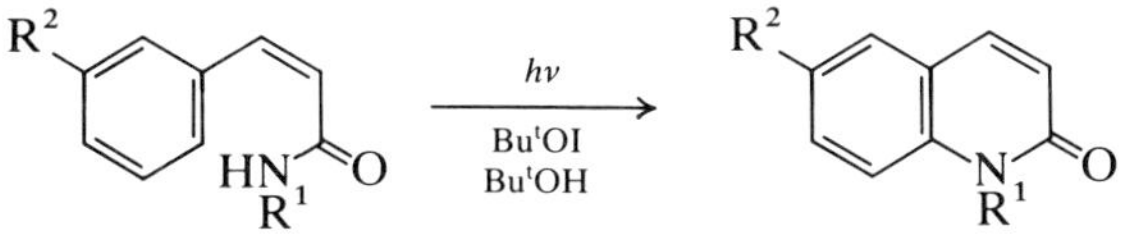

$R^2$ HN $R^1$ O (226) $h\nu$ $Bu^tOI$ $Bu^tOH$ $R^2$ N $R^1$ O (227)

$R^1 = R^2 = H$ (64%)
$R^1 = Me, R^2 = H$ (24%)

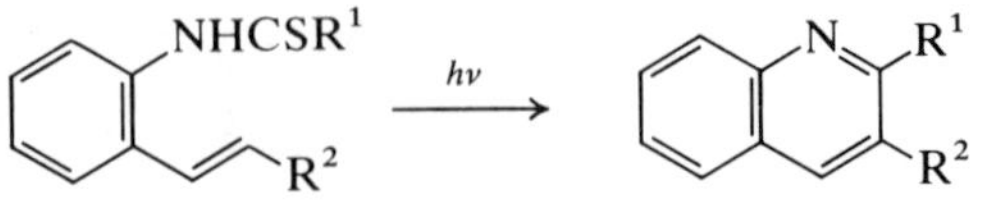

(228) → hν → (229)

$R^1$ = Me, Ph
$R^2$ = Me, Ph, CN

Merck). The fraction eluted with benzene afforded a pale-yellow oil (1.1 g), BP 138–143 °C (bath temperature) at 1 mmHg which was a mixture of the *cis* (**179**) and *trans* isomers (**178**) and separated by preparative TLC (silica gel 60 PF254, Merck) with chloroform as eluent: (**179**) (190 mg, 12%) oil, BP 138–140 °C (bath temperature) at 1 mmHg; and (**178**) (750 mg, 47%), MP 98–99 °C (n-hexane).

**Experimental 7.12** 16,17,19,20-Tetradehydro-17-methoxyyohimban-21-one [132]

To a solution of the enamide (**203**) (318 mg) in acetonitrile (150 ml) were added sodium borohydride (350 mg) and methanol (15 ml) successively at room temperature. After the sodium borohydride had dissolved, the resulting solution was cooled to 5–10 °C and irradiated for 30 min with a high-pressure mercury lamp (300 W, Eikosha PIH-300) through a Pyrex filter whilst dry nitrogen gas was bubbled through. The reaction mixture was then evaporated at room temperature under reduced pressure. Water was added to the residue to separate the colourless solid which was recrystallized from methanol to afford the lactam (**204**) (288 mg, 90%). MP 248–250 °C.

### 7.4.1.2 Preparation by ring transformation

The hydroxyoxindole (**230**) was photochemically transformed into a 4-hydroxyquinol-2-one (**231**) via homolytic ring cleavage and recyclization [144].

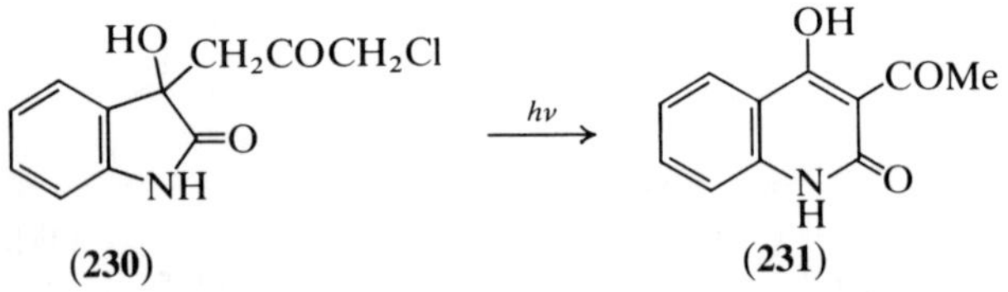

(230) → hν → (231)

Irradiation of the benzoisothiazoline dioxide (**232**) in the presence of $\beta$-chloroacrylic acid gave the tetrahydroquinoline (**233**) [145].

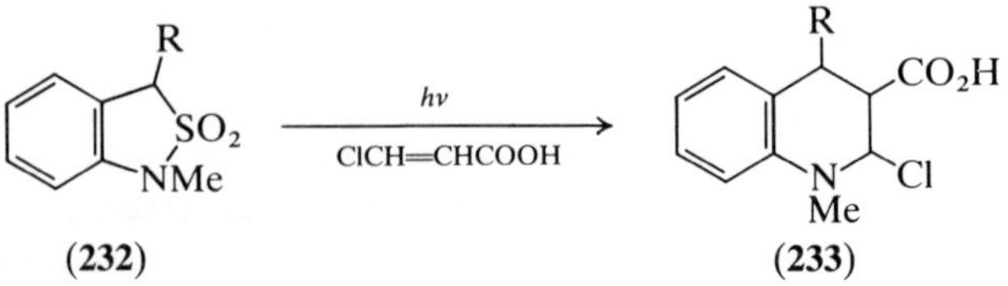

(232) → hν, ClCH=CHCOOH → (233)

Photochemical ring transformation of the spiro $\alpha$-aminoketone (**234**) gave the pyridone (**235**), which is related to the indole alkaloids [146].

NH$_2$ MeO$_2$C MeO NH MeO$_2$C MeO

(**234**)

$h\nu$

MeO$_2$C MeO

(**235**)

### 7.4.2 Other six-membered heterocycles

As with the photochemical synthesis of pyridines and their benzo analogues, other six-membered heterocycles, including those with two or more hetero atoms, have been prepared by photochemical methods involving photocyclization and photochemical ring transformations. The pyridodiazines (**236**) and (**237**) were prepared by photocyclization of the corresponding acylanilides [147].

Irradiation of the 3-pyridyltetrazolium salts (**238**) gave a mixture of 5,6-tetrazole fused derivatives of the system (**239**), which can be converted into the parent system by various reducing agents [148].

Pyrans have also been synthesized by photocyclization and photochemical ring transformation. Several groups have sought to displace the dienone–2*H*-pyran equilibrium in favour of the heterocycles. Thus irradiation of the dienone (**240**) gave the *trans* isomer (**241**) together with an equilibrium mixture of the pyran (**242**) and its ring-opened isomer, which consisted mainly of (**242**) [149].

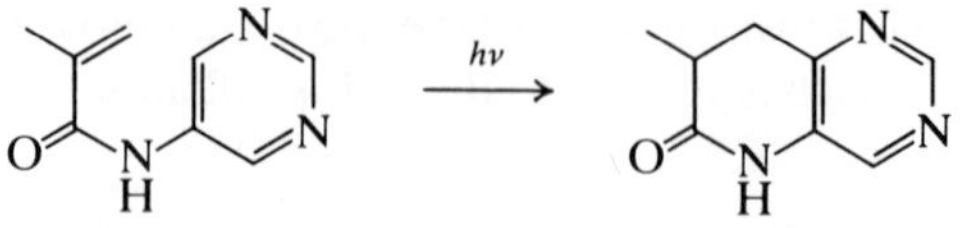

(236)

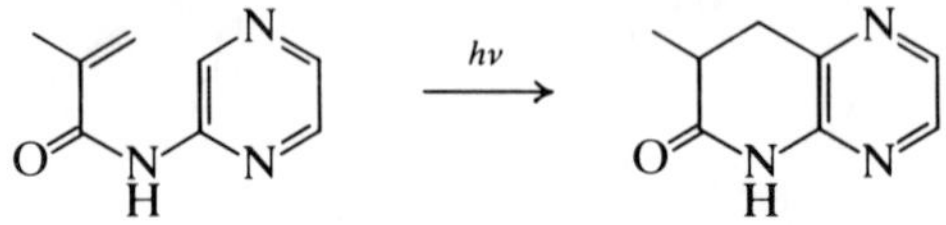

(237)

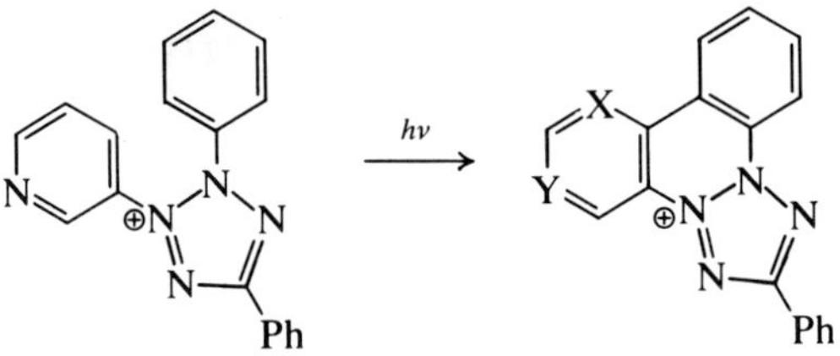

(238)

(239ab)
(**a**) X = N, Y = CH
(**b**) X = CH, Y = N

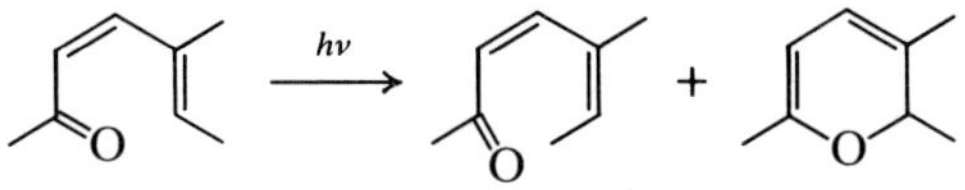

(240) (241) (242)

Brief irradiation of β-ionone (**243**) using a high-pressure mercury lamp with a Pyrex filter gave the fused pyran (**244**) in good yield on a preparative scale; without the filter the dienone (**245**) was produced in quantitative yield [150].

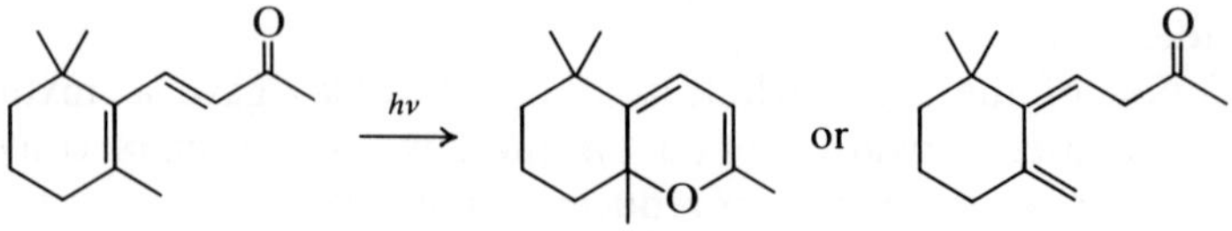

(243) (244) or (245)

Photochemical ring transformation of furans has provided a synthetic route to pyrans. Irradiation of the 5-aryl-3*H*-furan-2-ones (**246**, R = OMe or OAc) in benzene using a medium-pressure mercury lamp led to formation of the 2,3-dimethylchromones (**249**, R = OMe or OAc). When ethanol was used as the solvent, the analogous 3-ethoxycarbonylmethyl compounds (**250**) resulted. The primary process is photochemical cleavage of the O—CO bond and subsequent cyclization involving an intramolecular radical addition to the ester carbonyl group. The resulting diradical (**247**) collapses to the chromanone (**248**), which is either solvolysed in ethanol or decarboxylated in benzene [151].

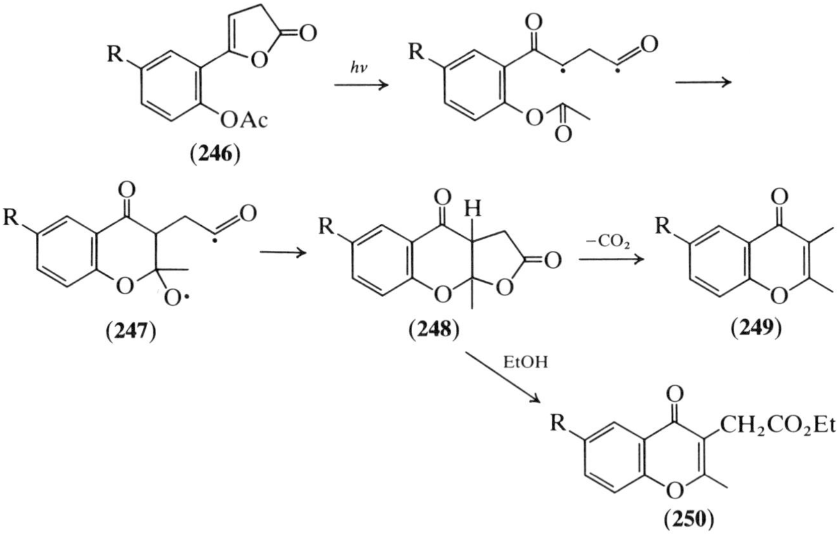

(**246**) (**247**) (**248**) (**249**) (**250**)

## 7.5 SEVEN- AND LARGER-MEMBERED HETEROCYCLES

Photochemical syntheses of seven- and larger-membered compounds have recently been reported. These are fundamentally divided into two photochemical methods: (i) photochemical cycloaddition followed by thermal ring expansions, and (ii) photocyclization.

### 7.5.1 Preparation of seven- and larger-membered heterocycles by photochemical cycloaddition followed by thermal ring expansion

A large number of heteropines have been prepared by photochemical cycloaddition of heterocyclic enones to C═C double bonds followed by ring expansion of the resulting cyclobutanes. The photochemical adducts (**252**), prepared by photocycloaddition of 4-hydroxy-2-quinolone and its 3-substituted congeners (**251**), were subjected to retro-aldol reactions in the presence of base to give the azocines (**253**) [152]. Similarly, benzo[*b*]thiophenes (**254**), benzo[*b*]furans (**255**) [154] and *N*-substituted indoles (**256**) [155] underwent photocycloaddition to appropriate alkenes, and then thermal or base-catalysed ring expansion led to the formation of a variety of heteropines as follows.

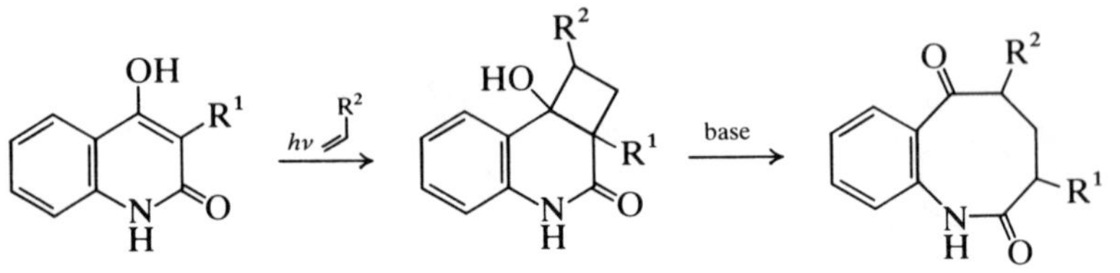

(**251**) (**252**) (**253**) [152]

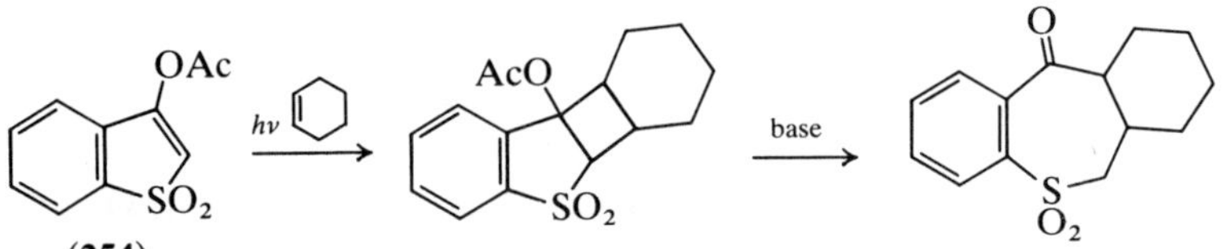

(**254**) [153]

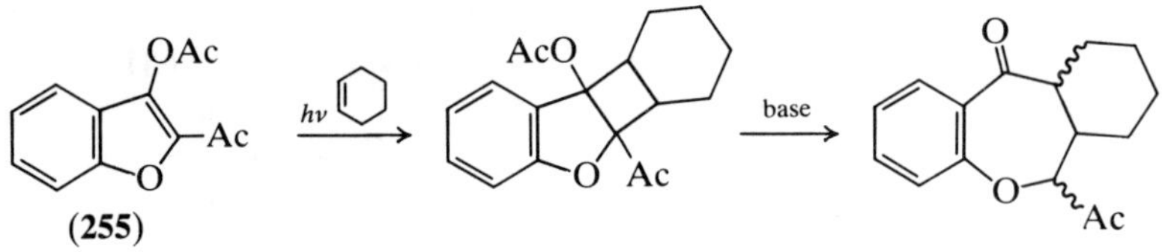

(**255**) [154]

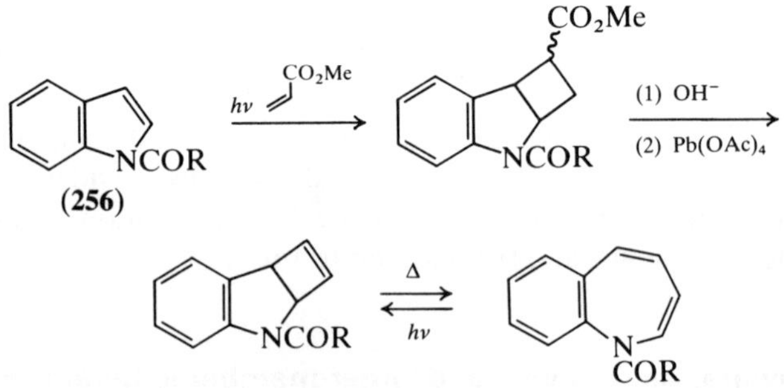

(**256**) [155]

2,3-Dioxopyrrolidine-4-carboxylates (**257**) gave azepinones (**258**) via a route involving photocycloaddition to alkenes followed by a retro-aldol reaction [156]. Photocyclization of 2-phenyl-4,5-dioxopyrroline-3-carboxylates (**259**) has been thoroughly studied; it gave the cyclobutanes (**260**), which were subjected to ring expansion under basic conditions or thermally to give azatropolones (**261**) [157, 158].

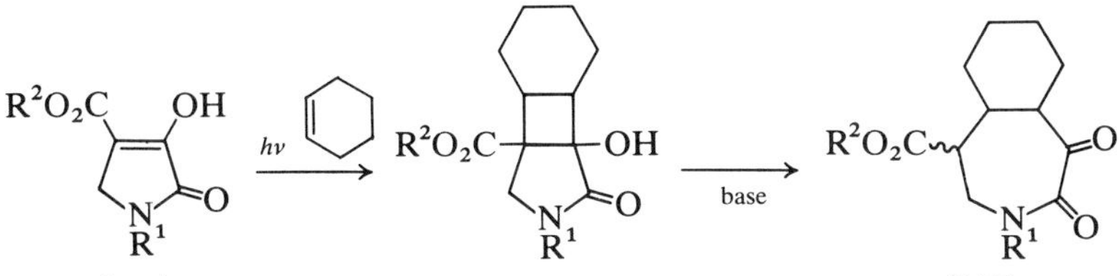

(257) (258)

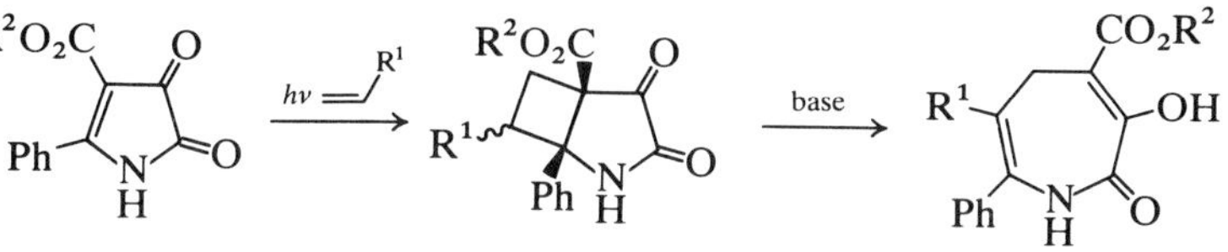

(259) (260) (261)

The photochemistry of imides has been investigated mainly by Kanaoka [159] and Mazzochi's groups [160] and provides an excellent synthetic route to nitrogen heterocycles. Irradiation of a series of cyclic aliphatic imides in the presence of alkenes gave the corresponding oxetanes in good yields, whereas attempts to carry out this reaction with the open-chain analogues have been fruitless [161, 162]. The reaction can occur intramolecularly to give tricyclic products such as (**263**), which can be conveniently converted to the seven-membered azepinone (**264**) with acid.

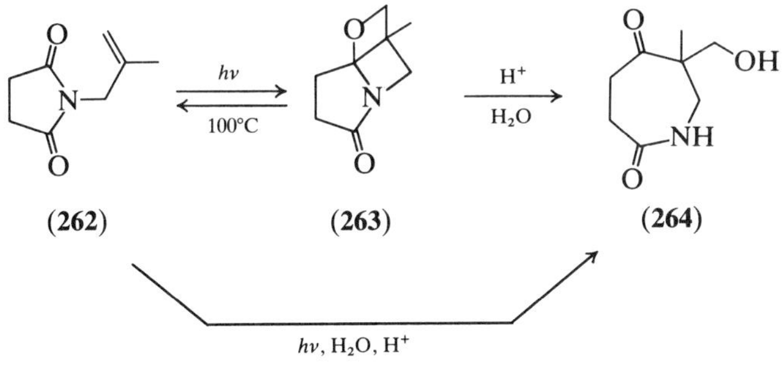

(262) (263) (264)

An interesting synthesis of azepine derivatives (**4**) involves photocycloaddition of the azide (**1**), followed by ring opening of the resulting aziridines (**3**) [2].

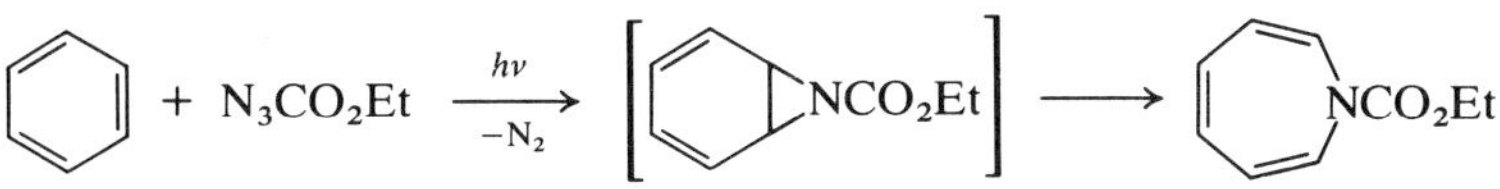

(1) (3) (4)

**Experimental 7.13** Photolysis of *N*-2-alkenyl alicyclic imides

Irradiation of (**262**) (0.02 M) in acetonitrile with a 120 W low-pressure mercury arc for about 120 h gave, after evaporation of the solvent, an oily product (**263**) almost quantitatively. After prolonged heating at about 100 °C, (**263**) was decomposed into the starting material. Irradiation of (**262**) in water acidified with a trace of hydrochloric acid also gave (**264**) in good yield.

## 7.5.2 Preparation of seven- and larger-membered heterocycles by photocyclization

### 7.5.2.1 From imides

*N*-Alkylated cyclic imides like (**265**) undergo rather efficient hydrogen abstraction from the $\beta$-position of the *N*-alkyl group to give the radical (**266**), which gives the azepinedione (**268**) via the azacyclobutane (**267**) [163]. A number of examples are given in Table 7.2. In all cases biradical closure is the dominant process, but is accompanied by 20–30% of the type-II cleavage product, e.g. succinimide from (**265**).

(**265**) $\xrightarrow{h\nu}$ (**266**) $\longrightarrow$ (**267**) $\longrightarrow$ (**268**)

TABLE 7.2

Irradiation of succinimides and glutarimides

| *n* | $R^1$ | $R^2$ | | $R^3$ | $R^4$ | Yield (%) |
|---|---|---|---|---|---|---|
| 2 | H | H | | H | H | 45 |
| 2 | H | H | | Me | H | 42 |
| 2 | H | Me | | H | H | 56 |
| 2 | H | H | | Et | H | 31 |
| 2 | H | H | | Me | Me | 33 |
| 2 | Me | Me | | H | H | 49 |
| 2 | H | | $(CH_2)_3$ | | H | 50 |
| 2 | H | | $(CH_2)_4$ | | H | 42 |
| 2 | H | | $(CH_2)_5$ | | H | 38 |
| 3 | H | H | | H | H | 37 |
| 3 | H | H | | Me | H | 52 |
| 3 | H | H | | Et | H | 33 |
| 3 | H | | $(CH_2)_3$ | | H | 28 |

O, $(CH_2)_n$, N, $R^1$, $R^2$, $R^3$, $R^4$, O $\xrightarrow{h\nu}$ O, $R^3$, $R^4$, $R^1$, $R^2$, $(CH_2)_n$, N, H, O

The photochemistry of the phthalimide system (**269**) was first described by Kanaoka's group [164]. *N*-Alkylphthalimides give a series of products from initial hydrogen abstraction. The results are outlined below. The product distribution is quite solvent-dependent: only the benzazepinedione (**272**) was observed in alcohol solvents, whereas the alkene (**271**) and phthalimide (**270**) were observed in acetonitrile and acetone [159, 164].

O, N, O (**269**) $\xrightarrow{h\nu}$ O, N, HO $\longrightarrow$ O, N, HO

β cleavage → O, NH, O (**270**)

H transfer → O, N, HO, H (**271**)

↓ O, NH, O (**272**)

The photocyclization of phthalimides with nitrogen as the *N*-alkyl substituents has been studied by several groups [159, 160]. Kanaoka's group investigated the cyclization of a series of compounds (**273**) with *n* varying

O, N–$(CH_2)_n$, :NMe, Ph, O (**273**) $\xrightarrow{h\nu}$ O, N–$(CH_2)_n$, $^{\ominus}$O, ·NMe, $^{\oplus}$Ph (**274**) $\longrightarrow$ O, N–$(CH_2)_n$, HO, NPh (**275**)

| *n* | Yield (%) |
|---|---|
| 5 | 15 |
| 6 | 10 |
| 10 | 8 |
| 12 | 9 |

from 1 to 12, resulting in the formation of new ring systems containing between 5 and 16 members [165]. These authors suggest that the reaction proceeds through the intermediacy of the radical anion and radical cation (**274**), generated by intramolecular electron transfer; proton transfer then gives a biradical, which subsequently closes.

In the course of a study of the photochemistry of amino acids, Kanaoka *et al.* [166] discovered the photochemical formation of the ring-closed product (**277**) from the methionine derivative (**276**, R = $CH_2CH_2SCH_3$).

(**277**) (**276**) (**278**)

This reaction was subsequently extended to a series of phthalimides (**279**), which give pairs of compounds (**280**) and (**281**), with (**280**) always the major product [167,168].

(**279**) (**280**) (**281**)

| $n$ | **280** (%) | **281** (%) |
|---|---|---|
| 5 | 78 | 6 |
| 6 | 58 | 10 |
| 8 | 45 | 3 |
| 9 | 29 | 0 |
| 10 | 26 | 4 |
| 12 | 25 | 4 |

This reaction is remarkably efficient, giving rings of up to 16 members in this parent series. In addition to the methylthio substituent, other types of phthalimides with *N*-alkyl substituents containing amide, ester and ether groups have also given macrocyclic compounds upon irradiation [160].

Some interesting examples of the products from these reactions are (**282**),

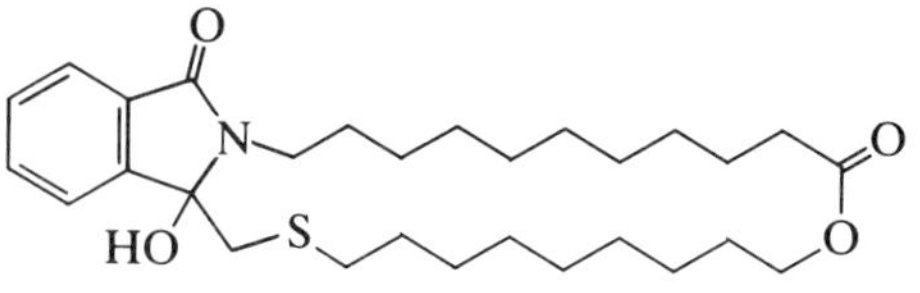

ring size = 27 (48%)
ring size = 25 (10%)

(**282**)

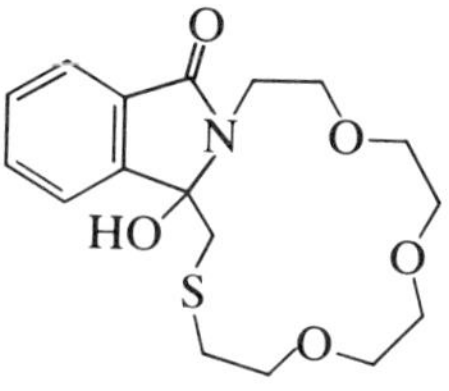

ring size = 15 (35%)
ring size = 13 (24%)

(**283**)

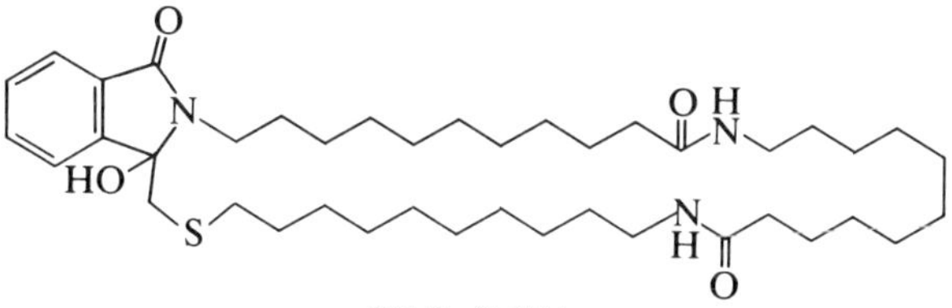

(**284**) (36%)

(**283**) and (**284**); in some cases these are the major products from closure at the methylene group adjacent to sulphur [168].

This reaction could proceed via the formation of an electron-transfer intermediate (**285**), which gives the biradical (**286**) after proton transfer. In fact, a weak charge-transfer band has been observed between methylbutyl sulphide and *N*-methylphthalimide [168].

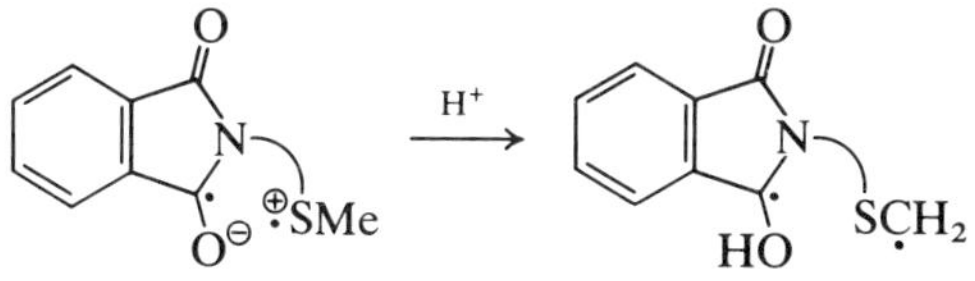

(**285**) $\xrightarrow{H^+}$ (**286**)

In the course of a study of the photochemical reactions between imides and olefinic compounds, Maruyama *et al.* [169, 170] have found a new photocyclization involving incorporation of the solvent used. Irradiation of the *N*-alkylphthalimides (**287**) in methanol gave the macrocyclic compounds (**288**) in good yields. Similar photocyclization of imides with $\beta$-styryl and 2-methyl-1-propenyl moieties proceeded smoothly to give the same type of macrocyclic products with incorporation of methanol [171].

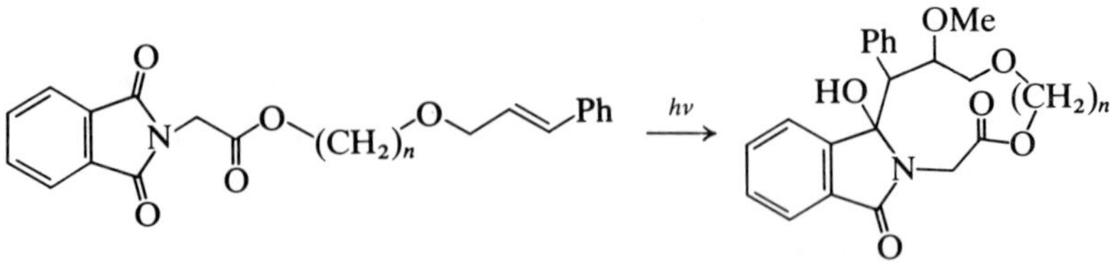

(**287abcd**)

(**288abcd**)
(**a**) $n = 2$ (68%)
(**b**) $n = 4$ (55%)
(**c**) $n = 5$ (65%)
(**d**) $n = 6$ (35%)

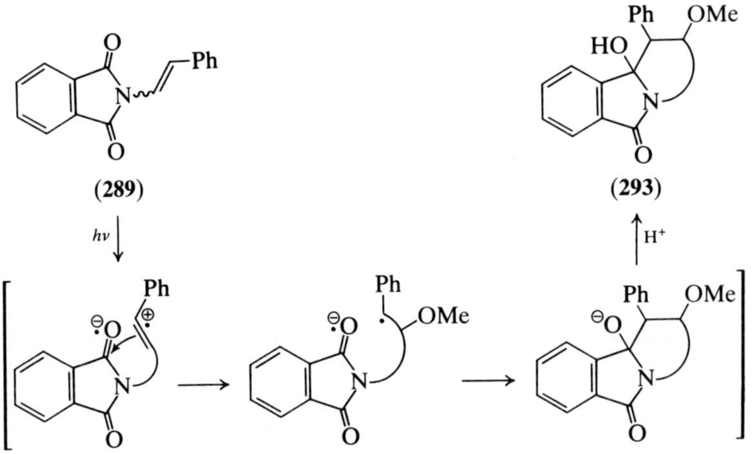

(**289**) (**293**) (**290**) (**291**) (**292**)

Maruyama and Kubo [171] explain the surprising efficiency of this reaction by the formation of an electron-transfer intermediate, (**289**) → (**290**) followed by anti-Markownikov addition of methanol, (**290**) → (**291**), radical cyclization, (**291**) → (**292**), and protonation, (**292**) → (**293**).

**Experimental 7.14** Photocyclization of imide [165]

A solution of (**273**, $n = 5$) (966 mg, 3 mmol) in acetone–petroleum ether (2.7 : 1, 1 l), in a water-cooled quartz immersion well was irradiated with a 500 W high-pressure mercury lamp (Eikosha, PIH-500) for 1.5 h under a nitrogen atmosphere. During irradiation, stirring of the reaction mixture was effected by the introduction of a stream of nitrogen at the bottom of the outer jacket. After removal of the solvent by evaporation, the residue was chromatographed over silica gel (Merck, silica gel 60, 70–230 mesh) with dichloromethane–acetone (10 : 1) as eluent to give the product (**275**, $n = 5$). The product was further purified by recrystallization from ethanol to give the pure product (145 mg, 15%), MP 192–194 °C.

### 7.5.2.2 From chloroacetamides

From extensive work on the photochemistry of aromatic compounds, including indoles, containing a chloroacetamide group, Yonemitsu *et al.* [172, 173] have found a new photochemical synthesis of many-membered nitrogen heterocycles. Irradiation of the chloroacetamide (**294**) with an *m*-phenol structure in a protic solvent gave the benzazepinone (**295**) [174]. The corresponding *p*-isomer (**296**) gave two products (**297**) and (**300**), of which the latter is a dimeric compound formed via two intermediates (**298**) and (**299**) [175].

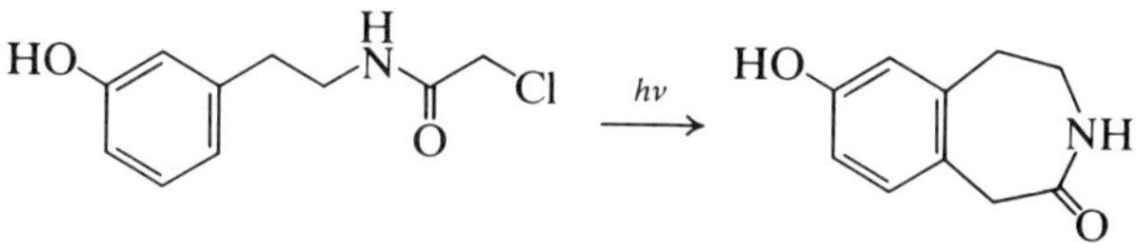

(**294**) (**295**) (70%)

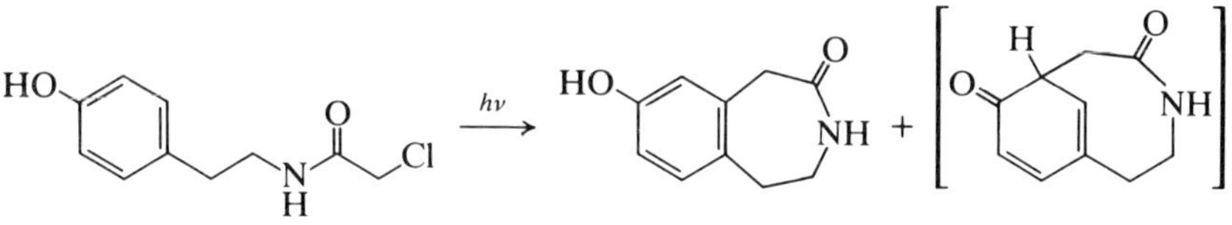

(**296**) (**297**) (5%) (**298**)

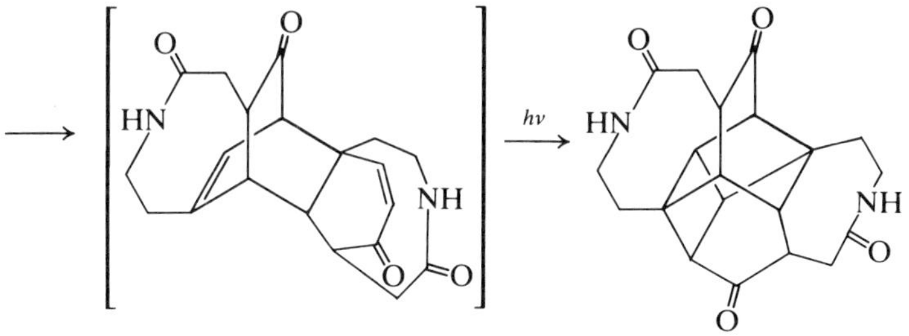

(**299**) (**300**) (40%)

Photocyclizations of the *p*- and *m*-methoxy substituted chloroacetamides (**301**) and (**303**) gave azaazulene (**302**) [176] and a mixture of (**304**), (**305**) and (**306**) [177, 178] respectively, via the routes shown.

When the reaction was carried out in a protic solvent the yields of (**304**) and (**305**) increased and those of (**306**) and (**308**) decreased. In a protic solvent facile electron transfer from the intramolecular exciplex followed by elimination of chloride ion gave a species that cyclized to form the products (**304**) and (**305**). On the other hand, in an aprotic solvent the biradical (**307**)

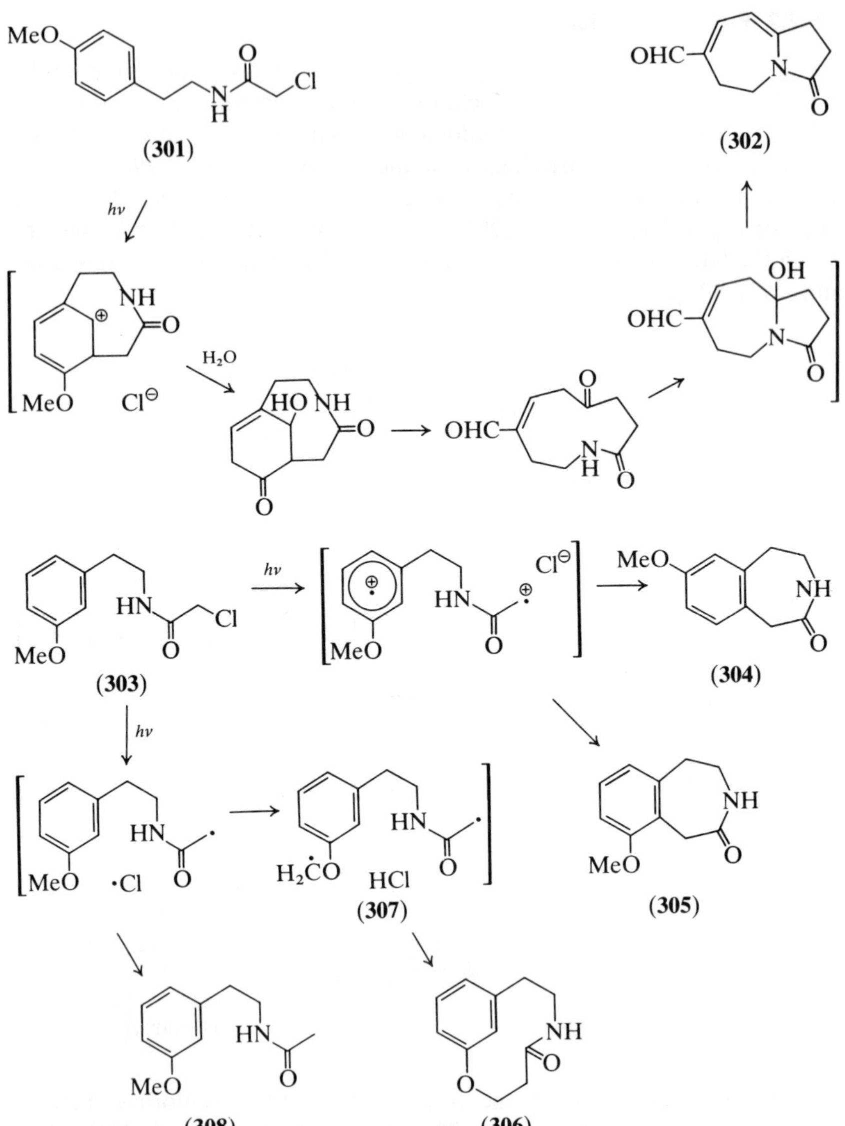

MeO O N H Cl (**301**)

OHC N O (**302**)

$h\nu$

NH ⊕ =O MeO $Cl^{\ominus}$

$H_2O$

HO NH =O O → OHC O N H O

OH OHC N O

HN Cl MeO O (**303**) $h\nu$ → ⊕ HN ⊕ $Cl^{\ominus}$ MeO O → MeO NH O (**304**)

$h\nu$

HN MeO ·Cl O → HN $H_2\dot{C}O$ HCl O (**307**)

NH MeO O (**305**)

HN MeO O (**308**)

NH O O (**306**)

was formed preferentially, and its recombination then gave the products (**306**) and (**308**).

Bryce-Smith *et al.* [179] reported the photocyclization of the phenyl-alkylamines (**309**) in methanol, leading to the formation of the bridged many-membered heterocycles (**310**).

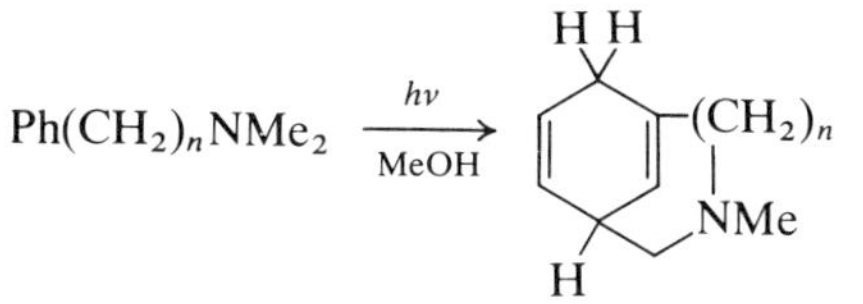

$Ph(CH_2)_nNMe_2 \xrightarrow[\text{MeOH}]{h\nu}$ (310)

(309ab)

(310)
(a) $n = 3$
(b) $n = 4$

### 7.5.2.3 From acylanilides and related compounds

As an extension of the photochemical $\alpha$-cleavage of carbonyl compounds, followed by a recombination reaction, photolysis of *N*-phenyllactams has provided a synthetic route to macrocyclic nitrogen heterocycles. Irradiation of the *N*-phenyllactams (**311**) led to cleavage of the N—CO bond to give the biradicals (**312**), which recombined at the *ortho* position to yield the cyclized intermediates (**313**); these then isomerized to the macrocyclic aminoketones (**314**) [180, 181]. Other examples of this route are shown below [182, 183].

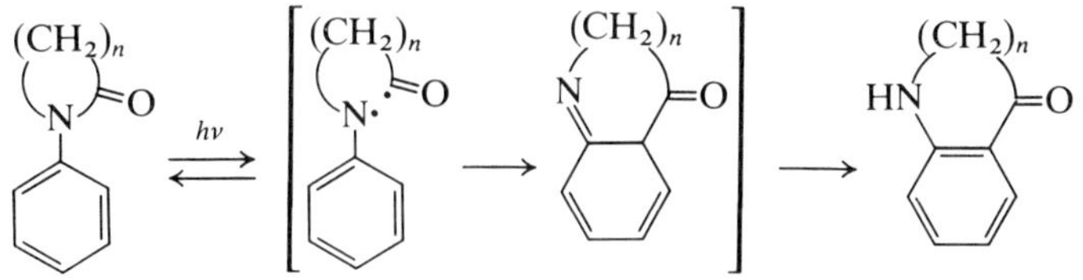

(311) (312) (313) (314abc)
(a) $n = 5$ (60%)
(b) $n = 6$ (83%)
(c) $n = 11$ (80%)

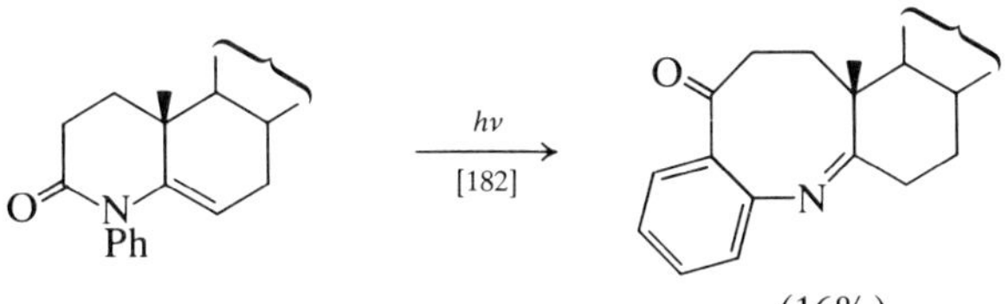

$h\nu$ [182]

(16%)

$h\nu$ [183]

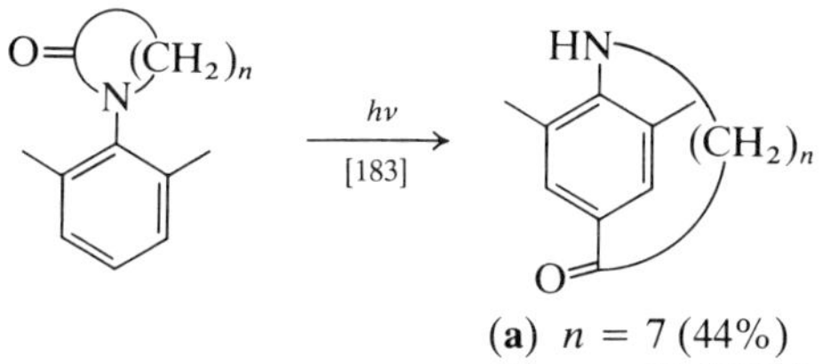

(a) $n = 7$ (44%)
(b) $n = 11$ (43%)

Photosensitized oxidation has also been used for the synthesis of medium and many-membered heterocyclic compounds [184, 185].

(95%) [184]

(14%) [185]

## REFERENCES


1. W. Lwawski and T. W. Mattingly, *J. Am. Chem. Soc.* **87**, 1947 (1965).
2. K. Hafner and C. Konig, *Angew. Chem. Int. Ed. Engl.* **2**, 96 (1963).
3. H. van der Plas, *Ring Transformations of Heterocycles*, Vol. 1, p. 277. Academic Press, New York, 1973.
4. J. K. Crandall and W. W. Conover, *J. Org. Chem.* **39**, 63, (1974).
5. H. Quast and L. Bieber, *Angew. Chem. Int. Ed. Engl.* **14**, 428 (1975).
6. W. Adam, J. Liu and O. Rodrigues, *J. Org. Chem.* **38**, 2269 (1973).
7. R. C. White, *Tetrahedron Lett.* **21**, 1021 (1980).
8. A. Krantz and J. Laureni, *J. Am. Chem. Soc.* **103**, 486 (1981).
9. J. S. Splitter and M. Calvin, *J. Org. Chem.* **23**, 651 (1958).
10. *US Patents*, 2 722 520 (1955); 3 448 121 (1969); 3 536 808 (1970).
11. D. R. Julian, in *Photochemistry of Heterocyclic Compounds* (ed. O. Buchardt), p. 602. Wiley, New York, 1976.
12. G. Jones, in *Organic Photochemistry* (ed. A. Padwa), Vol. 5, p. 1. Marcel Dekker, New York, 1981.
13. N. C. Yang and W. Eisenherdt, *J. Am. Chem. Soc.* **93**, 1277 (1971).

14. H. A. J. Carless and A. K. Mitra, *Tetrahedron Lett.* **1977**, 1411.
15. S. H. Schroeter and C. M. Orlando, *J. Org. Chem.* **34**, 1181 (1969).
16. S. Toki, K. Shima and H. Sakurai, *Bull. Chem. Soc. Jpn* **38**, 1806 (1966).
17. K. Shima and H. Sakurai, *Bull. Chem. Soc., Jpn* **39**, 1806 (1966).
18. G. Jones, H. M. Gilow and J. Low, *J. Org. Chem.* **44**, 2949 (1979).
19. T. Matsuura, A. Banda and K. Ogura, *Tetrahedron* **27**, 1211 (1971).
20. J. J. Beereboom and M. S. von Wittenau, *J. Org. Chem.* **30**, 1231 (1965).
21. R. Srinivasan, *J. Am. Chem. Soc.* **82**, 775 (1960).
22. R. Bishop and N. K. Hamer, *Chem. Commun.* **1969**, 804.
23. G. Buchi, C. G. Inman and E. S. Lipinsky, *J. Am. Chem. Soc.* **76**, 4327 (1954).
24. D. R. Arnold, in *Advances in Photochemistry* (ed. A. Noyes, S. G. Hammond and J. N. Pitts), Vol. 6. Interscience, New York, 1968.
25. G. Jones, S. B. Schwartz and M. T. Marton, *J. Chem. Soc. Chem. Commun.* **1973**, 374.
26. G. Adams, C. Bibby and R. Grigg, *J. Chem. Soc. Chem. Commun.* **1972**, 491.
27. H. A. J. Carless, *J. Chem. Soc. Chem. Commun.* **1974**, 982.
28. R. Rossi, *Synthesis* **1978**, 413.
29. C. A. Henrick, *Tetrahedron* **33**, 1845 (1977).
30. G. Jones, M. A. Acquandre and M. A. Carmody, *J. Chem. Soc. Chem. Commun.* **1975**, 206.
31. See Ref. [12], p. 77.
32. S. H. Schroeter, *J. Org. Chem.* **34**, 1188 (1969).
33. A. Zamojski and T. Kozluk, *J. Org. Chem.* **42**, 1089 (1984).
34. S. L. Schreber and K. Satake, *J. Am. Chem. Soc.* **106**, 4186 (1984).
35. D. R. Morton and R. A. Morge, *J. Org. Chem.* **43**, 2093 (1978).
36. J. Berger, M. Yoshioka, M. P. Zink, H. R. Wolf and O. Jengen, *Helv. Chim. Acta* **63**, 154 (1980).
37. P. Jost, P. Chaquin and J. Kossanyi, *Tetrahedron Lett.* **1980**, 465.
38. D. Bichan and M. Winnik, *Tetrahedron Lett.* **1974**, 3857.
39. M. L. J. Mihailovic, L. J. Lorenc and V. Pavlovik, *Tetrahedron Lett.* **33**, 441 (1977).
40. W. Adam and L. Szendrey, *Chem. Commun.* **1971**, 1299.
41. J. C. Arnould, A. Enger, A. Feigenbaum and J. P. Pete, *Tetrahedron* **35**, 2501 (1979).
42. S. G. Cohen, A. Parola and G. H. Parsons, *Chem. Rev.* **73**, 141 (1973).
43. E. H. Gold, *J. Am. Chem. Soc.* **93**, 2793 (1971).
44. H. Aoyama, S. Suzuki, T. Hasegawa and Y. Omote, *Chem. Commun.* **1979**, 899.
45. H. Aoyama, H. Hasegawa and Y. Omote, *J. Am. Chem. Soc.* **101**, 5343 (1979).
46. W. Kirmse and L. Horner, *Chem. Ber.* **89**, 2759 (1956).
47. H. Aoyama, M. Sakamota, K. Yoshida and T. Omote, *J. Heterocycl. Chem.* **20**, 1099 (1983).
48. T. Kato and Y. Nakamura, *Heterocycles* **16**, 135 (1981).
49. L. A. Paquette and G. Slomp, *J. Am. Chem. Soc.* **85**, 765 (1963).
50. E. J. Corey and J. Streith, *J. Am. Chem. Soc.* **86**, 950 (1964).
51. J. Brennan, *J. Chem. Soc. Chem. Commun.* **1981**, 880.
52. T. Kametani, T. Mochizuki and T. Honda, *Heterocycles* **18**, 89 (1982).
53. C. Kaneko, M. Sato and N. Katagiri, *J. Synth. Org. Chem. Jpn* **44**, 1058 (1986).
54. C. Kaneko, K. Shiba, H. Fujii and Y. Momose, *J. Chem. Soc. Chem. Commun.* **1980**, 1177.
55. H. Fujii, K. Schiba and C. Kaneko, *J. Chem. Soc. Chem. Commun.* **1980**, 537.
56. C. Kaneko, T. Naito and A. Saito, *Tetrahedron Lett.* **25**, 1591 (1984).
57. C. Kaneko, N. Katagiri, M. Sato, M. Moto, T. Sakamoto, S. Saikawa, T. Naito and A. Saito, *J. Chem. Soc. Perkin Trans. 1* **1986**, 1283.

58. N. Katagiri, M. Sato, T. Naito and C. Kaneko, *Tetrahedron Lett.* **25**, 5665 (1984).
59. M. Sato, N. Yoneda and C. Kaneko, *Chem. Pharm. Bull.* **34**, 621 (1986).
60. N. Katagiri, T. Haneda and C. Kaneko, *Nucleic Acids Res. Symp. Ser.* No. 16, 113 (1985).
61. A. Ohno, Y. Ohnishi and G. Tsuchihashi, *Tetrahedron Lett.* **1969**, 161.
62. A. Ohno, Y. Ohnishi and G. Tsuchihashi, *J. Am. Chem. Soc.* **91**, 5038 (1969).
63. K. Oda, M. Machida and Y. Kanaoka, *Synthesis* **1986**, 768.
64. M. Machida, K. Oda, E. Yoshida, S. Wakao, K. Ohno and Y. Kanaoka, *Heterocycles* **23**, 1615 (1985).
65. M. Machida, K. Oda, E. Yoshida and Y. Kanaoka, *Tetrahedron* **42**, 4619 (1986).
66. M. Machida, K. Oda and Y. Kanaoka, *Chem. Pharm. Bull.* **33**, 3552 (1985).
67. K. Oda, M. Machida, K. Aoe, Y. Nishikata, Y. Sato and Y. Kanaoka, *Chem. Pharm. Bull.* **34**, 1411 (1986).
68. T. Wolff and R.W. Schmidt, *J. Am. Chem. Soc.* **102**, 6098 (1980).
69. K. H. Grellmann, G.M. Sherman and H. Linschitz, *J. Am. Chem. Soc.* **85**, 1881 (1963).
70. H. Linschitz and K. H. Grellmann, *J. Am. Chem. Soc.* **86**, 303 (1964).
71. A. G. Schultz and C. K. Sha, *Tetrahedron* **36**, 1757 (1980).
72. O. L. Chapman, G. L. Eian, A. Bloom and J. Clardy, *J. Am. Chem. Soc.* **93**, 2918 (1971).
73. O. L. Chapman and G. L. Eian, *J. Am. Chem. Soc.* **90**, 5329 (1968).
74. J. F. Nicoud and H. B. Kagan, *Isr. J. Chem.* **15**, 78, (1976/1977).
75. A. G. Schultz and I.-C. Chiu, *J. Chem. Soc. Chem. Commun.* **1978**, 29.
76. H. Iida, Y. Yuasa and C. Kibayashi, *J. Org. Chem.* **44**, 1236 (1970).
77. J. P. Ferris and F. R. Antonucci, *J. Am. Chem. Soc.* **96**, 2010 (1974).
78. R. R. Bard and J. F. Bunnett, *J. Org. Chem.* **45**, 1546 (1980).
79. R. Beugelmans, B. Boudet and L. Quinterop, *Tetrahedron Lett.* **21**, 1943 (1980).
80. J. F. Wolfe, M. C. Sleevi and R. R. Goehring, *J. Am. Chem. Soc.* **102**, 3646 (1980).
81. A. Padwa, M. Dharan, J. Smolanoff and S. I. Wetmore, *J. Am. Chem. Soc.* **95**, 1945 (1973).
82. N. Gakis, H. Heimgartner and H. Schmid, *Helv. Chim. Acta* **57**, 1403 (1974).
83. A. Padwa, P. H. Carlson and A. Ku, *J. Am. Chem. Soc.* **100**, 3494 (1978).
84. A. Padwa, J. Smolnoff and A. Tremper, *J. Am. Chem. Soc.* **97**, 4682 (1975).
85. F. Bellamy, P. Martz and J. Streith, *Heterocycles* **3**, 217 (1975).
86. G. G. Spence, E. C. Taylor and O. Buchardt, *Chem. Rev.* **70**, 231 (1979).
87. F. Bellamy and J. Streith, *Heterocycles* **4**, 1391 (1976).
88. C. Kaneko and R. Kitamura, *Heterocycles*, **6**, 111 (1977).
89. C. Kaneko and R. Kitamura, *Heterocycles* **6**, 117 (1977).
90. C. Kaneko, H. Fujii, S. Kowai, A. Yamamoto, K. Hashida, T. Kimata, R. Hayashi and H. Somei, *Chem. Pharm. Bull.* **28**, 1157 (1980).
91. J. C. Swenton, T. J. Ikele and B. H. Williams, *J. Am. Chem. Soc.* **92**, 3103 (1970).
92. J. Bratt, B. Iddon, A. G. Mack, H. Suschitzky, J. A. Taylor and B. Wakefield, *J. Chem. Soc. Perkin Trans. 1,* **1980**, 648.
93. W. A. Henderson and A. Zweig, *Tetrahedron Lett.* **1969**, 625.
94. J. A. Elix and D. P. Murphy, *Aust. J. Chem.* **28**, 1559 (1975).
95. A. G. Schultz, R. D. Lucci, W. Y. Fu, M. H. Berger, J. Erhardt and W. K. Hagmann, *J. Am. Chem. Soc.* **100**, 2150 (1978).
96. A. G. Schultz and M. B. Detar, *J. Am. Chem. Soc.* **98**, 3564 (1976).
97. A. Cox, F. R. Kemp, R. Lapouyade, P. de Mayo, J. Joussat-du-Bien and R. Bonneau, *Can. J. Chem.* **53**, 2386 (1975).
98. F. B. Mallory and C. W. Mallory, in *Organic Reactions* (ed. W. G. Dauben), Vol. 30, p. 1. Wiley, New York, 1984.
99. H. A. Carles and D. J. Haywood, *J. Chem. Soc. Chem. Commun.* **1980**, 657.
100. G. R. Lappin and J. S. Zanucci, *Chem. Commun.* **1969**, 1113.
101. F. B. Mallory and C. S. Wood, *Tetrahedron Lett.* **1965**, 2643.

102. P. Hugelschofer, J. Kalvoda and K. Schaffner, *Helv. Chim. Acta* **43**, 1322 (1960).
103. G. M. Badger, C. P. Joshua and G. E. Lewis, *Tetrahedron Lett.* **1964**, 3711.
104. H.-H. Perkampus and B. Behjati, *J. Heterocycl. Chem.* **11**, 511 (1974).
105. T. Onaka, Y. Kanda and M. Natsume, *Tetrahedron Lett.* **1974**, 1179.
106. S. Prabhakar, A. M. Labo and M. R. Tavares, *J. Chem. Soc. Chem. Commun.* **1978**, 884.
107. R. E. Doolittle and C. K. Bradsher, *J. Org. Chem.* **31**, 2616 (1966).
108. J. W. McFarland and H. L. Howes, *J. Med. Chem.* **12**, 1079 (1969).
109. P. S. Mariano, E. Krochmal and A. Leone, *J. Org. Chem.* **42**, 1122 (1977).
110. J. Dejong and J. H. Boyer, *Chem. Commun.* **1971**, 961.
111. I. Ninomiya, *Heterocycles* **2**, 105 (1974).
112. G. R. Lenz, *Synthesis* **1978**, 489.
113. I. Ninomiya and T. Naito, *Heterocycles* **15**, 1433 (1981).
114. I. Ninomiya and T. Naito, in *The Alkaloids* (ed. A. Brossi), Vol. XXII, p. 189. Academic Press, New York, 1983.
115. I. Ninomiya and T. Naito, *J. Synth. Org. Chem. Jpn* **42**, 225 (1984).
116. A. L. Campbell and G. R. Lenz, *Synthesis* **1987**, 421.
117. I. Ninomiya, S. Yamauchi, T. Kiguchi, A. Shinohara and T. Naito, *J. Chem. Soc. Perkin Trans. 1*, **1974**, 1747,
118. P. G. Cleveland and O. L. Chapman, *J. Chem. Soc. Chem. Commun.* **1967**, 1064.
119. I. Ninomiya, T. Naito and T. Kiguchi, *J. Chem. Soc. Perkin Trans. 1*, **1973**, 2257.
120. I. Ninomiya, T. Naito, T. Kiguchi and T. Mori, *J. Chem. Soc. Perkin Trans. 1*, **1973**, 1696.
121. I. Ninomiya, T. Kiguchi, O. Yamamoto and T. Naito, *J. Chem. Soc. Perkin Trans. 1*, **1979**, 1723.
122. I. Ninomiya, T. Kiguchi, S. Yamauchi and T. Naito, *J. Chem. Soc. Perkin Trans. 1*, **1980**, 197.
123. B. S. Tyagarajan, N. Kharasch, H. B. Lewis and W. Wolf, *Chem. Commun.* **1967**, 614.
124. Y. Kanaoka and K. Itoh, *J. Chem. Soc. Chem. Commun.* **1973**, 647.
125. S. V. Kessar, G. Singh and P. Balakrishnan, *Tetrahedron Lett.* **1974**, 2269.
126. T. Naito, Y. Tada, Y. Nishiguchi and I. Ninomiya, *J. Chem. Soc. Perkin Trans. 1*, **1985**, 487.
127. I. Ninomiya, C. Hashimoto, T. Kiguchi and T. Naito, *J. Chem. Soc. Perkin Trans. 1*, **1985**, 941.
128. T. Kiguchi, C. Hashimoto and I. Ninomiya, *Heterocycles* **23**, 1925 (1985).
129. T. Kiguchi, C. Hashimoto and I. Ninomiya, *Heterocycles* **23**, 1377 (1985).
130. I. Ninomiya, C. Hashimoto and T. Kiguchi, *Heterocycles* **22**, 1035 (1984).
131. T. Kiguchu, C. Hashimoto and I. Ninomiya, *Heterocycles* **22**, 43 (1984).
132. T. Naito, Y. Hirata, O. Miyata and I. Ninomiya, *J. Chem. Soc. Perkin. Trans 1*, **1988**, 2219.
133. O. Miyata, Y. Hirata, T. Naito and I. Ninomiya, *Heterocycles* **22**, 2719 (1984).
134. O. Miyata, Y. Hirata, T. Naito and I. Ninomiya, *Heterocycles* **22**, 1041 (1984).
135. T. Naito, N. Kojima, O. Miyata and I. Ninomiya, *J. Chem Soc. Perkin Trans. 1*, **1985**, 1611.
136. T. Naito, N. Kojima, O. Miyata and I. Ninomiya, *Heterocycles* **24**, 2117 (1986).
137. T. Naito, O. Miyata and I. Ninomiya, *Heterocycles* **26**, 1739 (1987).
138. T. Naito, N. Kojima, O. Miyata and I. Ninomiya, *Chem. Pharm. Bull.* **34**, 3530 (1986).
139. T. Naito, Y. Tada and I. Ninomiya, *Heterocycles* **16**, 1141 (1981).
140. T. Naito, K. Katsumi, Y. Tada and I. Ninomiya, *Heterocycles* **20**, 799 (1983).
141. T. Kametani, N. Takagi, M. Toyota, T. Honda and K. Fukumoto, *J. Chem. Soc. Perkin Trans. 1*, **1981**, 2830.
142. S. A. Glover and A. Goosen, *J. Chem. Soc. Perkin Trans 1*, **1977**, 1348.
143. P. de Mayo, L. K. Sydnes and G. Wenska, *J. Chem. Soc. Chem. Commun.* **1979**, 499.

144. C. M. Foltz and T. Kondo, *Tetrahedron Lett.* **1970**, 3163.
145. M. Lancaster and D. J. H. Smith. *J. Chem. Soc. Chem. Commun.* **1980**, 471.
146. T. Kametani, Y. Hirai, M. Kajiwara, T. Takahashi and K. Fukumoto, *Chem. Pharm. Bull.* **23**, 2634 (1975).
147. M. Ogata and H. Matsumoto, *Chem. Pharm. Bull.* **20**, 2264 (1972).
148. J. W. Barton and R. B. Walker, *Tetrahedron Lett.* **1975**, 569.
149. A. F. Kluge and C. P. Lillya, *J. Org. Chem.* **36**, 1988 (1971).
150. S. Kurata, T. Kusumi, Y. Inoue and H. Kakisawa, *J. Chem. Soc. Perkin Trans. 1*, **1976**, 532.
151. R. Mactinez-Utrilla and M. A. Miranda, *Tetrahedron* **37**, 2111 (1981).
152. T. Naito and C. Kaneko, *Chem. Pharm. Bull.* **28**, 3150 (1980).
153. N. V. Kirby and S. T. Reid, *J. Chem. Soc. Chem. Commun.* **1980**, 3150.
154. J. H. M. Hill and S. T. Reid, *J. Chem. Soc. Chem. Commun.* **1983**, 501.
155. M. Ikeda, K. Ohno, T. Uno and Y. Tamura, *Tetrahedron Lett.* **1980**, 3403.
156. S. T. Reid and D. DeSilva, *Tetrahedron Lett.* **24**, 1949 (1983).
157. T. Sato, Y. Horiguchi and Y. Tsuda, *Heterocycles* **16**, 355 (1981).
158. T. Sano, Y. Horiguchi, S. Kanbe and Y. Tsuda, *Heterocycles* **16**, 363 (1981).
159. Y. Kanaoka, *Acc. Chem. Res.* **11**, 407 (1978).
160. P. H. Mazzochi, in *Organic Photochemistry* (ed. A. Padwa), Vol. 5, p. 421. Marcel Dekker, New York, 1981.
161. T. Kanaoka, K. Yoshida and Y. Hatanaka, *J. Org. Chem.* **44**, 664 (1979).
162. K. Maruyama, T. Ogawa and Y. Kubo, *Chem. Lett.* **1978**, 1107.
163. Y. Kanaoka, and Y. Hatanaka, *J. Org. Chem.* **41**, 400 (1976).
164. Y. Kanaoka, Y. Migita, K. Koyama, Y. Sato, H. Nakai and T. Mizoguchi, *Tetrahedron Lett.* **1973**, 1193.
165. M. Machida, H. Kakechi and Y. Kanaoka, *Heterocycles* **7**, 273 (1977).
166. M. Sato, H. Nakai, T. Mizoguchi, M. Kawanishi and Y. Kanaoka, *Chem. Pharm. Bull.* **21**, 1164 (1973).
167. Y. Sato, H. Nakai, T. Mizoguchi and Y. Kanaoka, *Tetrahedron Lett.* **1976**, 1889.
168. Y. Sato, H. Nakai, T. Mizoguchi, Y. Hatanaka and Y. Kanaoka, *J. Am. Chem. Soc.* **98**, 2349 (1976).
169. K. Maruyama, Y. Kubo, M. Machida, K. Oda, Y. Kanaoka and K. Furuyama, *J. Org. Chem.* **43**, 2303 (1978).
170. K. Maruyama and Y. Kubo, *Chem. Lett.* **1978**, 851.
171. K. Maruyama and Y. Kubo, *J. Am. Chem. Soc.* **100**, 7772 (1978).
172. O. Yonemitsu, P. Cerutti and B. Witkop, *J. Am. Chem. Soc.* **88**, 3941 (1966).
173. S. Naruto and O. Yonemitsu, *Tetrahedron Lett.* **1975**, 3399.
174. S. Naruto, O. Yonemitsu, N. Kataoka and K. Kimura, *J. Am. Chem. Soc.* **93**, 4053 (1971).
175. T. Iwakuma, H. Nakai, O. Yonemitsu and B. Witkop, *J. Am. Chem. Soc.* **96**, 2564 (1974).
176. O. Yonemitsu, T. Tokuyama, M. Chaykovsky and B. Witkop, *J. Am. Chem. Soc.* **90**, 776 (1968).
177. Y. Okuno and O. Yonemitsu, *Chem. Pharm. Bull.* **23**, 1039 (1975).
178. Y. Okuno and O. Yonemitsu, *Tetrahedron Lett.* **1974**, 1169.
179. D. Bryce-Smith, A. Gilbert and G. Klunkin, *J. Chem. Soc. Chem. Commun.* **1973**, 330.
180. M. Fischer, *Tetrahedron Lett.* **1968**, 4295.
181. M. Fischer, *Chem. Ber.* **102**, 342 (1969).
182. R. P. Gandhi, M. Singh, Y. P. Sachdeva and S. M. Mukherji, *Tetrahedron Lett.* **1973**, 661.
183. M. Fischer, *Tetrahedron Lett.* **1969**, 2281.
184. I. Saito, M. Imuta and T. Matsuura, *Chem. Lett.* **1972**, 1197.
185. W. Adams and J.-C. Liu, *J. Am. Chem. Soc.* **94**, 1206 (1972).

# – 8 –

# Cage Compounds

As shown in the photochemical synthesis of cubane (**8**) [1, 2], photochemical reactions have proved to be very useful in the synthesis of various types of highly strained compounds and cage compounds that are difficult to prepare by normal thermal reactions [3]. As examples, the photochemical syntheses of the highly symmetrical hydrocarbons cubane (**8**) [1, 2], dodecahedrane (**11**) [4–6] and pentaprismane (**15**) [7] are described here.

## 8.1 SYNTHESIS OF CUBANE

The monoketal (**2**) of the bromocyclopentadienone dimer (**1**), prepared from cyclopentenone in four steps, was irradiated in benzene, leading to

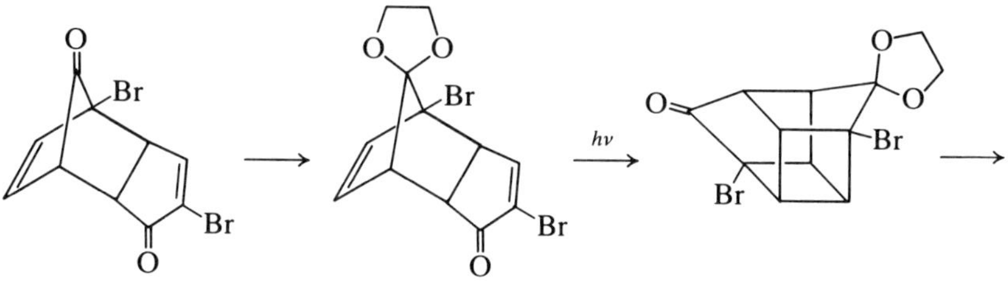

(**1**) (**2**) $h\nu$ (**3**) (95%)

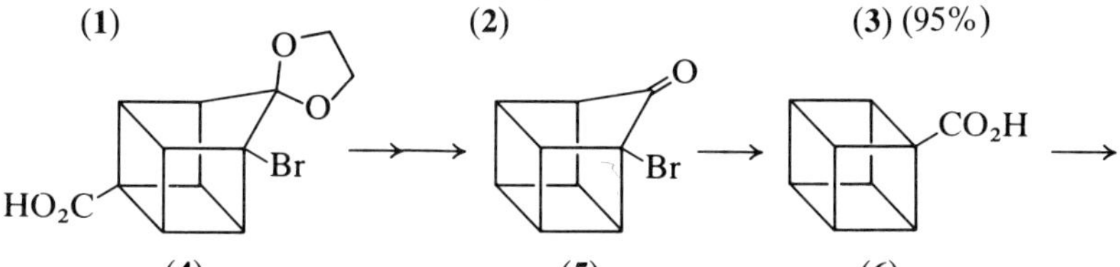

(**4**) (**5**) (**6**)

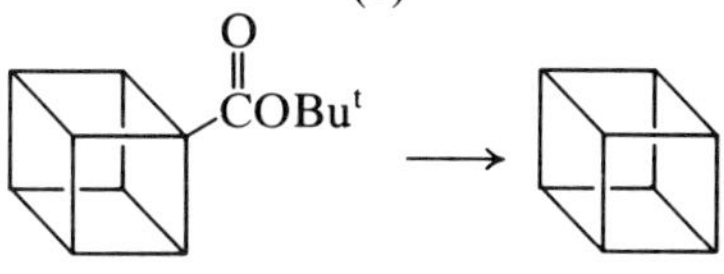

(**7**) (**8**) cubane

a [2 + 2] photocycloaddition that gave the cage compound (**3**). This was then subjected to a Favorskii rearrangement in the presence of potassium hydroxide to give the acid (**4**). After decarboxylation of (**4**) by radical fragmentation, another Favorskii rearrangement of the resulting bromoketone (**5**) gave the cubane carboxylic acid (**6**), which was converted into the desired cubane (**8**) by thermal decomposition of the corresponding t-butyl ester (**7**) [1, 2].

## 8.2 SYNTHESIS OF DODECAHEDRANE

Intramolecular photochemical $\delta$-hydrogen abstraction followed by cyclobutane formation from the resulting 1,4-biradical (Norrish type II reaction) were crucial steps in an elegant synthesis of dodecahedrane in which the same reactions were used repeatedly. Oxidative dimerization of sodium cyclopentadienide with iodine gave 9,10-dihydrofulvalene, which was subjected to a Diels–Alder reaction with dimethyl acetylene-dicarboxylate to give the diester (**9**) as a starting compound. Compound (**9**)

$CO_2Me$
$CO_2Me$
(**9**)

$PhOH_2C$ CHO
(**10**)

(1) $h\nu$
(2) Li, $NH_3$
(3) $H^+$

HO $CH_2OH$

O CHO

O

$h\nu$

(**11**)
dodecahedrane

was then converted into the aldehyde (**10**) via a 17-step reaction sequence. The desired dodecahedrane (**11**) was prepared from (**10**) via a nine-step reaction sequence in which two of the three C—C bond formations were achieved by photochemical hydrogen abstraction in a seven-membered transition state [4–6].

## 8.3 SYNTHESIS OF PENTAPRISMANE

Pentaprismane (**15**) is composed of ten identical methine units arranged at the corners of a regular pentagonal prism. Eaton *et al.* [7] have succeeded in synthesizing this compound via a route involving bridgehead functionalization of homopentaprismanone and ultimate Favorskii contraction. Homopentaprismanone (**12**) was prepared by two photochemical [2 + 2] cycloaddition reactions. Functionalization of (**12**) at the carbonyl bridgehead gave the α-tosyloxyketone (**13**), which was subjected to a Favorskii rearrangement to yield pentaprismanecarboxylic acid (**14**), which was decarboxylated via thermolysis of its t-butyl ester to give the desired pentaprismane (**15**).

MeO OMe Cl Cl Cl Cl O O $\xrightarrow[\text{[95\%]}]{h\nu}$ MeO OMe Cl Cl Cl Cl O O ⟶ MeO OMe $\xrightarrow[\text{[90\%]}]{h\nu}$

O (**12**) ⟶ O OTs (**13**) ⟶ $CO_2H$ (**14**) ⟶ (**15**) pentaprismane

## REFERENCES

1. P. E. Eaton and T. W. Cole, *J. Am. Chem. Soc.* **86**, 3157 (1964).
2. P. E. Eaton and T. W. Cole, *J. Am. Chem. Soc.* **86**, 962 (1964).
3. R. C. Cookson, E. Crundwell and J. Hudec, *Chem. Ind.* **1958**, 1003.
4. L. A. Paquette and D. W. Balogh, *J. Am. Chem. Soc.* **104**, 774 (1982).
5. L. A. Paquette, R. J. Ternansky and D. W. Balogh, *J. Am. Chem. Soc.* **104**, 4502 (1982).
6. R. J. Ternansky, D. W. Balogh and L. A. Paquette, *J. Am. Chem. Soc.* **104**, 4503 (1982).
7. P. E. Eaton, Y. D. Or and S. J. Branca, *J. Am. Chem. Soc.* **103**, 2134 (1981).

# –9–

# Spiro Compounds

Structurally interesting spirocyclic systems have been synthesized mainly by three types of photochemical reactions: (i) photorearrangement of cross-conjugated dienones; (ii) photocycloadditions; and (iii) combination of two photochemical reactions.

## 9.1 PREPARATION BY PHOTOREARRANGEMENT OF CROSS-CONJUGATED DIENONES

The photochemical transformations of most cross-conjugated dienones are quite complex, involving rearrangements via several different pathways. However, almost all cross-conjugated cyclohexadienones display analogous modes of rearrangement. In aqueous acidic media cross-conjugated cyclohexadienones of the type (**1**) are converted principally to one or more hydroxyketone photoadducts [1].

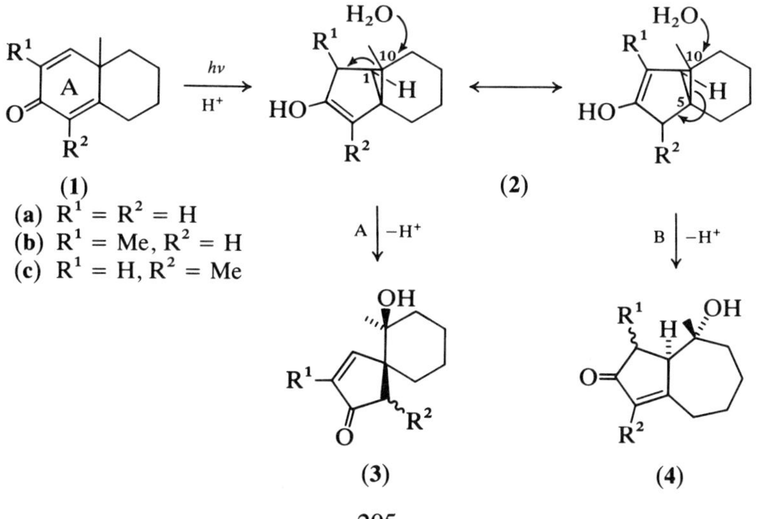

(**1**)
(**a**) $R^1 = R^2 = H$
(**b**) $R^1 = Me, R^2 = H$
(**c**) $R^1 = H, R^2 = Me$

(**2**)

(**3**) (**4**)

The formation of hydroxyketones from cross-conjugated cyclohexadienones like (**1**) apparently involves the cyclopropane intermediate (**2**), which serves as a common precursor to both the spiro (**3**) and 5/7-fused (**4**) photoproducts [2, 3]. The spiro product (**4**) is formed by nucleophilic attack of solvent at C-10 from the front side, followed by cleavage of the rear cyclopropyl bond (path A); whereas rearside attack of the solvent and cleavage of the front cyclopropyl bond gives the 5/7-fused hydroxyketone (**4**) (path B). In the absence of A-ring substituents, the two paths occur with approximately equal facility. The 4-methyl derivative (**1c**) underwent preferential path-B cleavage owing to the inductive effect exerted by the methyl group, which causes localization of the positive charge in (**2c**).

On the other hand, the 2-methyl dienone (**1b**) was found to give the spiro ketone (**3b**) in excellent yield, with the formation of no detectable amount of the alternative hydroazulenone product (**4b**) [4].

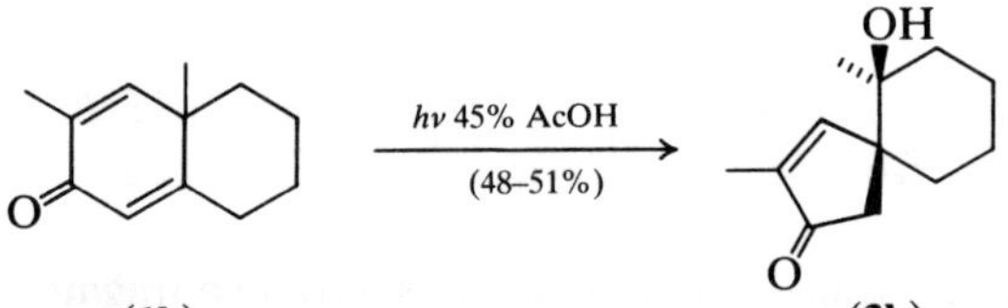

$h\nu$ 45% AcOH (48–51%)

(**1b**) (**3b**)

From the results of a number of investigations of the photorearrangement of cross-conjugated dienones, it has been found that the reaction is a powerful tool for the synthesis of spirocyclic systems [1]. Photochemical rearrangement of bicyclohexenones to spiro compounds has provided an efficient and simple synthetic route to spirocyclosesquiterpenes, as exemplified by the stereoselective total synthesis of ($\pm$)-$\alpha$-vetispirene (**6**) [5]. Photochemical rearrangement of the methoxydienone (**5**) was a key step in the construction of the spiro[4,5]decane ring system.

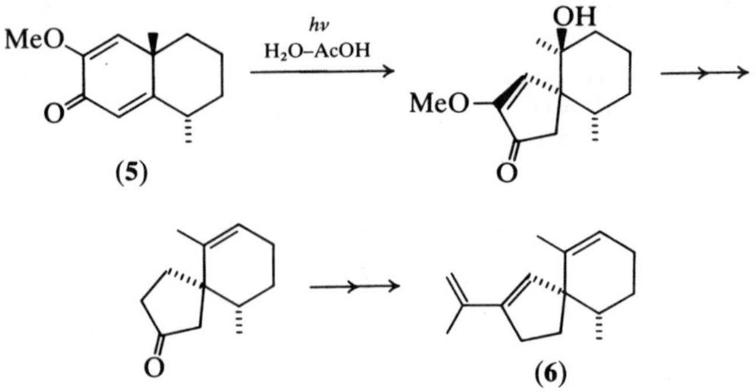

$h\nu$ $H_2O$–AcOH

(**5**) (**6**)

## 9.2 PREPARATION BY PHOTOCYCLOADDITION

As an extension of [2 + 2] photocycloaddition, [3 + 2] photocycloaddition has received attention as a new synthetic method for polycyclic systems. This reaction is also known as *meta*-photocycloaddition between alkenes and arenes [6,7].

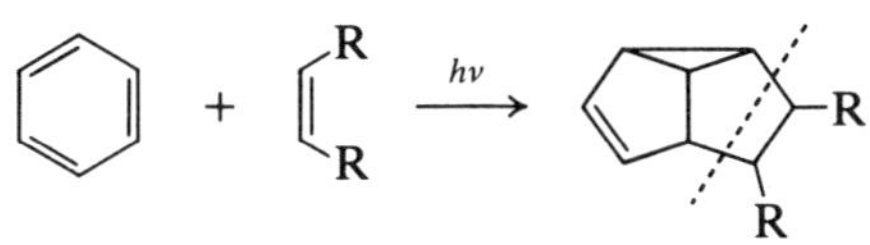

Recently, Wender and Howbert [8] have applied an intramolecular version of the reaction to the synthesis of (±)-α-cedrene (**7**), which was accomplished in just three steps, including a photochemical process.

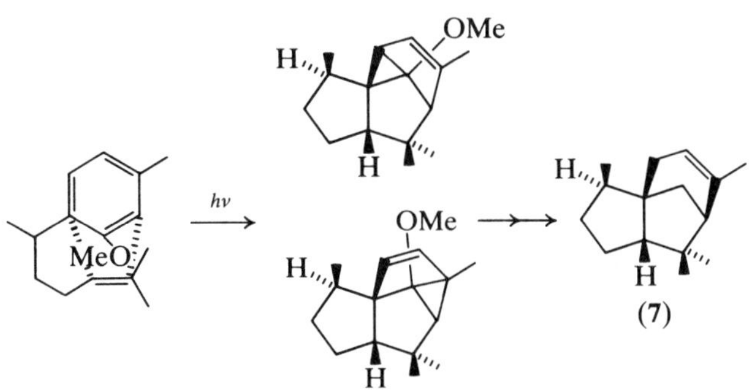

(**7**)

## 9.3 PREPARATION BY COMBINATION OF TWO PHOTOCHEMICAL REACTIONS

Fetizon *et al.* [9] have succeeded in the stereoselective synthesis of a spirosesquiterpene, (±)-α-acoradiene (**11**), by applying two photochemical reactions: [2 + 2] photocycloaddition and a subsequent Norrish type II reaction. Irradiation of the enone (**8**) led to a stereoselective intramolecular [2 + 2] cycloaddition that gave the cyclobutane (**9**). The carbonyl group of (**9**) was then excited, resulting in intramolecular hydrogen abstraction of the methyl group followed by ring opening of the cyclobutane to give the key intermediate (**10**) in 55% yield from the enone (**8**). The intermediate (**10**) had already been converted into α-acoradiene (**11**).

$h\nu$, $-50°C$; $h\nu$, $0°C$; [85%]

**(8)** **(9)**

**(10)** **(11)**

($\pm$)-$\alpha$-acoradiene

## REFERENCES


1. P. J. Kropp, in *Organic Photochemistry* (ed. O. L. Chapman), Vol. 1, p. 1. Marcel Dekker, New York, 1967.
2. P. J. Kropp and W. F. Erman, *J. Am. Chem. Soc.* **85**, 2456 (1963).
3. C. Ganter, E. C. Utzinger, K. Schaffner, D. Arigoni and O. Jeger, *Helv. Chim. Acta* **45**, 2403 (1962).
4. P. J. Kropp, *J. Am. Chem. Soc.* **86**, 4053 (1964).
5. D. Caine, A. A. Boucugnani, S. T. Chao, J. B. Dawson and P. F. Ingwalson, *J. Org. Chem.* **41**, 1539 (1976).
6. K. K. Wilzbach and L. Kaplan, *J. Am. Chem. Soc.* **88**, 2066 (1966).
7. D. Bryce-Smith, A. Gilbert and B. H. Orger, *J. Chem. Soc. Chem. Commun.* **1966**, 512.
8. P. A. Wender and J. J. Howbert, *J. Am. Chem. Soc.* **103**, 688 (1981).
9. D. D. K. Manh, J. Ecoto, M. Fetizon, H. Colin, and J.-D. Masa, *J. Chem. Soc. Chem. Commun.* **1981**, 953.

# – 10 –

# Equipment for and Techniques in Photochemical Synthesis

Most photochemical research has been carried out using two main approaches. One approach is fundamental research on the nature of photochemistry by physical chemists and physical organic chemists with the aim of elucidating the detailed mechanism of photochemical processes and the nature of reactive species formed in the process. The other approach is the application of photochemical reactions to the synthesis of organic molecules with the aim of finding a more useful, convenient, or new synthetic route to known molecules, or of establishing new syntheses of previously unattainable and often complicated molecules.

Research into reaction mechanisms obviously requires highly elaborate equipment, such as monochromatic-light sources, a filter train, a beam collimator, and instruments for measuring the intensity of the incident light and the amount of light absorbed by the molecule under study. On the other hand, synthetic organic chemists require light sources that have a high intensity particularly at the wavelengths absorbed by the reacting molecule, and vessels and instruments for easy handling of a sizeable amount of the study compound at one time. In keeping with the purpose of this series of books, this final chapter is devoted to the practical techniques, and equipment and reaction conditions required for the photochemical synthesis of compounds.

Many useful books on photochemistry by a number of noted organic photochemists have been published [1–6]. All these books contain at least one chapter which deals with techniques, apparatus and reaction conditions in detail. Therefore, the present chapter provides useful and practical information for novices in the field of photochemistry who wish to know what to do when applying photochemical means to the synthesis of a compound and how a molecule behaves when irradiated.

## 10.1 LIGHT SOURCES AND LIGHT FILTERS

Two major instrumental conditions crucial to the success of photochemical reactions are the choice of the light source and the light filter.

### 10.1.1 Light sources

The first source of irradiation energy used for a photochemical reaction in a laboratory, was sunlight (Fig. 10.1). However, the exposure times required were often long, for example more than one month under tropical summer sunshine, and it was very difficult to duplicate the reaction conditions.

In addition, sunlight has serious disadvantages as a radiation source: it has low intensity in the short wavelength region (the most important disadvantage), its angle of incidence cannot be varied but does depend on the time of day, season, latitude, etc., and, in addition, special care is required to remove its thermal effect.

However, these problems have been completely solved by modern technology by the development of a number of lamps that are commercially available at relatively low cost. The availability of many types of lamp to any laboratory in the world has stimulated photochemical research and the publication of literature on this subject.

A good light source for photochemical research must satisfy the conditions of accurate and reproducible work within a convenient timespan. From these points of view, mercury-vapour arc lamps are now the sources of choice for most photochemical research in solution. These lamps are very useful practically since they cover a range of wavelengths from the UV region

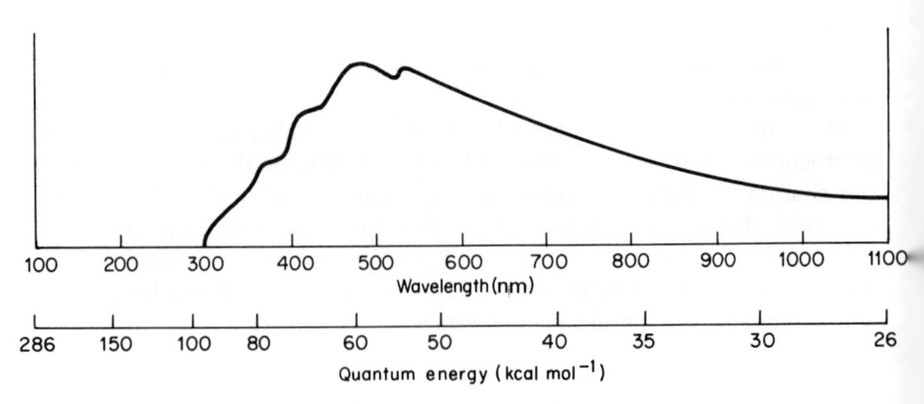


FIG. 10.1. Relative spectral-energy distribution of sunlight [1].

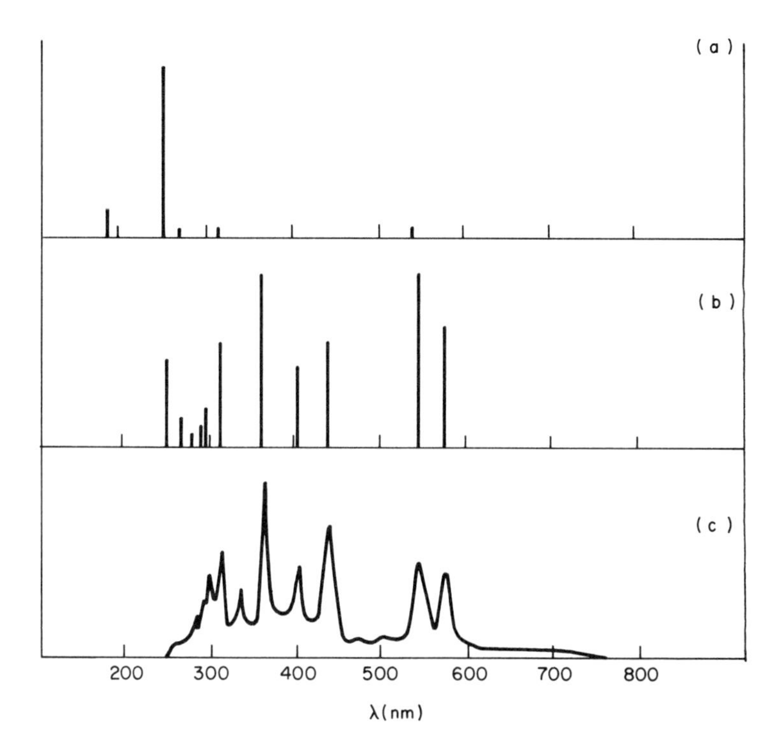


FIG. 10.2. The spectral emission of mercury arc lamps [2]. (a) Low-pressure arc; (b) medium-pressure arc; (c) super-high-pressure arc.

at 200 nm (143 kcal mol$^{-1}$, 599 kJ mol$^{-1}$) to the visible region at 750 nm (38 kcal mol$^{-1}$, 159 kJ mol$^{-1}$).

There are three basic types of "mercury arc lamps": the low-pressure or resonance lamp, the medium-pressure arc, and the high-pressure mercury arc lamps. Each of these types has different characteristics, thus making each lamp suitable for specific experimental needs (Fig. 10.2).

The low-pressure mercury arc lamp has a mercury-vapour pressure of 0.005–0.1 Torr, emits light rich in 253.7-nm radiation (see Table 10.1), operates at room temperature and, if ultrapure quartz such as Suprazil quartz is used for the lamp, it is rich in 184.9-nm radiation which is the most powerful from this type of light source.

The emission from the low-pressure arc, which is particularly intense at 253.7 and 184.9 nm, corresponds to the transitions:

$$Hg(^1P_1) \rightarrow Hg(^1S_0) + h\nu$$

TABLE 10.1

Energy distribution in low- and medium-pressure mercury arcs [3]

| Wavelength (Å) | Relative energy | |
|---|---|---|
| | Low-pressure mercury arc* | Medium-pressure mercury arc† |
| 13 673 | — | 15.3 |
| 11 287 | — | 12.6 |
| 10 140 | — | 40.6 |
| 5770–5790 | 10.14 | 76.5 |
| 5461 | 0.88 | 93.0 |
| 4358 | 1.00 | 77.5 |
| 4045–4078 | 0.39 | 42.2 |
| 3650–3663 | 0.54 | 100.0 |
| 3341 | 0.03 | 9.3 |
| 3126–3132 | 0.60 | 49.9 |
| 3022–3028 | 0.06 | 23.9 |
| 2967 | 0.20 | 16.6 |
| 2894 | 0.04 | 6.0 |
| 2804 | 0.02 | 9.3 |
| 2753 | 0.03 | 2.7 |
| 2700 | — | 4.0 |
| 2652–2655 | 0.05 | 15.3 |
| 2571 | — | 6.0 |
| 2537 | 100.00 | 16.6‡ |
| 2482 | 0.01 | 8.6 |
| 2400 | — | 7.3 |
| 2380 | — | 8.6 |
| 2360 | — | 6.0 |
| 2320 | — | 8.0 |
| 2224 | — | 14.0 |

* Hanovia Lamp Division, Engelhard Industries, Newark, NJ, SC-2537 lamp.
† Hanovia's Type A, 673 A, 550-W lamp.
‡ Reversed radiation.

Thus low-pressure arc lamps are used in the study of mercury sensitized reactions and are often useful in cases where direct photolysis at 253.7 nm is desired in mercury free systems. The emission range of this type of lamp means that it can be used without having to pay special attention to the temperature of the reaction solution.

The medium-pressure mercury-arc lamps operate at a pressure of 1–10 atmos and at relatively high temperatures. Therefore a few minutes of warm-up time are required before the lamp reaches operational stability. This type of lamp emits photochemically useful light at wavelengths below 300 nm. The emission from these lamps consists of a chain of weak but nearly equal spectral lines, with the intensities of the 253.7-nm and 184.9-nm lines diminishing as the pressure of lamps increases (Table 10.1). This type of lamp in combination with a monochromator or suitable filter system provides the best intense source of a variety of near-monochromatic frequencies which is often required in photochemical studies.

The high-pressure arc lamps, sometimes called "super-high-pressure lamps", operate at 200 atmos and emit a broad spectrum of many mercury lines (see Fig. 10.2). The high operating pressure increases the number of emission lines and broadens the principal ones, thus giving a continuum of lines but weakening the emission below 280 nm. This type of lamp is very useful in instances where light sources of high intensity in the visible region or the near-UV region are required. Because of small dimensions of the arc, the light can be easily focused by optical means. On the other hand, temperature control by a cooling system is essential for success in reactions using high-pressure are lamps.

Mercury vapour lamps, therefore, are a highly versatile means of irradiation in preparative organic synthesis. Other types of lamps, such as xenon–mercury and phosphor-coated lamps, are also useful for general irradiation procedures.

### 10.1.2 Light filters

In preparative organic photochemistry it is often necessary to use monochromatic light or a selected range of wavelengths emitted by an arc in order to prevent primary photoproducts from undergoing further secondary photoreactions, which can render a photochemical reaction almost useless. The choice of lamp for a reaction depends on two main factors. The first is the need for overlap in the spectral ranges of the lamp and the compound to be irradiated. This condition can be achieved in most cases by the use of a medium-pressure lamp which adequately covers the UV region ranging from 250 to 450 nm. However, a greater degree of selectivity in the wavelength of the irradiating light is often required, for example in the reaction of a particular compound by irradiating one of the absorption bands of the molecule, or if the product of the reaction is sensitive to a wavelength different from the one used to excite the starting compound. This selectivity can be achieved by using a monochromatic source (such as a mercury arc lamp), a diffraction grating, or less expensively, a system of filters.

#### 10.1.2.1 Solid filters

The simplest filters used for photochemical reactions are quartz, Vycor, Corex, or Pyrex (Table 10.2) reaction vessels. These are undoubtedly the most useful solid filters for preparative work in which an immersion-type photoreactor is used. The walls of the immersion well act as both cut-off and band-pass filters. Approximate values of the wavelength transmission of these solid materials are given in Table 10.2.

TABLE 10.2

The wavelengths (nm) of irradiation that give 20%, 50% and 90% transmission through a 1-mm filter of some glasses [4]

| Glass | Transmission | | |
|---|---|---|---|
| | 20% | 50% | 90% |
| Quartz | <200 | <200 | 240 |
| Vycor | 200 | 220 | 280 |
| Corex | 270 | 290 | 360 |
| Pyrex | 290 | 300 | 360 |

A quartz arc lamp placed within the reaction solution in an ordinary glass vessel, or a quartz-enclosed arc lamp placed outside a quartz reaction vessel are the most commonly used combinations of apparatus for photochemical processes.

#### 10.1.2.2 Liquid filters

A large number and wide variety of liquid-filter systems have been developed as alternatives to solid filters. The transmission data of some liquid filters are given in Table 10.3.

Liquid filters are solutions prepared from readily available chemicals and can be divided into two classes: (i) solutions which slightly improve the use of a glass filter; and (ii) combinations of solutions that provide narrow band-pass irradiation suitable for quantum-yield measurements. The first type of solution filters, as shown in Table 10.3, are short- and long-wavelength cut-off filters and serve the same purpose as glass filters. Some examples of the second type of filter are listed in Table 10.4.

Many of the solutions used in these combinations are themselves light sensitive and care must be taken to avoid their overirradiation.

TABLE 10.3

Short- and long-wavelength cut-off filter solutions*

| Wavelength of cut-off (nm) | Chemical composition |
|---|---|
| <250 | $Na_2WO_4$ |
| <305 | $SnCl_2$ in HCl (0.1 M in 2 : 3 HCl–$H_2O$) |
| <330 | 2-M $Na_3VO_4$ |
| <355 | $BiCl_3$ in HCl |
| <400 | KH phthalate + $KNO_2$ (in glycol at pH 11) |
| <460 | 0.1-M $K_2CrO_4$ (in $NH_4OH$–$NH_4Cl$ at pH 10) |
| >360 | 1-M $NiSO_4$ + 1-M $CuSO_4$ (in 5% $H_2SO_4$) |
| >450 | $CoSO_4$ + $CuSO_4$ |

* Ref. 4, p. 493.

TABLE 10.4

Some filter solution combinations*

| Wavelength (nm) | Solution 1 | Solution 2 | Solution 3 |
|---|---|---|---|
| 232–268 | 2-M $NiSO_4$† | 0.25-M $CoSO_4$† | $9.0 \times 10^{-3}$ M 2,7-dimethyl-3,6-diazacyclohepta-1,6-diene perchlorate§ |
| 260–300 | 2-M $NiSO_4$‡ | 0.8-M $CoSO_4$‡ | $1.23 \times 10^{-3}$-M $BiCl_3$‖ |
| 255–305 | 2-M $NiSO_4$‡ | 2.0-M $CoSO_4$† | $2.0 \times 10^{-4}$-M $BiCl_3$‖ |
| 250–325 | 2-M $NiSO_4$‡ | 0.8-M $CoSO_4$‡ | $2.46 \times 10^{-4}$-M $BiCl_3$‖ |
| 290–350 | 2-M $NiSO_4$‡ | 0.8-M $CoSO_4$‡ | 0.1-M $CuSO_4$‡ |
| 282–356 | 2-M $NiSO_4$‡ | 1.0-M $CoSO_4$‡ | 0.1-M $CuSO_4$‡ |
| 300–350 | 1.71-M $NiSO_4$‡ | 1.0-M $CoSO_4$† | 0.0133-M $SnCl_2$‖¶ |
| 310–355 | 2-M $NiSO_4$‡ | 0.8-M $CoSO_4$‡ | $2.2 \times 10^{-2}$-M $SnCl_2$‖¶ |
| 310–375 | 0.5-M $NiSO_4$** | 2-M $CoSO_4$† | 1.0-M $CuSO_4$‡ |
| 328–388 | 2-M $NiSO_4$† | 1.0-M $CoSO_4$† | 0.01-M $NaVO_3$†† |
| 330–440 | 0.14-M $NiSO_4$† | 1.0-M $CoSO_4$† | 0.33-M $SnCl_2$‖¶ |
| 335–450 | 1.0-M $CoSO_4$‡ | 0.023-M $CuSO_4$‡ | 0.004-M $KVO_3$‡‡ |
| 370–450 | 1.0-M $CoSO_4$† | 1.0-M $CoSO_4$† | 0.10-M $NaVO_3$‡‡ |
| 375–470 | $2.5 \times 10^{-3}$-M $FeCl_3$ | 0.20-M $CoSO_4$† | Saturated (*ca.* 1.25-M) $CuSO_4$† |

* Ref. 4, p. 493.
† In 10% $H_2SO_4$.
‡ In 5% $H_2SO_4$.
§ This solution decomposes during a 3-h photolysis.
‖ In 2 : 3 HCl : $H_2O$.
¶ $SnCl_2$ solution decomposes after contact with air for 24 h.
** In 1.5% $H_2SO_4$.
†† In 0.1-M NaOH.
‡‡ In 10% HCl.

## 10.2 SOLVENTS FOR PHOTOCHEMICAL REACTIONS

The choice of solvent for a particular photochemical reaction is very important and often plays a crucial role in the success of the reaction. The factors associated with the solvent that affect photochemical reactions are: (i) the choice of solvent, (ii) the concentration of the reaction solution, and (iii) the purity of the solvent.

### 10.2.1 Solvent selection

Many solvents can be used in photochemical reactions. The spectral transmissions of some commonly used solvents are given in Table 10.5. From these data, a suitable solvent with an appropriate optical transparency at the wavelength of the compound to be irradiated can be chosen.

TABLE 10.5

The wavelength (nm) transmission characteristics of some solvents*

| | Transmission | |
|---|---|---|
| Solvent | 10% | 100% |
| Acetone | 329 | 366 |
| Acetonitrile | 190 | 313 |
| Benzene | 280 | 366 |
| Carbon tetrachloride | 265 | 313 |
| Cyclohexane | 205 | 254 |
| Diethyl ether | 215 | 313 |
| Dimethyl sulphoxide | 262 | 366 |
| Ethanol | 205 | 313 |
| Hexane | 195 | 254 |
| Propan-2-ol | 205 | 313 |
| Tetrahydrofuran | 233 | 366 |

* Measured for a 1-cm path length of pure solvent. Ref. 4, p. 507.

Hydrocarbons such as cyclohexane are the most widely used solvents because they are relatively trouble-free, in that they have no low-lying excited states that interfere with the reaction of the solute and, in particular, they are chemically inactive. Furthermore, hydrocarbons can be easily purified and are often available in an impurity-free grade of material. Ethers and alcohols are also useful solvents. However, special care must be taken to ensure that ethers are free of peroxides and that alcohols are free of acids since these impurities, which are often present in these solvents, could lead to undesirable side reactions such as explosion or polymerization.

Acetonitrile has been employed as the solvent of choice in some of the recently developed photocyclization reactions, although why the particular solvent is suitable for one particular photochemical reaction is not well understood.

### 10.2.2 Solvent purity

It is important that the impurities which are inherent in some solvents, such as peroxides in ethers and acids in alcohols, usually originating from the synthetic precursors, are removed before irradiation.

The most hazardous impurity that is commonly present in solvents is oxygen. Therefore, the exclusion of air (oxygen) from a photochemical reaction is essential, particularly when a triplet species is involved. Furthermore, when non-oxidative conditions are required for a non-oxidative photochemical reaction, the presence of even a trace of oxygen will change the course of the reaction to oxidative, thereby changing the structure of the photoproduct; often dehydrogenated products are obtained. The concentrations of oxygen present in some non-degassed solvents are given in Table 10.6.

Even a very low concentration of dissolved oxygen is sufficient to quench a triplet reaction, which thus prevents the desired photochemical reaction

TABLE 10.6

The concentration of oxygen in some non-degassed solvents*

| Solvent | Oxygen concentration (mol ter$^{-1}$) |
|---|---|
| Acetone | 0.0024 |
| Benzene | 0.0019 |
| Carbon tetrachloride | 0.0026 |
| Cyclohexane | 0.0023 |
| Diethyl ether | 0.0040 |
| Ethanol | 0.0021 |
| Hexane | 0.0031 |
| Propan-2-ol | 0.0021 |

* Ref. 4, p. 507.

itself. Therefore, it is essential to degass the solvent prior to irradiation which requires special care not only in handling the apparatus but also in purifying the solvent. Technically, degassing by a vacuum technique at $-10$ to 1 Torr is effective. However, quantitative studies have shown that this method is not adequate to give a completely degassed solvent. The most convenient method and the one most often used is to bubble an inert gas such as nitrogen, argon, or helium through the reaction system. For most purposes, nitrogen bubbling should suffice. However, even nitrogen of the highest quality supplied in large cylinders contains an appreciable amount of oxygen and water. Thus it is desirable to remove these impurities by passing the gas through a scrubbing chain. The more recent availability of helium has made it much easier to carry out photochemical reactions under completely degassed conditions.

When oxidative conditions are required for a photochemical process, bubbling air through the reaction medium offers a very convenient and simple procedure. However, in many cases the use of a non-degassed solvent provides the same oxidative results.

### 10.2.3 Concentration of the solute

The role of the solvent in a chemical reaction is to provide a homogeneous system, in which the required reaction can proceed to a maximum whilst the unwanted side reactions are suppressed to a minimum. Dilution of a neat compound prevents the accumulation of high concentrations of excited species which if produced, for example, near the walls of the reaction vessel can lead to considerable side reactions including polymerization. Dilution also allows for reasonable penetration of light into the reaction solution and so reasonable reaction rates can be achieved.

The optimum concentration of a reactant can only be found by repeated trial-and-error type experiments. Therefore it is advisable to check the optimum concentrations reported for related reactions. If the reaction solution is too dilute then most of the light passes through the solution, while if the solution is too concentrated then side reactions such as undesirable dimerization occur very readily. However, a number of excellent results have been reported for the photodimerization of alkenes by the irradiation of a neat sample.

## 10.3 OTHER FACTORS

### 10.3.1 Influence of reaction temperature

Since the emission from arc lamps covers the range from UV to visible light, heat is inevitably evoluted from the light in the visible region. When using medium- and high-pressure arc lamps care must be taken to control the thermal effect of the irradiation. In particular, the power of high-pressure mercury lamps can be in the range of 200 W to 1 or 2 kW and when such lamps are used for irradiation the temperature of the reaction mixture rises very quickly and may cause the reactant to decompose. Therefore, special care should be taken to reduce the temperature of the reaction solution by circulating a cooling solution around the vessel or lamp or by using a fan to blow air away from the reaction vessel.

### 10.3.2 Sensitizers for photochemical reactions

A sensitizer is a compound (S), often organic, which absorbs energy at the specific wavelength being used in a photochemical reaction to form the excited molecule (S*). Once the sensitizer has absorbed this energy it then liberates it and it is transferred to the acceptor compound (A), often a reactant. The acceptor molecule thus becomes excited (A*), and the sensitizer returns to its ground state (S). The activated molecule (A*) then undergoes a photochemical reaction to form the product (B). Thus the sensitizer acts as an energy donor. Some examples of compounds commonly used for this purpose, are listed in Table 10.7.

$$\mathrm{S} \longrightarrow \mathrm{S}^* \xrightarrow{\text{stilbene}} \text{dimeric stilbene} + \mathrm{S}$$

TABLE 10.7

Triplet sensitizers and triplet quenchers [5]

| Triplet sensitizers | $E_T$ (kcal $mol^{-1}$) | Triplet quenchers | $E_T$ (kcal $mol^{-1}$) |
|---|---|---|---|
| Acetone | 78–90 | Dichloroethylene | ~80 |
| Benzene | 84.5 | Naphthalene | 60.9 |
| Acetophenone | 73.6 | *trans*-Piperylene | 58.8 |
| Benzophenone | 68.5 | Biacetyl | 54.9 |
| Naphthalene | 60.9 | 1,3-Cyclohexadiene | ~54 |
| Biacetyl | 54.9 | | |
| Eosine | 42.4 | | |
| Rose bengal | 39.4 | | |

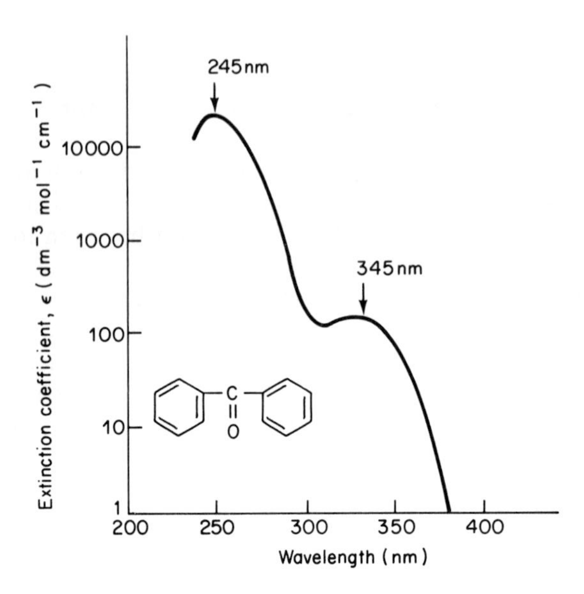


FIG. 10.3. The absorption spectrum of benzophenone in ethanol [6].

For example, benzophenone is a well-known sensitizer which absorbs radiation in the UV range (see Fig. 10.3). It has two maximum absorptions at 245 nm and 345 nm (488 kJ $mol^{-1}$ and 347 kJ $mol^{-1}$ respectively) which are absorbed to form the activated species (S*). In the presence of stilbene (A), which has an absorption maximum at 313 nm but does not absorb energy so smoothly from the irradiated light, the energy is first absorbed by benzophenone and is then smoothly transferred to the stilbene upon collision to give the activated species (A*). Further reaction is thus triggered such as the well-known dimeric cycloaddition of stilbene which gives a cyclobutane type product.

The use of sensitizers has been found to be very effective in a number of photochemical reactions. Some examples are: the photosensitized *cis–trans* dimerization of stilbenes (benzophenone as sensitizer); the photosensitized valence isomerization in the photosensitized dimerization of maleic anhydride (benzophenone); and the photosensitized oxidation of isopropyl alcohol and anthracene (benzophenone).

This energy transfer has even been observed in cases where the amount of energy absorbed by the sensitizer is lower than that required for the activation of the acceptor. The mechanism of energy transfer via a sensitizer has thus been a topic of heated discussion in photochemistry.

## 10.4 APPARATUS FOR PHOTOCHEMICAL REACTIONS

Preparative organic photochemistry is generally carried out in one of two ways: (i) the immersion-well method where the reaction solution surrounds the lamp; or (ii) the external irradiation method where the reaction solution is surrounded by a battery of lamps.

### 10.4.1 Immersion-well method

The immersion-well method is very effective in capturing the lamp's output and, therefore, is very economical. A typical example of an immersion-well apparatus for irradiation is shown in Fig. 10.4. This type of reactor is highly efficient since the lamp is placed inside the reaction solution to be irradiated, thereby assuring the maximum capture of the light emitted by the lamp.

The lamps used in this method are contained in a double-walled immersion-well structure made of either quartz or high-grade borosilicate

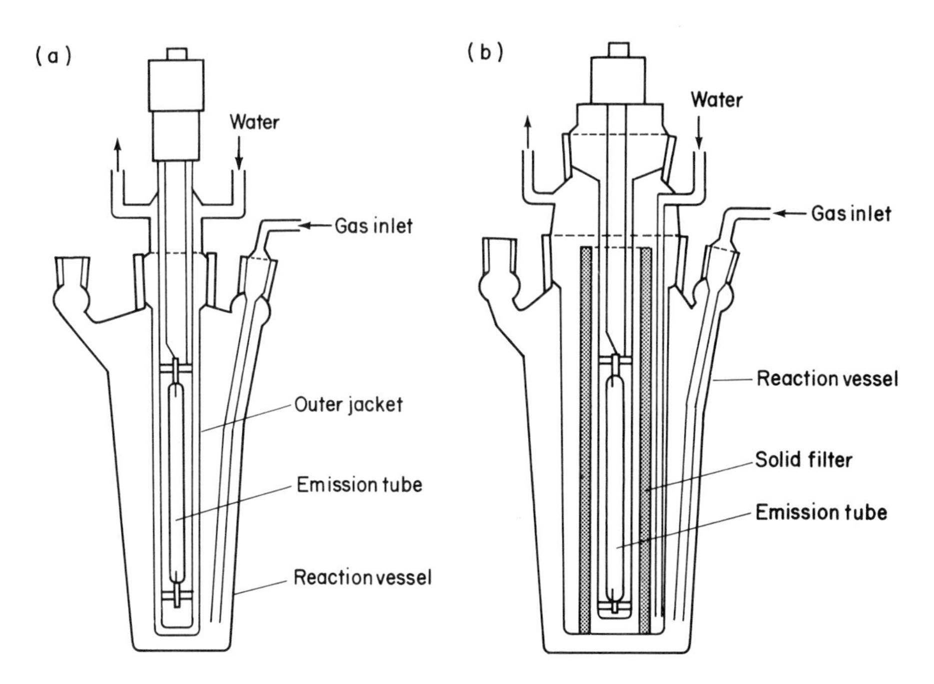


FIG. 10.4. Cross-sections of (a) EHB-W and (b) EMB-WI type photochemical reactors. (Reproduced by permission of EIKOSHA Co. Ltd, Japan.)

glass, which are transparent to UV light. The well is fitted with a standard taper joint which allows it to be used with a variety of reactor sizes. The well is usually made of Pyrex or quartz. Pyrex is less expensive than quartz but, only transmits light of wavelengths above 300 nm, whilst quartz is transparent up to 200 nm. Quartz immersion wells often come equipped with filter sleeves which fit around the lamp, thus permitting the cut-off of short-wavelength radiation.

The double-walled structure of the wells allows the reaction solution to be insulated from the heat of the medium- or high-pressure mercury-arc lamp by the circulation of a cooling solution (water) between them. The outer vessel of the lamp is usually made of Pyrex and can vary in size from 100 ml to a few litres; the number of outlets and inlets attached to it also varies. Therefore, this type of apparatus can be set up by the combination of a variety of parts depending on the reaction conditions, for example irradiation under aerobic or anaerobic condition, the size of experiment (the amount of reactants), and the reaction temperature (low or high).

Efficient stirring of the solution to be irradiated, which is the most important condition particularly in case of large-scale experiments, is effectively handled either by the use of a mechanical stirrer or by gas bubbling (nitrogen or helium). Nitrogen gas bubbling has the advantage of removing the remaining oxygen in the reaction mixture by blowing, but at the same time has the disadvantage in that solvent and volatile products are lost during the course of experiment.

### 10.4.2 External-irradiation method

The alternative mode of irradiation is with an external lamp arrangement. In this type of reactor, the solution to be irradiated is prepared in a quartz beaker-type flask and placed in the centre of a battery of lamps set in a circle within a radiation chamber. Irradiation is usually at room temperature.

Temperature control can only be achieved by a circulation fan placed underneath the apparatus table. The disadvantage of this method is that temperature control is difficult and may not be adequate if a strict regime is required.

This type of multilamp reactor can, therefore, be used with low- and medium-pressure arc lamps which do not yield too much heat. Furthermore, the efficiency of light capture by the reaction vessel is rather poor compared with the immersion method.

### 10.4.3 Thin-film irradiation

This method has found application in the irradiation of small volumes of reactant or for the irradiation of concentrated solutions where irradiated light is required to penetrate only a fraction of a millimeter.

## REFERENCES

1. A. Schönberg, G. O. Schenck and O. A. Neumuller, *Preparative Organic Photochemistry*, p. 473, Springer-Verlag, New York, 1968.
2. W. M. Horspool, *Aspects of Organic Photochemistry*, p. 38. Academic Press, London, 1976.
3. J. G. Calvert and J. N. Pitts, *Photochemistry*, p. 696. Wiley, New York, 1967.
4. W. M. Horspool, in *Synthetic Organic Photochemistry* (ed. W. M. Horspool), p. 492. Plenum Press, New York, 1984.
5. T. Matsuura, *Organic Photochemistry* (in Japanese), p. 38. Kagaku-Dohjin, Kyoto, 1970.
6. M. Imoto, *Introductory Organic Electronic Theory* (in Japanese), p. 268. Tokyo Kagaku-Dohjin, Tokyo, 1987.

# Index of Compounds and Methods

V



W



X



Y